Principles of Bloodstain Pattern Analysis

Theory and Practice

CRC SERIES IN PRACTICAL ASPECTS OF CRIMINAL AND FORENSIC INVESTIGATIONS

VERNON J. GEBERTH, BBA, MPS, FBINA *Series Editor*

Practical Homicide Investigation: Tactics, Procedures, and Forensic Techniques, Third Edition
Vernon J. Geberth

The Counterterrorism Handbook: Tactics, Procedures, and Techniques, Second Edition
Frank Bolz, Jr., Kenneth J. Dudonis, and David P. Schulz

Forensic Pathology, Second Edition
Dominick J. Di Maio and Vincent J. M. Di Maio

Interpretation of Bloodstain Evidence at Crime Scenes, Second Edition
William G. Eckert and Stuart H. James

Tire Imprint Evidence
Peter McDonald

Practical Drug Enforcement, Second Edition
Michael D. Lyman

Practical Aspects of Rape Investigation: A Multidisciplinary Approach, Third Edition
Robert R. Hazelwood and Ann Wolbert Burgess

The Sexual Exploitation of Children: A Practical Guide to Assessment, Investigation, and Intervention, Second Edition
Seth L. Goldstein

Gunshot Wounds: Practical Aspects of Firearms, Ballistics, and Forensic Techniques, Second Edition
Vincent J. M. Di Maio

Friction Ridge Skin: Comparison and Identification of Fingerprints
James F. Cowger

Footwear Impression Evidence, Second Edition
William J. Bodziak

Principles of Kinesic Interview and Interrogation, Second Edition
Stan Walters

Practical Fire and Arson Investigation, Second Edition
David R. Redsicker and John J. O'Connor

The Practical Methodology of Forensic Photography, Second Edition
David R. Redsicker

Practical Aspects of Interview and Interrogation, Second Edition
David E. Zulawski and Douglas E. Wicklander

Investigating Computer Crime
Franklin Clark and Ken Diliberto

Practical Homicide Investigation Checklist and Field Guide
Vernon J. Geberth

Bloodstain Pattern Analysis: With an Introduction to Crime Scene Reconstruction, Second Edition
Tom Bevel and Ross M. Gardner

Practical Aspects of Munchausen by Proxy and Munchausen Syndrome Investigation
Kathryn Artingstall

Quantitative-Qualitative Friction Ridge Analysis: An Introduction to Basic and Advanced Ridgeology
David R. Ashbaugh

Practical Criminal Investigations in Correctional Facilities
William R. Bell

Officer-Involved Shootings and Use of Force: Practical Investigative Techniques
David E. Hatch

Sex-Related Homicide and Death Investigation: Practical and Clinical Perspectives
Vernon J. Geberth

Global Drug Enforcement: Practical Investigative Techniques
Gregory D. Lee

Practical Investigation of Sex Crimes: A Strategic and Operational Approach
Thomas P. Carney

Principles of Bloodstain Pattern Analysis: Theory and Practice
Stuart James, Paul Kish, and T. Paulette Sutton

Series Editor's Note

This book is part of a series entitled *Practical Aspects of Criminal and Forensic Investigation*. This series was created by Vernon J. Geberth, New York City Police Department Lieutenant Commander (Retired), who is an author, educator, and consultant on homicide and forensic investigations.

This series, written by authors who are nationally recognized experts in their respective fields, has been designed to provide contemporary, comprehensive, and pragmatic information to the practitioner involved in criminal and forensic investigations.

Principles of Bloodstain Pattern Analysis

Theory and Practice

Stuart H. James
Paul E. Kish
T. Paulette Sutton

Taylor & Francis Group

Boca Raton   London   New York   Singapore

A CRC title, part of the Taylor & Francis imprint, a member of the Taylor & Francis Group, the academic division of T&F Informa plc.

Cover design by Alexei Pace (with photos from Herbert Leon MacDonell and Detective Jeffrey A. Strobach).

Published in 2005 by
CRC Press
Taylor & Francis Group
6000 Broken Sound Parkway NW, Suite 300
Boca Raton, FL 33487-2742

© 2005 by Taylor & Francis Group, LLC
CRC Press is an imprint of Taylor & Francis Group

No claim to original U.S. Government works
Printed in the United States of America on acid-free paper
10 9 8 7 6 5 4

International Standard Book Number-10: 0-8493-2014-3 (Hardcover)
International Standard Book Number-13: 978-0-8493-2014-9 (Hardcover)
Library of Congress Card Number 2004062860

This book contains information obtained from authentic and highly regarded sources. Reprinted material is quoted with permission, and sources are indicated. A wide variety of references are listed. Reasonable efforts have been made to publish reliable data and information, but the author and the publisher cannot assume responsibility for the validity of all materials or for the consequences of their use.

No part of this book may be reprinted, reproduced, transmitted, or utilized in any form by any electronic, mechanical, or other means, now known or hereafter invented, including photocopying, microfilming, and recording, or in any information storage or retrieval system, without written permission from the publishers.

For permission to photocopy or use material electronically from this work, please access www.copyright.com (http://www.copyright.com/) or contact the Copyright Clearance Center, Inc. (CCC) 222 Rosewood Drive, Danvers, MA 01923, 978-750-8400. CCC is a not-for-profit organization that provides licenses and registration for a variety of users. For organizations that have been granted a photocopy license by the CCC, a separate system of payment has been arranged.

Trademark Notice: Product or corporate names may be trademarks or registered trademarks, and are used only for identification and explanation without intent to infringe.

Library of Congress Cataloging-in-Publication Data

James, Stuart H.
   Principles of bloodstain pattern analysis: theory and practice / Stuart H. James, Paul Erwin Kish, T. Paulette Sutton.
      p. cm. -- (Practical aspects of criminal & forensic investigation)
   Includes bibliographical references and index.
   ISBN 0-8493-2014-3 (alk. paper)
   1. Bloodstains. 2. Forensic serology. 3. Crime scene searches. I. Kish, Paul Erwin. II. Sutton, T. Paulette. III. Title. IV. Series: CRC series in practical aspects of criminal and forensic investigations.

HV8077.5.B56J35 2005
363.25'62--dc22   2004062860

Taylor & Francis Group
is the Academic Division of T&F Informa plc.

Visit the Taylor & Francis Web site at
http://www.taylorandfrancis.com

and the CRC Press Web site at
http://www.crcpress.com

Foreword

The discipline of Bloodstain Pattern Analysis has advanced to the level that if "blood evidence" exists investigators and prosecutors want and should have an interpretation. To do less would be nothing short of ignoring valuable physical evidence.

The evolution of this evidence has been driven by discovery and fueled by DNA. By using methodical, scientific principles, crime scene information can be discovered. This information will re-create events that support or refute witness statements and will assist the court in determining intent and help the jury to visualize the final moments. The language of bloodstain pattern analysis will provide information that cannot be gained by other means. Let the physical evidence speak for itself and tell the stories of self-defense or intent to commit murder.

As a bloodstain pattern analyst I relish in the opportunity to gain further insight into this discipline, sometimes as confirmation that a different approach can gain the same result or that this new approach provides a more in-depth review of my case. In any event I seek to articulate this evidence so that those concerned will gain the same confidence that I have.

The logic of bloodstain pattern analysis can sometimes be a trap. Just when you think that you know how to interpret patterns, they give you a different twist. Given classic patterns it is not difficult to determine their origin; however, we know that crime scenes are not predictable. The evidence is more often a combination or a portion of the patterns we know. This subject requires a complete understanding of all aspects of the case. The ability to apply training, knowledge, and common sense is key to a successful interpretation of a bloodletting scene. If you want to fully understand the intricacies of bloodstains you need only refer to the table of contents in this book. Where, when, and how do we begin? Is blood a unique substance? How do we recognize one pattern or event from another? Is it possible to alter a bloodstain pattern? What equipment will we need? How do we get this evidence to court? What if the bloodstains have been entirely or partially cleaned up? Is there an error rate, and how do we explain our results to a jury?

I believe that this book will meet your requirements and I would recommend it to readers at every level. Those who are researching at a college or university level will benefit as much as those who are working in the field of forensic science, or otherwise involved in the criminal justice system. The completeness of this book is a testament to the authors. Those involved in the field of bloodstain pattern analysis, even to the slightest degree, will recognize their names.

The authors are forensic scientists who have gained the respect of the bloodstain analysis community through years of work, not only through painstaking analysis of bloodletting evidence but also through expert courtroom testimony. As you look at these names you will also see teachers—people who take the time and care to share their knowledge. Years of expertise have come together to write one of the most comprehensive texts on the subject. The reader need only turn these pages with a mild interest to become both fascinated and educated.

Pat Laturnus
Bloodstain Pattern Analyst
Instructor–Ontario Police College
Ontario, Canada
Retired–Royal Canadian Mounted Police

Preface

Principles of Bloodstain Pattern Analysis: Theory and Practice presents an in-depth text of bloodstain pattern analysis that emphasizes a modern thought process of a taxonomic classification of bloodstains based on their physical characteristics of size, shape, and distribution as well as the specific mechanisms that produce them. The concept of a multidisciplinary approach using scene and laboratory examinations in conjunction with forensic pathology, forensic serology, and chemical enhancement techniques is also presented to the reader. The technical content and quality of this book is increased dramatically with color images of bloodletting injuries, bloodstains, and crime scenes. Case studies are presented within individual chapters, and the book is complimented by two chapters that discuss details of legal issues as they pertain to bloodstain pattern analysis. The coauthors, Stuart H. James, Paul Erwin Kish, and T. Paulette Sutton, form a nucleus of bloodstain pattern analysts, each with many years of experience in casework and teaching of basic and advanced bloodstain pattern analysis courses.

The chapters that present the scientific principles and practical application of bloodstain pattern analysis represent the combined efforts of the authors. They have invited other respected and qualified forensic scientists and attorneys to contribute chapters in their specialties, which has enhanced the quality and scope of the text. The chapters are arranged in a logical order with Chapter 1, "Introduction to Bloodstain Pattern Analysis," discussing the evolution of bloodstain pattern analysis from significant historical events to its current level as a discipline within the forensic sciences.

Chapter 2, "Medical and Anatomic Aspects of Bloodshed," contributed by forensic pathologist Dr. Ronald K. Wright, MD, JD, provides an overview of the circulation of blood within the body and the rate of bleeding and blood volume loss resulting from medical conditions and specific injuries. He has also included an interesting section that describes postmortem lividity and identifiable patterned abrasions and bruises on the skin. The authors agreed that this chapter was essential to the bloodstain pattern analyst for the understanding of mechanisms of bleeding within and outside the body of the victim.

Chapter 3, "Biological and Physical Properties of Human Blood," explains the cellular components of human blood and their functions within the body supplemented with excellent color images. The physical properties of blood relative to its behavior as a liquid outside the body is essential for the bloodstain analyst to appreciate in terms of bloodstain formation. The principles of physics pertaining to surface tension, adhesion and cohesion, relative density, viscosity, and the non-Newtonian behavior of blood are defined.

Chapter 4, "Physical Properties of Bloodstain Formation," applies the physical properties of blood to the shape of an airborne drop or droplet of blood and the size and shape of resultant bloodstains on horizontal and nonhorizontal surfaces. This chapter also describes the taxonomic classification of bloodstains that will be the emphasis in subsequent chapters of bloodstain pattern analysis.

Chapter 5, "Passive Bloodstains," is a detailed chapter that describes a major category of bloodstains. This chapter is supplemented with numerous color images that afford the reader many examples of this important taxonomic classification of bloodstains.

Chapter 6, "Formation of Spatter and Spatter Associated with a Secondary Mechanism," discusses the physical characteristics of spatter formation relative to stain size, shape, quantity, location, and distribution relative to its taxonomic classification. Emphasis is given to mechanisms associated with the creation of spatter patterns and the important ability of the bloodstain analyst to recognize the overlapping of stain sizes created by different mechanisms and the caution to be exercised in making a determination.

Chapter 7, "Impact Spatter Mechanisms," describes the production of spatter produced as the result of beating, stabbing, gunshot, explosion, and power tool events. Factors that limit the production of impact spatter are emphasized.

Chapter 8, "Spatter Associated with a Projection Mechanism," presents the characteristics of bloodstains and patterns produced by arterial, expiratory, and cast-off mechanisms. The uniqueness of the patterns as well as the overlap of stain size within a pattern that can be confused with impact spatter mechanisms is demonstrated.

Chapter 9, "Altered Bloodstain Patterns," is a chapter that applies to all categories of bloodstains because physical or biological alterations such as drying, clotting, dilution, insect activity, and effects of fire are frequently encountered in casework. Recognition of these alterations cannot be overemphasized. Void patterns and sequencing of bloodstain patterns are also included in this chapter.

Chapter 10, "Determination of the Area of Convergence and Area of Origin of Bloodstain Patterns," is a chapter devoted to the reconstruction of bloodstain events using the principles of mathematics, including trigonometry. Various methods of the determination of areas of convergence and origin are explained.

Chapter 11, "Directional Analysis of Bloodstain Patterns with a Computer," contributed by Dr. Alfred L. Carter with case studies provided by Craig C. Moore and Mike Illes, describes the BackTrack program devised by Dr. Carter for the measurement of bloodstains and the determination of areas of convergence and origin.

Chapter 12, "Documentation and Examination of Bloodstain Evidence," details the methods of preparing sketches, diagrams, photographs, and videos of bloodstain evidence at the scene as well as at the laboratory, where the examination of bloodstained clothing is crucial prior to testing for DNA analysis.

Chapter 13, "Examination of Bloodstain Patterns at the Scene," describes additional detail for proper processing of crime scenes and sample collection, including vehicles for bloodstain pattern analysis. An important feature of this chapter is the examination of the body of the victim for bloodstain patterns.

Chapter 14, "Presumptive Testing and Species Determination of Blood and Bloodstains," contributed by Robert Spalding, describes the various presumptive tests for blood, including their applications and limitations. Mr. Spalding also explains the confirmatory tests for blood and determination of species of origin.

Chapter 15, "The Detection of Blood Using Luminol," contributed by Dale Laux, explains the luminol reaction, its preparation, its use, and interpretation of patterns revealed by this chemi-luminescent reagent. The principles of luminol photography with excellent images are provided as well as the effect of the use of luminol on the subsequent analysis of bloodstains.

Chapter 16, "Chemical Enhancement of Latent Bloodstain Impressions," contributed by Martin Eversdijk, is an important addition to this text. Martin methodically describes selected methods for the chemical dye enhancement of trace impressions of blood augmented with excellent images of bloody enhanced fingerprints, palm prints, and footwear.

Chapter 17, "Approaching the Bloodstain Pattern Case," explains the importance of the examination of all relevant information and the logical sequential approach to case evaluation for the bloodstain analyst. The chapter emphasizes the utilization information from all involved forensic disciplines for the proper synthesizing of the available data and the scientific formulation and testing of hypotheses.

Chapter 18, "Report Writing," is an expansion of an excellent chapter by Paul E. Kish that appeared in the text, *Scientific and Legal Applications of Bloodstain Pattern Interpretation*. It is a logical format for the writing of a bloodstain pattern analysis report from introduction to final conclusions.

Chapter 19, "Legal and Ethical Aspects of Bloodstain Pattern Evidence," contributed by Carol Henderson and Brittan Mitchell, is updated from the text, *Scientific and Legal Applications of Bloodstain Pattern Interpretation*. This chapter details the issues of the admissibility, weight of the bloodstain evidence, and the qualifications of the bloodstain analyst who is called on to provide expert testimony in court. Methods of direct and cross-examination are discussed as well as ethical issues of the attorney and the expert.

Chapter 20, "Bloodstain Pattern Analysis: Postconviction and Appellate Application," contributed by Marie Saccoccio also has been updated from the text, *Scientific and Legal Applications of Bloodstain Pattern Interpretation*. She discusses the role of the postconviction attorney and the appeal process with case examples.

Finally, the Appendices supplement the text with scientific data including trigonometric tables, metric equivalents, and scene and laboratory checklists and biohazard safety precautions. Additionally, court decisions relating to bloodstain pattern analysis and presumptive blood testing are included. Numerous references are provided for bloodstain pattern analysis and the related topics.

The goal of the authors and contributors was to provide an up-to-date bloodstain pattern analysis text for crime scene investigators, forensic laboratory personnel, forensic pathologists, and prosecutors and defense attorneys within the criminal justice system. As Professor Andre A. Moenssens stated in an article entitled "Novel Scientific Evidence in Criminal Cases," published in the *Journal of Criminal Law and Criminology* in 1993:

> Attorneys have the responsibility to learn about the scientific evidence that they wish to admit. If the attorneys who are questioning and cross-examining the expert witnesses have a working knowledge of bloodstain pattern analysis, they will be better equipped to distinguish between those experts with sufficient qualifications and those without them. Also, attorneys will be able to more capably critique an expert's testimony and limit or eliminate conclusions that are speculative or over-stated. Unless attorneys know what questions to ask during cross-examination, much of what the witness testifies to will go unchallenged.

Stuart H. James

Paul E. Kish

T. Paulette Sutton

The Authors

Stuart H. James of James and Associates Forensic Consultants, Inc. is a graduate of Hobart College where he received a BA degree in biology and chemistry in 1962. He received his MT(ASCP) in medical technology from St. Mary's Hospital in Tucson, Arizona in 1963. Graduate courses completed at Elmira College included homicide investigation, bloodstain pattern analysis, and forensic microscopy. He has completed more than 300 hours of continuing education and training in death investigation and bloodstain pattern analysis. A former crime laboratory supervisor in Binghamton, New York, he has been a private consultant since 1981.

Mr. James has instructed in Forensic Science at the State University of New York and Broome Community College in Binghamton, New York. Additionally, he has lectured on the subjects of bloodstain pattern analysis and forensic science throughout the United States and abroad. He has instructed basic and advanced bloodstain pattern analysis courses with Paul E. Kish in Pontiac, Michigan, Appleton Wisconsin, Suffolk University in Boston, Massachusetts, the Henry C. Lee Institute at the University of New Haven in West Haven, Connecticut, the Centre of Forensic Sciences in Toronto, Canada, the Politie LSOP Institute for Criminal Investigation and Crime Science in Zutphen, The Netherlands and the University of Newcastle upon Tyne in the United Kingdom.

He has been consulted on homicide cases in 46 States and the District of Columbia as well as in Australia, Canada, Germany, South Korea, and the US Virgin Islands and has provided expert testimony in many of these jurisdictions in state, federal, and military courts.

Mr. James was a co-author of the text entitled, *Interpretation of Bloodstain Evidence at Crime Scenes*. He was also the Editor of *Scientific and Legal Applications of Bloodstain Pattern Interpretation* both of which were published in 1998. He is a co-editor with Jon J. Nordby of the text entitled *Forensic Science: An Introduction to Scientific and Investigative Techniques* published in 2003. Mr. James is a fellow in the American Academy of Forensic Sciences and a distinguished member of the International Association of Bloodstain Pattern Analysts (IABPA) as well as the current editor of the quarterly IABPA News. He is also a member of the Scientific Working Group for Bloodstain Pattern Analysis (SWGSTAIN).

Paul Erwin Kish is a consulting bloodstain pattern analyst in Corning, New York, as well as, an adjunct Instructor in the Criminal Justice Program at Elmira College. He holds a B.S. degree in Criminal Justice and an M.S. degree in Education from Elmira College. He has been consulted on homicide cases in 30 states within the United States and 7 countries while presenting expert testimony in 16 states, the District of Columbia, and Canada.

Mr. Kish is an internationally known lecturer on the subject of bloodstain pattern analysis. He has taught Basic Bloodstain Pattern Analysis courses throughout the United States, as well as, in Canada, The Netherlands, United Kingdom, and Sweden. He has educated over 850 students from 17 countries during these weeklong basic bloodstain pattern analysis courses. He has taught Advanced Bloodstain Pattern Analysis courses in the United States, The

Netherlands, and United Kingdom. He has lectured extensively throughout the United States at forensic and law related conferences and seminars.

He has authored various articles on the topic of bloodstain pattern analysis including contributions to the texts *Scientific and Legal Applications of Bloodstain Pattern Interpretation* (CRC Press, 1998); *Forensic Science: An Introduction to Scientific and Investigative Techniques* (CRC Press, 2003). Mr. Kish is an associate editor of the International Association of Bloodstain Pattern Analysts News having served as Editor for two years. He is a fellow in the American Academy of Forensic Sciences, a member of the Scientific Working Group for Bloodstain Pattern Analysis (SWGSTAIN), the International Association of Bloodstain Pattern Analysts, Canadian Society of Forensic Science, and International Association for Identification.

T. Paulette Sutton is an Associate Professor of Clinical Laboratory Sciences and Assistant Director of Forensic Services at the University of Tennessee Health Science Center in Memphis. She holds a B.S. in Medical Technology from the University of Tennessee and a M.S. in Operations Management Engineering from the University of Arkansas. She has practiced as a forensic serologist and bloodstain pattern analyst at the University of Tennessee, Memphis for the past 28 years and is also a member of the crime scene team for the Shelby County Medical Examiner's Office (Memphis). Sutton has been consulted on cases in many states and federal jurisdictions and has given expert testimony in criminal, as well as civil cases, in for both prosecution and defense. She has taught basic and advanced Bloodstain Analysis courses, as well as developing and teaching a course devoted entirely to the documentation of bloodstain pattern evidence. Ms. Sutton has lectured extensively on bloodstain pattern analysis and has received the Lecturer of Merit and the Distinguished Faculty awards from the National College of District Attorneys.

Ms. Sutton has contributed to the texts *Interpretation of Bloodstain Evidence at Crime Scenes*, 2nd edition (CRC Press, 1998); *Scientific and Legal Applications of Bloodstain Pattern Interpretation* (CRC Press, 1998); *Forensic Science: An Introduction to Scientific and Investigative Techniques* (CRC Press, 2003) and authored didactic and laboratory exercise manuals *Bloodstain Pattern Analysis in Violent Crimes*. She has authored numerous articles on forensic serology and on bloodstain pattern analysis and serves as an associate editor of the International Association of Bloodstain Pattern Analysts News. Sutton is a member of the International Association for Identification, the International Association of Bloodstain Pattern Analysts and the Scientific Working Group for Bloodstain Pattern Analysis (SWGSTAIN).

Contributors

Alfred L. Carter is a retired Professor of Physics at Carleton University in Ottawa, Canada. He has done extensive work in the computer analysis of bloodstains and developed software for this application. He is a member of the International Association of Bloodstain Pattern Analysts. He has consulted with the Royal Canadian Mounted Police in Ottawa, Canada and has taught computer analysis of bloodstains in both Canada and the United States.

Martin Eversdijk began his career with the Dutch Police in 1986 and joined the Forensic Scene of Crime Department in the city of Amstelveen in 1993. He became a staff member/trainer at the Institute for Criminal Investigation and Crime Science, a national training center for forensic scene of crime officers in the Netherlands. He conducts research and teaches in the areas of bloodstain pattern analysis, blood searching techniques and blood enhancement techniques. Mr. Eversdijk is a member of the International Association of Bloodstain Pattern Analysts and the FBI Scientific Working Group for Bloodstain Pattern Analysis (SWGSTAIN).

Carol Henderson is Director of the National Clearing House for Science, Technology and the Law and a Visiting Professor of Law at Stetson University College of Law in Gulfport, Florida. She is Vice President of the American Academy of Forensic Sciences and has served on the Board of Directors and as Chairman of the Jurisprudence Section. She is co-editor of the *Encyclopedia of Forensic and Legal Medicine* published by Elsevier and a co-author of the 4th Edition of *Scientific Evidence in Civil and Criminal Cases* with Moenssens, Starrs and Inbau published by The Foundation Press in 1995. She is the author of more than 40 articles on scientific evidence and ethics.

Mike Illes has been a member of the Ontario Provincial Police for the past 18 years. He has been involved in the field of forensics for 14 years and is the Unit Commander of the Central Region Forensic Identification Unit located in Peterborough, Ontario. In 1994 Sergeant Illes received training in Basic Bloodstain Pattern Recognition in the United States. He completed the RCMP Forensic Bloodstain Pattern Analyst program and RCMP Math and Physics Course in 1995. Sergeant Illes is internationally known for his work in Bloodstain Pattern Analysis. He has done bloodstain crime scene work in Canada (Ontario and Newfoundland), the United States and Holland. He has given expert opinion evidence in the Canadian and Netherlands Court systems. Sergeant Illes has conducted forensic lectures and training courses in Canada, the United States and the Netherlands. Sergeant Illes sits on the Trent University/Sir Sandford Fleming College Forensic Degree Program Advisory Committee and is an instructor for that program. He is also a member of the FBI Scientific Working Group for Bloodstain Pattern Analysis (SWGSTAIN).

Dale Laux began his forensic career in 1980 after graduating from The Ohio State University with a Masters of Science degree in developmental biology. He studied limb regeneration in salamanders and owes his scientific reasoning to his advisor, Dr. Roy Tassava, who taught him how to think and rationalize. Dale has spent his entire career with the Ohio Bureau of Criminal Identification and Investigation (BCI), a division of the Attorney General's Office. He has witnessed and been a part of the evolution of forensic biology from ABO and genetic markers to DNA. He has authored or co-authored 12 papers, presented 20 papers, has given numerous workshops on the use of luminol, and has lectured on a wide variety of forensic topics. He is a member and past president of the Midwestern Association of Forensic Scientists (MAFS) and a Fellow of the American Academy of Forensic Sciences. He was named Ohio Peace Officer of the Year in 1988, received the Superintendent's Award from BCI in 2004, and was recently given the Distinguished Service Award from MAFS, their highest honor. He dedicates his chapter on luminol to his wife Denise, for her encouragement and support, and to their sons, David and Kevin, for their assistance in his experiments.

Brittan L. Mitchell is a Law and Science Fellow for the National Clearing House for Science, Technology and the Law. She graduated *magna cum laude* from Stetson College of Law and is an active member of the Florida Bar. She also received a B.S. degree in Accounting from Florida State University and a B.S. in Economics from Brigham Young University.

Marie Elena Saccoccio is an alumna of New England School of Law and Yale Law School and is a practicing attorney in Cambridge, Massachusetts concentrating on appellate/post-conviction matters. She is admitted to the bars of Colorado, Florida and Massachusetts where she works as a sole practitioner. She is the former judicial law clerk to the Honorable Wade Brorby of the United States Court of Appeals for the Tenth Circuit.

Robert P. Spalding served in the FBI for 28 years as an investigative Special Agent in Cleveland and at the Forensic Science Research and Training Center (FSRTC) in Washington, DC and the FBI Academy. In 1993 he was assigned to the newly formed Evidence Response Team (ERT) Unit where he taught crime scene investigation to FBI field office personnel. In this assignment, he also taught bloodstain pattern analysis to field FBI Evidence Response Team Personnel at the FBI Academy. Spaulding received his B.S. and M.S. degrees from the University of Maine in 1965 and 1968 respectively. He is currently a member of several professional organizations and is the owner of Spalding Forensics, LLC, a consulting and training firm in Centreville, Virginia, specializing in casework involving bloodstain patterns and crime scene reconstruction.

Ronald K. Wright is a forensic pathologist in private practice. He has a B.S. in Biology and Chemistry from Southwest Missouri University, an M.D. from St. Louis University School of Medicine and a J.D. from the University of Miami School of Law. He has practiced as a forensic pathologist and toxicologist for more than 30 years, holding positions of Deputy Chief Medical Examiner of the State of Vermont and Miami-Dade County Florida and Chief Medical Examiner of Broward County, Florida.

Acknowledgments

The authors express their thanks and appreciation to the contributing authors, Dr. Ronald Wright, Dr. Al Carter, Robert P. Spalding, Dale Laux, Martin Eversdijk, Carol Henderson, Brittan Mitchell, Mike Illes, and Marie Saccoccio for their excellent chapters. We also appreciate the efforts of Alexei Pace for providing excellent diagrams that supplement the text as well as the design of the cover, Lisa DiMeo of Arcana Forensics in San Diego, California for her review of some technical aspects of the book and for providing photographs, and Lisa Elrod of Memphis Tennessee for her invaluable consultation with regard to many of the photographic images.

The staff at CRC Press in Boca Raton, Florida including Life Sciences Publisher, Barbara Norwitz, Senior Acquisitions Editor Becky McEldowney, Editorial Project Development Manager, Helena Redshaw, Project Editor Suzanne Lassandro, and Editorial Assistant Susan Longo deserve special recognition for their guidance and assistance throughout the development of this book.

The authors also thank the following individuals and organizations for their support and providing photographs that have enhanced the quality of this book.

Abbott Laboratories, Abbott Park, Illinois
Larry Barksdale, Lincoln Police Department, Lincoln, Nebraska
Joe Cocco, Pennsylvania State Police
Greg Coughlin, New Castle County Police, New Castle, Delaware
Dazor Manufacturing Company, St. Louis, Missouri
Anthony Dinardo, New Castle County Police, New Castle, Delaware
Philippe Esparanca, French Gendarmerie Laboratory, France
Jan Ewanow, Binghamton Police Department, Binghamton, New York
Dr. Jerry Francisco, MD, Retired Medical Examiner, Memphis, Tennessee
Cynthia Gardner, MD, Hamilton County Coroner's Office, Cincinnati, Ohio
Dennis Kunkel Microscopy, Inc., Kailua, Hawaii
Herbert Leon MacDonell, Laboratory of Forensic Science, Corning, New York
Martin County Sheriffs Office, Stuart, Florida
Paul Mergenthaler, New Castle County Police, New Castle, Delaware
Linda Netzel, Kansas City Crime Laboratory, Kansas City, Missouri
Steve Nichols, University of Tennessee, Memphis, Tennessee
Pinellas County Sheriffs Office, Clearwater, Florida
Rex Sparks, Des Moines Police Department, Des Moines, Iowa
Jeffrey A. Strobach, Wausau Police Department, Wausau, Wisconsin
Steven Symes, Mercyhurst College, Erie, Pennsylvania
Todd Thorne, Kenosha Police Department, Kenosha, Wisconsin

David Williams, Rochester Police Department, Rochester, New York
Jim Wallace, University of Tennessee, Memphis, Tennessee
United States Attorneys Office, Washington, DC

Additionally, the continuing patience, support and understanding of Victoria Wendy Hilt of Fort Lauderdale, Florida and Pam Kish of Corning, New York are truly appreciated.

Stuart H. James

Paul E. Kish

T. Paulette Sutton

Contents

Foreword	vii
Preface	ix
The Authors	xiii
Contributors	xv
Acknowledgments	xvii

1 Introduction to Bloodstain Pattern Analysis — 1

Introduction	1
Objectives of Bloodstain Pattern Analysis	1
Scientific Approach to Bloodstain Pattern Analysis	2
Historical Development	3
International Association of Bloodstain Pattern Analysts	7
Classification of Bloodstains	7
Low-Velocity Impact Blood Spatter	7
Medium-Velocity Impact Blood Spatter	7
High-Velocity Impact Blood Spatter	7
Acceptance within the Scientific Community	8
Scientific Working Group for Bloodstain Pattern Analysis	8
Education and Training in Bloodstain Pattern Analysis	9
Conclusion	10

2 Medical and Anatomical Aspects of Bloodshed — 11

Introduction	11
The Circulation of Blood	11
Rate of Bleeding	13
Blood Volume	13
Causes of Blood Volume Loss	15
Blood Loss Resulting from Medical Conditions	15
Bleeding from the Skin	15
Bleeding from the Mouth and Nose	16
Bleeding from the Respiratory System	20
Bleeding from the Gastrointestinal Tract	20
Blood Loss Resulting from Injuries	21
Abrasions	21
Lacerations	21
Incised Wounds	24

Stab Wounds	26
Gunshot Wounds	27
Contact Wounds	29
Intermediate-Range Gunshot Wounds	30
Distant or Long-Range Gunshot Wounds	31
Gunshot Exit Wounds	31
Postmortem Aspects of Blood in the Body	32
Bloodstain Patterns within the Body	34
Patterns Associated with Postmortem Lividity	34
Patterns Associated with Postmortem Marbling of the Skin	35
Petechial Hemorrhaging (Tardieu Spots)	35
Patterns Associated with Bruises	36
Patterns Associated with Abrasions	37
Conclusion	39

3 Biological and Physical Properties of Human Blood — 41

Biological Functions of Human Blood	41
Composition of Blood	42
Volume and pH	42
Plasma	42
White Blood Cells (Leucocytes)	44
Red Blood Cells (Erythrocytes)	45
Hemoglobin	45
Platelets (Thrombocytes) and Clotting Factors	45
Physical Properties of Blood	48
Viscosity	48
Blood — A Non-Newtonian Fluid	49
Poiseuille's Equation	49
Surface Tension	50
Adhesion, Cohesion, and Capillarity	53
Relative Density	53
Effects of Drugs and Alcohol on Bloodstains	55
Blood Use in Experiments	56

4 Physical Properties of Bloodstain Formation — 59

Drop Formation and Travel	59
The Shape of a Falling Drop	61
Volume of a Drop of Blood	63
Drop Volume as a Function of Source	63
Distance Fallen	65
Classification of Bloodstain Patterns	67
Passive	68
Spatter	68
Altered	69

5 Passive Bloodstains — 71

Introduction — 71
Free-Falling Drops on Horizontal Surfaces — 71
Free-Falling Drops on Angular Surfaces — 73
Drip Patterns — 76
Drip Trails — 79
Splashed Bloodstain Patterns — 82
Flow Patterns — 84
Blood Pools — 87
Saturation — 87
Transfer Bloodstains — 88
Recognizable Blood Transfer Stains — 92
Conclusion — 92

6 Formation of Spatter and Spatter Associated with a Secondary Mechanism — 99

Introduction — 99
Physical Properties of Blood Relative to Spatter Formation — 99
Physical Characteristics of Spatter Patterns — 100
    Size Range and Quantity of Spatters — 100
    Distribution and Location of Spatter — 102
    Shape of Spatters — 103
    Effect of Target Surface Texture on Spatter Appearance — 106
    Mechanisms Associated with the Creation of Spatter Patterns — 107
Significance of Spatter Identification — 108
Spatter Associated with a Secondary Mechanism — 108
Satellite Spatters on Horizontal Surfaces — 110
Satellite Spatters on Vertical Surfaces — 112

7 Impact Spatter Mechanisms — 119

Introduction — 119
Impact Spatter Associated with Beating and Stabbing Events — 119
Impact Spatter Associated with Gunshot, Explosions, and Power Tools — 131
Back Spatter — 136
Forward Spatter — 138
Detection of Blood in the Barrels of Firearms — 147
Caveats — 148

8 Spatter Associated with a Projection Mechanism — 149

Introduction — 149
Arterial Mechanisms — 149
Bloodstain Patterns Produced by Medical Conditions of the Venous System — 158

	Expiratory Mechanisms	160
	Expiratory Bloodstain Experiments	167
	Cast-Off Mechanisms	169

9   Altered Bloodstain Patterns   179

Introduction	179
Diffusion and Capillarity of Bloodstains	179
Aging of Blood	180
Drying of Blood	182
Clotting of Blood	188
Diluted Bloodstains	194
Removal of Bloodstains	196
Effects of Fire and Soot on Bloodstains	200
Fly and Other Insect Activity	206
Void Patterns	210
Sequencing of Bloodstains	212
Conclusion	216

10   Determination of the Area of Convergence and Area of Origin of Bloodstain Patterns   217

Introduction	217
A Pattern	217
Area of Convergence	217
Area of Origin	219
Basic Trigonometry of a Right Triangle	220
Angle of Impact	221
Measuring Bloodstains	223
Selection of Stains	224
Methods of Area of Origin Determination	225
Documentation	226
Alternative Methods	233
Trigonometric Method	233
Graphic Method	234
"Using" Surfaces	234

11   Directional Analysis of Bloodstain Patterns with a Computer   241

Introduction	241
Some Basic Theory	241
Locating the Blood Sources	242
An Example of Virtual Paths Compared to Actual Flight Paths	243
The Two Impact Angles (α and γ) Depend on the Shape of the Bloodstain	244
Using Ellipses to Analyze Bloodstains	246

	More about Ellipses	248
	Laboratory Study of a Double Blow with the BackTrack Software	249
	Summary	253
	A Case Study by Detective Craig C. Moore of the Niagara Regional Police Service, Ontario, Canada	254
	Two Case Studies by Sgt. Mike Illes of the Ontario Provincial Police, Peterborough Detachment	257
	Case #1 Multiple impacts in a limited space	257
	Case #2 Expired bloodstaining	259
	Credits	261
12	**Documentation and Examination of Bloodstain Evidence**	**263**
	Introduction	263
	Methods of Documenting Bloodstain Patterns	264
	Notes	264
	Sketches and Diagrams	265
	Photographs	267
	Basic Photographic Equipment List	274
	Video	274
	Documentation of the Crime Scene	275
	Indoor Crime Scenes	275
	Outdoor Crime Scenes	281
	Documentation and Examination of Motor Vehicles	282
	Documentation and Examination of Bloodstains in the Laboratory	282
	Examination of Bloodstained Clothing	283
	Collection and Preservation of Bloodstain Evidence	294
13	**Evaluation of Bloodstain Patterns at the Scene**	**297**
	Introduction	297
	Processing Scenes	297
	General Processing for Bloodstain Pattern Analysis	300
	Processing for Data	304
	Doors and Locks	305
	Refrigerator	306
	Knife Blocks	307
	Medicines/Medicine Cabinets	309
	Sinks and Faucets	310
	Trash Cans	311
	Temperature	314
	Lighting	315
	Transient Evidence	316
	Sequencing	316

Vehicles as Crime Scenes	319
Negative Checklist for Vehicles	323
Collisions	323
Seatbelts	323
Airbags	323
Effects of Motion and Impact	327
Processing the Body for Bloodstain Patterns	329
Documentation of Bodies at the Scene	329
Use of Scales	330
Body Removal	330
Alterations in Transport	333
Mouth and Nose of the Decedent	334
Feet or Shoes of the Decedent	334
Nonambulatory Movement	336
Posture at Time of Injury	336
Claims of Self-Defense	339
Time Lapse	339
Sequencing Injuries	339
Victim Viability	341
Manner of Death	342
Sample Collection	343
Suspect Clothing	343
Separating Contributors	344
Linking an Accused	344

14 Presumptive Testing and Species Determination of Blood and Bloodstains   349

Introduction	349
The Analysis of Blood in Forensic Serology	350
Presumptive Tests for Blood	351
Catalytic Color Tests	351
Benzidine (Adler Test)	352
Phenolphthalein (Kastle-Meyer Test)	353
o-Tolidine	353
Leucomalachite Green	354
Tetramethylbenzidine	355
Comparing the Tests	356
Tests Using Chemiluminescence and Fluorescence	357
Luminol	358
Fluorescein	360
Other Methods	361
Confirmatory Tests for Blood	361
Teichmann Test	362
Takayama Test	363
Species Origin Determination for Blood	364
Serum Protein Analysis	364
General Methods	364

Ring Precipitin Test	365
Ouchterlony Double Diffusion Test	365
Crossed-Over Electrophoresis	366
Nonserum Protein Analysis	367
Antihuman Hemoglobin	367
Other Techniques	368

15 The Detection of Blood Using Luminol — 369

Introduction	369
Discovery	369
The Luminol Reaction	371
Preparation of Luminol	374
Interpretation of Luminescence	376
Use of Additional Presumptive Tests	377
Interpretation of Patterns	377
Effect of Substrates	378
Collection of Samples	378
Effects of Luminol on the Subsequent Analysis of Bloodstains	379
Luminol Photography	380
Report Writing	383
Testimony	384
Case Studies	384
Case 1	384
Case 2	386
Case 3	387

16 Chemical Enhancement of Latent Bloodstain Impressions — 391

Introduction	391
Preliminary Procedures	392
Fixing Blood Impressions	392
Fixative with 2% Aqueous Solution of 5-Sulphosalicylic Acid	393
Preparation of the Fixing Solution 2% Aqueous 5-Sulphosalicylic Acid	393
Fixing Impressions with Methanol	394
General Methods for Application of the Fixing Solutions	394
Method 1	394
Method 2	394
Method 3	395
General Methods for the Application of the Chemical Dye Reagents	395
Method 1	395
Method 2	395
Method 3	395
Rinsing Procedure	395
Preparation of the Rinse Solution for Water-Based Chemical Dyes	395
Application of the Rinse Solution for Methanol-Based Chemical Dyes	396

Preparation of the First and Second Rinse Solutions for Methanol-Based Chemical Dyes	396
Rinse Solution 1	396
Rinse Solution 2	396
General Methods for the Application of the Rinse Solutions	396
Method 1	396
Method 2	397
Method 3	397
Caveats	397
Amido Black Water-Based	398
Advantages	398
Disadvantages	398
Preparation of Amido Black Water-Based Without Fixative	400
Preparation of Amido Black Water-Based with Fixative	400
Method of Use	401
Caveats	402
Amido Black Methanol-Based	402
Advantages	402
Disadvantages	402
Preparation of Methanol-Based Amido Black	403
Method of Use	403
Amido Black Water-Ethanol Based	403
Preparation of Amido Black Water-Ethanol Based	403
Preparation of the Rinse Solution for Amido Black Water-Ethanol Based	404
Method of Use	404
Caveats	404
Hungarian Red Water-Based	405
Advantages	406
Disadvantages	406
Method of Use	406
Caveats	406
Aqueous Leucocrystal Violet	408
Advantages	408
Disadvantages	409
Preparation of ALCV	409
Method of Use	409
Caveats	409
Coomassie Brilliant Blue	412
Advantages	412
Disadvantages	412
Preparation of Coomassie Brilliant Blue Methanol Based	412
Coomassie Brilliant Blue Water-Methanol Based	413
Preparation of Coomassie Brilliant Blue Water-Methanol Based	413
Preparation of the Rinse Solution for Water-Methanol Based Coomassie Brilliant Blue	414
Method of Use	414
Caveats	414

Crowle's Double Stain	414
Advantages	415
Disadvantage	415
Preparation of Crowle's Double Stain	415
Method of Use	415
Caveats	416
Titanium Dioxide Suspension	416
Advantages	417
Disadvantages	417
Preparation of Titanium Dioxide Suspension	417
Method of Use	417
Caveats	419
General Notes of Precaution	419

17 Approaching the Bloodstain Pattern Case   421

Introduction	421
Sequence for Approaching Bloodstain Cases	422
Obtain a Brief Background of the Case	422
Examination of the Crime Scene	423
Initial Review of Crime Scene and Evidence Photographs	424
Examination of Autopsy or Hospital Photographs	424
Examination of Autopsy Protocol or Hospital Records	425
Examination of Crime Scene Sketches/Diagram(s)	425
Examination of Crime Scene and Evidence Photographs	426
Examination of Physical Evidence	427
Sequencing Evidence Examinations	428
Serology Reports	429
Crime Scene Investigator's Notes and Reports	429
Responding Officer's Notes	430
Emergency Medical Technicians' Notes or Depositions	430
Additional Forensic Reports	430
Synthesizing the Data	431
Data Synthesis for Bloodstain Pattern (A)	431
Formulation and Testing of Hypotheses	431
Data Synthesis for Bloodstain Pattern (A)	431
Hypothesis for Bloodstain Pattern (A)	431
Testing of Hypothesis of Bloodstain Pattern (A)	431
Formulation of a Scenario	432
Factors to Consider When Analyzing a Bloodstain Pattern Case	432

18 Report Writing   433

Introduction	433
Scene-Specific Reports	434
Evidence-Specific Laboratory Reports	434
Final Report	434

Report Content	434
Report Phrases	435
Introduction of the Report	436
Sample Introduction	436
Body of the Report	437
Crime Scene	438
Physical Evidence	438
Special Issues	438
Conclusion(s) of the Report	440
The Finished Report	441

19 Legal and Ethical Aspects of Bloodstain Pattern Evidence   443

Introduction	443
Legal Issues	443
Admissibility	443
Weight of the Evidence	449
Qualifications of Experts	450
Burden of Proof	454
Chain of Custody	454
Discovery	455
Direct and Cross-Examination of the Expert	459
Attorney's Goals and Methods of Direct Examination	459
Demonstrative Evidence	460
Attorneys' Goals and Methods of Cross-Examination	462
Advice to the Expert — Guidelines for Deposition and Trial Testimony	464
Ethical Issues	466
Expert's Ethics	466
Attorneys' Ethics in Dealing with Experts	467
Expert Witness Malpractice	470
Conclusion	478

20 Bloodstain Pattern Analysis: Postconviction and Appellate Application   479

Introduction	479
A Bit of History	479
The Postconviction Attorney's Role—The Daunting Task	483
The Appeal	483
Admissibility	484
Expert Qualifications	484
The Collateral Attack	486
Perjurious Expert Testimony—Newly Discovered Evidence	487
Prosecutorial Misconduct—Denial of Due Process	487

Ineffective Assistance of Counsel	489
Conclusion	496

References   497

Bloodstain Pattern Analysis	497
Medical and Medical–Legal Aspects of Bloodshed	499
Presumptive Testing and Species Determination of Blood and Bloodstains	499
General References	499
Specific References	500
Luminol	502
Chemical Enhancement of Latent Blood Impressions	504

Appendix A: Trigonometric Tables—Sine and Tangent Functions   505

Appendix B: Metric Measurements and Equivalents   507

Appendix C: Scene and Laboratory Checklists   511

Appendix D: Biohazard Safety Precautions   517

Universal Precautions Policy	517
Human Immunodeficiency Virus	518
Hepatitis	518
Basic Rules for Biohazard Safety	519
Minimizing Biohazard Contamination at the Crime Scene	519
Guidelines for the Selection of the Proper Personal Protective Equipment	520
Precautions for Evidence Collection	520
Disposables	520
Personal Protective Equipment Removal	521
Selection of Personal Protective Equipment	521
References	522

Appendix E: Court Decisions Relating to Bloodstain Pattern Analysis   523

Appendix F: Court Decisions Relating to Presumptive Blood Tests   527

Presumptive Testing Cases	527
Luminol Cases	527
Benzidine Cases	528
Orthotolidine/Ortho-Tolidine/o-Tolidine Cases	529

　　　　Leucomalachite/Leuco-Malachite Green Cases　　　　529
　　　　Hemastix® Case　　　　529
　　　　Phenolphthalein Cases　　　　530

Glossary of Terminology　　　　531

Index　　　　533

Introduction to Bloodstain Pattern Analysis 1

Introduction

Blood is one of the most significant and frequently encountered types of physical evidence associated with the forensic investigation of death and violent crime. The identification and individualization of human bloodstains have progressed over the past 100 years since the ABO grouping system was discovered by Landsteiner in 1901. The techniques for the individualization of human blood in forensic science relied on the ABO system for many years. The development of the characterization of the red cell isoenzymes and serum genetic markers in the late 1970s dramatically increased the individualization of human blood. The work of Sir Alec Jeffreys in the development of DNA profiling in 1985 was a milestone in forensic science. Since then the techniques of DNA analysis in forensic cases has rapidly evolved through PCR (polymerase chain reaction) and STR (short tandem repeat) techniques and afforded the forensic scientist a powerful tool for the individualization of human blood. Bloodstains collected from a scene of violent death where bloodshed has occurred and blood samples collected from clothing of the victim and the accused can now provide a link between an assailant and a victim to a high degree of scientific certainty.

The identification and individualization of human blood is cojoined with the discipline of bloodstain pattern analysis (BPA). BPA focuses on the analysis of the size, shape, and distribution of bloodstains resulting from bloodshed events as a means of determining the types of activities and mechanisms that produced them. This information coupled with DNA individualization and wound interpretation from the autopsy examination of the victim by the forensic pathologist provides a basis for the reconstruction of the bloodshed events. The scientific analysis of bloodstain pattern evidence has proved crucial in numerous cases where the manner of death is questioned and the issue of homicide, suicide, accident, or natural death must be resolved in a criminal or civil litigation or proceeding.

Objectives of Bloodstain Pattern Analysis

BPA is a discipline that uses the fields of biology, physics, and mathematics. BPA may be accomplished by direct scene evaluation and/or careful study of scene photographs (preferably color photographs with measuring device in view) in conjunction with detailed examination of clothing, weapons, and other objects regarded as physical evidence. Details of hospital records, postmortem examination, and autopsy photographs also provide useful information and should be included for evaluation and study. In cases where a scene investigation is not possible and photographs must be relied on, detailed sketches, diagrams, reports of crime scene investigators, and laboratory reports should be available for review.

Relative to the reconstruction of a crime scene, BPA may provide information to the investigator in many areas.

- Areas of convergence and origin of the bloodstains
- Type and direction of impact that produced bloodstains or spatter
- Mechanisms by which spatter patterns were produced
- Assistance with the understanding of how bloodstains were deposited onto items of evidence
- Possible position of victim, assailant, or objects at the scene during bloodshed
- Possible movement and direction of victim, assailant, or objects at the scene after bloodshed
- Support or contradiction of statements given by accused and/or witnesses
- Additional criteria for estimation of postmortem interval
- Correlation with other laboratory and pathology findings relevant to the investigation

The goal of the reconstruction of the crime scene using BPA is to assist the overall forensic investigation with the ultimate questions that must be addressed, which include, but are not limited to, the following:

- What event(s) occurred?
- Where did the event(s) occur?
- When and in what sequence did they occur?
- Who was there during each event?
- Who was not there during each event?
- What did not occur?

Scientific Approach to Bloodstain Pattern Analysis

The approach to BPA must adhere to the scientific method and rely on the principles of biology, physics, and mathematics. Education in these areas is highly recommended. A combination of training through formal instruction, personal experimentation, and experience with actual casework is necessary before an individual acquires adequate proficiency in the analysis of bloodstain patterns to answer these types of questions. Contemporaneous experiments to duplicate specific patterns should be considered relative to a given case to support an analysis or conclusion. Some conservative speculation is permissible during the initial investigative stages of a case. However, final opinions, contents of a written report, and ultimate court testimony must be based on scientific fact with *no speculation*. All potential explanations should be explored thoroughly and recognized and acknowledged by the analyst. Analysis of bloodstains should be correlated with postmortem and laboratory findings in an investigation. For example, when an arterial spurt pattern is observed, the autopsy report should indicate a cut or breached artery in the victim. In those cases where an assailant as well as the victim produces bloodshed or where there are multiple victims, the individualization of the bloodstains by the forensic laboratory is critical. It is important to stay within the realm of that which can be proven scientifically and not to overinterpret bloodstain evidence. This axiom particularly applies when the number of bloodstains is limited because a single or few small bloodstains do not often lend themselves to useful valid analysis. Conclusions based on crime scene photographs should be conservative when the investigator has not had the opportunity to examine the crime scene personally and must rely on the photographic documentation of others.

Historical Development

The study of bloodstain patterns and the consideration of the physical processes in which the distribution of these patterns can reconstruct details of activities at scenes of death and violent crime have recently emerged as a recognized forensic skill. Historically, bloodstain analysis has suffered through a long period of neglect, and as a result investigators in death cases frequently have not appreciated the very obvious information available from this forensic tool. Significant contributions have been made in this discipline as documented in the following timeline.

1895 Dr. Eduard Piotrowski   The earliest known significant study in bloodstain interpretation that has been documented and preserved was done by Dr. Eduard Piotrowski, assistant at the Institute for Forensic Medicine in Krakow, Poland. His work, entitled *Uber Entstehung, Form, Richtung und Ausbreitung der Blutspuren nach Hiebwunden des Kopfes*, was published in Vienna in 1895 (Figure 1.1). Piotrowski recognized that "It is of the highest importance to the field of forensic medicine to give the fullest attention to bloodstains found at the

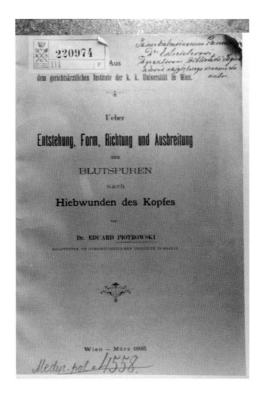

Figure 1.1   The 1895 Eduard Piotrowski published study, *Concerning Origin, Shape, Direction and Distribution of the Bloodstains following Head Wounds Caused by Blows.*

scene of a crime because they can throw light on a murder and provide an explanation for the essential moments of the incident." Through the efforts of Herbert Leon MacDonell of Corning, New York, the Historian for the International Association of Bloodstain Pattern Analysts, this work has been translated from the German text and reprinted in German and English as *Concerning Origin, Shape, Direction and Distribution of the Bloodstains following Head Wounds Caused by Blows*. This work is completely reproduced with color plates of the extensive bloodstain experiments performed by Dr. Piotrowski. According to MacDonell, "No one preceded Piotrowski in designing meaningful scientific experiments to show blood dynamics with such imagination, methodology and thoroughness. He had an excellent knowledge of the scientific method and a good understanding of its practical application to bloodstain pattern interpretation."

1900 Dr. Paul Jeserich Subsequent significant work involving the study of bloodstain patterns at a crime scene is documented by Dr. Paul Jeserich, a forensic chemist in Berlin who examined homicide scenes during the first decade of the 20th century.

1939 Dr. Victor Balthazard The French scientist Dr. Victor Balthazard and his associates conducted original research and experimentation with bloodstain trajectories and patterns and presented a paper at the 22nd Congress of Forensic Medicine entitled *Etude Des Gouttes De Sang Projete*. This was translated from French to English as *Research on Blood Spatter*.

1955 Dr. Paul Kirk Dr. Paul Kirk of the University of California at Berkeley prepared an affidavit regarding his findings based on bloodstain evidence to the Court of Common Pleas in the case of the *State of Ohio vs. Samuel Sheppard*. This was a significant milestone in the recognition of bloodstain evidence by the legal system. Dr. Kirk was able to establish the relative position of the attacker and victim at the time of the administration of the beating.

1971 Herbert Leon MacDonell The further growth of interest and use of the significance of bloodstain evidence is a direct result of the scientific research and practical applications of bloodstain theory by Herbert Leon MacDonell of Corning, New York (Figure 1.2). Through the assistance of a Law Enforcement Assistance Administration (LEAA) grant, MacDonell conducted research and performed experiments to re-create and duplicate bloodstain patterns observed at crime scenes. This resulted in his publication of the first modern treatise on bloodstain analysis, entitled *Flight Characteristics and Stain Patterns of Human Blood* (Figure 1.3).

1973 First Formal Bloodstain Training Course Given MacDonell established a training program for basic bloodstain pattern interpretation and conducted his first Bloodstain Institute in Jackson, Mississippi in 1973. His second publication, *Laboratory Manual on the Geometric Interpretation of Human Bloodstain Evidence*, was used by students. Since that time he and others have conducted numerous basic and advanced bloodstain analysis courses throughout the United States and abroad and trained hundreds of police and crime scene investigators, forensic scientists, and crime laboratory personnel.

1982 *Bloodstain Pattern Interpretation* Published MacDonell expanded his original work in a publication entitled *Bloodstain Pattern Interpretation*.

1983 Formation of the IABPA MacDonell conducted the first advanced class for BPA, and the participants organized the International Association of Bloodstain Pattern Analysts (Figure 1.4).

Figure 1.2 Professor Herbert Leon MacDonell.

Figure 1.3 Copy of *Flight Characteristics and Stain Patterns of Human Blood*, written by Professor Herbert Leon MacDonell, that was published in 1971.

Figure 1.4 The International Association of Bloodstain Pattern Analysts logo.

Experiments and Practical Exercises in Bloodstain Pattern Analysis **Published** Terry L. Laber and Barton P. Epstein cowrote a laboratory manual entitled *Experiments and Practical Exercises in Bloodstain Pattern Analysis* for use in their basic bloodstain analysis courses.

1989 *Interpretation of Bloodstain Evidence at Crime Scenes* **Published** The first edition of the book *Interpretation of Bloodstain Evidence at Crime Scenes*, written by Dr. William G. Eckert and Stuart H. James, was published. This work included numerous case studies involving BPA.

1990 *Bloodstain Pattern Analysis—Theory and Practice* **Published** Ross Gardner and Tom Bevel cowrote a laboratory manual entitled *Bloodstain Pattern Analysis—Theory and Practice.*

1993 *Bloodstain Pattern Analysis in Violent Crimes* **Published** T. Paulette Sutton at the University of Tennessee in Memphis produced a comprehensive manual entitled *Bloodstain Pattern Analysis in Violent Crimes.*

1993 *Bloodstain Pattern Interpretation* **Updated to** *Bloodstain Patterns* MacDonell updated his original bloodstain pattern manual *Bloodstain Pattern Interpretation* and renamed it *Bloodstain Patterns.*

1997 *Bloodstain Pattern Analysis with an Introduction to Crime Scene Reconstruction* **Published** Tom Bevel and Ross M. Gardner cowrote the text entitled *Bloodstain Pattern Analysis with an Introduction to Crime Scene Reconstruction.*

1998 *Scientific and Legal Applications of Bloodstain Pattern Analysis* **Published** Edited by Stuart H. James, *Scientific and Legal Applications of Bloodstain Pattern Analysis* was a compilation of chapters concerning bloodstain analysis and legal issues written by Dr. Alfred Carter, William Fischer, Carol Henderson, Paul Kish, Marie Saccoccio, and T. Paulette Sutton.

2001 *Blood Dynamics* **Published** Anita Wonder wrote the text entitled *Blood Dynamics* in 2001.

2004 Herbert Leon MacDonell Received Honorary Degree MacDonell received the honorary degree of Doctor of Science from the University of Rhode Island in recognition of his contributions to forensic science and BPA.

As a direct result of MacDonell's efforts, the evolution of bloodstain analysis in forensic science increased rapidly and others have made considerable contributions to the field in crime scene reconstruction, teaching, research, and publications.

Scientific articles pertaining to aspects of BPA are being seen more frequently in the scientific literature in well-known publications including the *Journal of Forensic Sciences, Forensic Science International,* the *American Journal of Forensic Medicine and Pathology,* the *Journal of the Canadian Society of Forensic Science,* and the *Journal of Forensic Identification.*

International Association of Bloodstain Pattern Analysts

As of 2004, the IABPA organization consisted of more than 750 members from throughout the United States and Canada as well as countries throughout the world including Great Britain, Denmark, Finland, Sweden, Norway, the Netherlands, New Zealand, Australia, Taiwan, Guam, and Columbia. The association publishes the *IABPA News*, which is devoted to current bloodstain topics, providing a schedule of training courses and exploring such issues as the curriculum for basic instructional courses in bloodstain interpretation, uniformity in bloodstain terminology, and research in the field. The annual IABPA conference agenda includes numerous case presentations and research topics by members and guest lecturers.

Classification of Bloodstains

The conventional method of classifying bloodstains was based on the correlation between the velocity of the force influencing the blood source or drop that governed the characteristics and size or diameters of the resulting bloodstains. Three basic categories of stain groups were used based on the concept that the size of the bloodstain being inversely proportional to the force applied to the static blood.

Low-Velocity Impact Blood Spatter

Low-velocity impact spatters (LVIS) are bloodstains created when the source of blood is subjected to a force with a velocity up to 5 ft/sec. Primary stains measure generally 4 mm in diameter or greater.

Medium-Velocity Impact Blood Spatter

Medium-velocity impact spatters (MVIS) are bloodstains created when the source of blood is subjected to a force with a velocity in the range of 5 to 25 ft/sec. The diameters of the resulting stains are in the size range of 1 to 3 mm, although smaller and larger stains may be present. Stains in this category were usually associated with beatings and stabbings.

High-Velocity Impact Blood Spatter

High-velocity impact spatters (HVIS) are bloodstains created when the source of blood is subjected to a force with a velocity of greater than 100 ft/sec. The diameters of the spatters are predominately less than 1 mm, although smaller and larger stains are often observed within the pattern. Stains in this category were usually associated with gunshot injuries. Other mechanisms that produced stains within the size range of the conventional medium- and high-velocity categories such as satellite spatter and expiratory bloodstains were not appreciated to the extent that misinterpretations could and did occur.

Many bloodstain analysts have chosen to discontinue this conventional terminology and classification for a more holistic approach to bloodstain classification. The issues that kindled the rethinking of the conventional classification of low-medium-high velocity were the overlapping of stain sizes between the medium- and high-velocity categories and the realization that mechanisms other than beatings, stabbings, and gunshots frequently produced stains with size ranges within these categories. The bloodstains and patterns are classified based on their physical features of size, shape, location, concentration, and distribution into *passive stains, spatter stains,* or *altered stains.* They are further classified relative to mechanisms that may produce stains with those characteristics with reference to pertinent scene, medical and case related facts, and history of the evidence. The analyst may then be able to establish the specific mechanism(s) by which the pattern was created.

Acceptance within the Scientific Community

The reliability of scientific evidence must be addressed within the forensic science community including the discipline of BPA. The courts make a preliminary determination of relevance and admissibility of scientific evidence as to whether it is relevant, competent, and properly applied to the facts of the case. So-called "pseudoscience" has no place in the judicial process. This is referred to as the "gatekeeping" function of the court. Some states apply the Frye standard, although many apply the Daubert standard. The interpretation of a Frye inquiry is to determine whether or not the procedure, evidence, or proposed testimony is relevant and has gained general acceptance within the scientific community. Daubert substitutes a reliability test for a relevancy test. Scientific knowledge must be derived from the scientific method supported by validation of expert testimony establishing a standard of evidentiary reliability. The factors for Daubert are outlined as follows:

- Has the scientific theory or technique been empirically tested?
- Has the scientific theory or technique been subjected to peer review and publication?
- What is the known or potential error rate?
- What are the expert's qualifications and stature in the scientific community?
- Can the technique and its results be explained with sufficient clarity and simplicity so that the court and the jury can understand its plain meaning?

Scientific Working Group for Bloodstain Pattern Analysis

The Federal Bureau of Investigation (FBI) Laboratory has actively organized and funded Scientific Working Groups (SWG) to establish professional forums composed of experts from local, state, federal, and international agencies as well as academic scientists and private practitioners to address issues within specific forensic disciplines. The Scientific Working Group on Bloodstain Pattern Analysis (SWGSTAIN) was established, and the first meeting was held at the FBI Academy in Quantico, Virginia in March 2002. Recognized bloodstain pattern analysts and practitioners in related fields from North America and Europe convened to discuss and evaluate methods, techniques, protocols, quality assurance, education, and research relating to BPA. The group meets biannually. Subcommittees were established in the following areas:

- **Education and Training.** This subcommittee addresses educational standards for students and instructors as well as standards for continuing education.
- **Legal.** This subcommittee addresses issues of scientific acceptance, peer review, publications, error rates, and existence of standards.
- **Quality Assurance.** This subcommittee addresses issues of report writing, proficiency testing, methodology, and standard operating procedures.
- **Research.** This subcommittee addresses issues of development of new methodologies, standards, resources, references, and validation.
- **Taxonomy and Terminology.** This subcommittee addresses taxonomy, glossary, and definitions.

Education and Training in Bloodstain Pattern Analysis

Bloodstain analysts represent a range of forensic scientists and crime scene investigators with diverse levels of education. The courts have accepted testimony from individuals with strong backgrounds in chemistry, biology, and physics. Many of these individuals possess degrees in science and/or forensic medicine. Many of these individuals are employed in crime laboratories or medical examiner offices that have crime scene responsibilities. Crime scene investigators, evidence technicians, and detectives who do not necessarily possess scientific backgrounds have also offered expert testimony. However, college-level courses in geometry, trigonometry, and basic physics are valuable assets for the understanding of BPA.

It is highly recommended that individuals enroll in a basic 40-hr course in BPA. These courses are usually taught over a 5-day period and occasionally as a semester course at a college or university by qualified instructors at numerous locations in the United States and abroad. They provide instruction in the theory and practical aspects of BPA, case presentations, and opportunities to perform laboratory experiments. Participation in laboratory experiments is crucial for understanding the dynamics of bloodstain pattern production and the mechanisms involved. Students create stain patterns similar to those present at crime scenes using various types of apparatuses. Patterns are created on cardboard and other surfaces that may be preserved and retained for future reference.

Advanced courses are also available following the successful completion of a basic course. The advanced courses are designed to review basic concepts and provide additional training in areas not explored in the basic courses. For example, an advanced course may concentrate on computer analysis; digital imaging; examination of bloodstained clothing and footwear; mock crime scenes; and, in some cases, mock trials. Students are encouraged to present cases to the group for peer review sessions. Successful completion of basic and advanced courses in BPA does not imply that an individual is a qualified bloodstain analyst. The formal education must be coupled with years of experience with crime scenes and evidence examinations along with regular attendance at scientific seminars or conferences. It is also important to stay abreast of the information in scientific journals and periodicals.

Membership in professional organizations is encouraged. Students who have successfully completed a basic course in BPA are qualified to apply for membership in the IABPA.

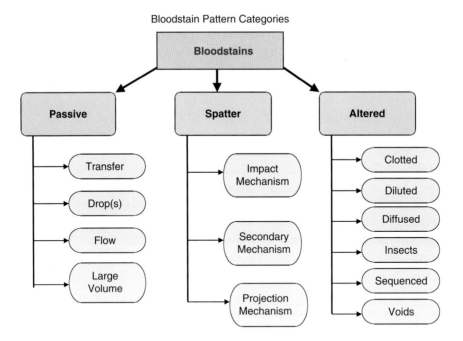

Figure 1.5 Bloodstain pattern categories.

Conclusion

The taxonomic approach to the classification of bloodstains and patterns is not entirely new. It has been discussed among bloodstain pattern analysts for a long time and has been described to varying extents in several texts published within the past ten years. The hierarchy of bloodstain categories that are discussed in this text combines the geometric characterization of the bloodstains with the events that caused the bloodstains (Figure 1.5). The objective is to strengthen the scientific process of BPA in actual casework through the ultimate test in the legal process and the courts. The variable level taxonomic classification is not complex but addresses the issue of the mechanisms that produce overlapping patterns with respect to the size of the stains. The structure of the categories will permit the analyst to begin with the primary bloodstain pattern categories—namely, passive, spatter, and altered—and proceed to specific subcategories based on case-driven facts.

Medical and Anatomical Aspects of Bloodshed

2

RONALD K. WRIGHT

Introduction

Man has been fascinated with blood since before antiquity. An appreciation that blood had importance in maintaining life has existed since recorded time. Most likely this appreciation predated the introduction of literacy. Certainly Galen described blood as one of the four humors of life in his writings in 164 A.D. He discovered that arteries contained blood, not an airlike substance called pneuma as had been previously believed. Biblical accounts are numerous and can be found in Genesis 4:10. The Qur'an describes the importance of blood in Surah 2:30. Accounts of ritual bloodletting associated with both food preparation and ritualistic killing of animals and humans can be found in the accounts of most civilizations.

The Circulation of Blood

In modern times, Harvey described the circulation of blood in 1628. He noted that the heart, a muscle with four hollow, blood-filled chambers and four valves, was responsible for the pumping of blood through the two large arteries leading from the heart and then to smaller vessels in the body as a whole and in the lungs. These smaller arteries in the body are called arterioles.

The pressure of blood in arteries is determined by the interaction of the pressure produced when the heart contracts and then the status of the amount of dilatation of the arterioles. Blood pressure is expressed in units of millimeters of mercury (mm Hg). This refers to the height of mercury in a tube when an occlusive cuff is inflated with air, which is required to stop blood from flowing (systolic), and the level at which no noise is heard from turbulent flow when the pressure is slowly reduced (diastolic). These two numbers provide a person's blood pressure: systolic is the pressure during the contraction of the heart (systole of the heart), and diastolic is the pressure when the heart is relaxed (diastole of the heart). Of necessity, the systolic pressure must be higher than the diastolic pressure. The normal blood pressure for humans is 115/70. The fluctuation in pressure from systolic to diastolic, each time the heart beats, produces a pulse or pulsatile pressure. Pressures with a systolic value of greater than 160 are described as systolic hypertension (high blood pressure). A diastolic pressure of greater than 90 is described as diastolic hypertension or, more generally, as hypertension or high blood pressure. Pressures below 100 systolic and/or below 50 diastolic are considered hypotensive. Hypotension is also called shock.

In the circulation of blood, blood travels through the arteries and arterioles and then into microscopic vessels called capillaries. The arterial walls are thick and contain smooth muscle, connective tissue, and elastic fibers so as to tolerate the pressure of the blood coursing through them. In the capillaries, the circulation is no longer driven by the high pressure of the heart. The heart remains the driving force, but the high pressure and the pulsatile nature of the flow stops. Figure 2.1 shows the arterial system of the human body. Following the flow through the capillaries, the blood flows to venules (small veins), then to large veins, and then back to the heart. The propulsive force for this part of circulation in small part comes from the pumping of the heart but in large part by a combination of gravity and the contraction of skeletal muscles through which the venules and veins run. Indeed, in humans the veins in the legs contain bicuspid valves, which act as a pump to direct blood back to the heart and prevent backflow and stasis in the extremities. The pressure of blood in the veins is determined primarily by the effect of gravity working on the column of fluid in the venous system.

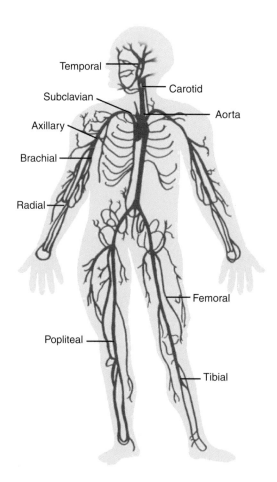

Figure 2.1 Schematic representation of the human arterial system. (Courtesy of Jim Wallace, University of Tennessee, Memphis.)

The venous system possesses thinner and less muscular walls with less elasticity as compared with the walls of the arteries because the flow is under reduced pressure. Central venous pressure is the pressure of the large central veins at the entrance to the heart (inferior and superior vena cava), is measured in centimeters of water (cm H_2O), and is normally less than 10. Ten cm H_2O of pressure is equal to 7.4 mm Hg pressure. As can be seen, the venous pressure is a small fraction of arterial pressure.

Rate of Bleeding

The rate of bleeding — that is, the volume of blood loss from inside the circulation system per unit of time, whether external or internal or both — is determined by the type of vessel that is bleeding (arterial, venous, or capillary), the diameter of the vessel, and the pressure of the blood in the vessel. Also of great importance in determining the rate of bleeding is the characteristic of the lesion in the vessel. As previously described, arteries have smooth muscle making up part of the blood vessel wall. Because of this, a completely transected artery will bleed less than an artery that is partially transected. With complete transection, the vessel retracts because of the contraction of the muscle wall, and this partially cuts off the flow of blood. With partial transection, the vessel cannot retract; therefore there is little reduction in the size of the hole and rate of bleeding.

Physical trauma or damage to arteries produces significant and frequently fatal consequences because blood is released rapidly under pressure. The mechanism of trauma is frequently associated with cutting and stabbing injuries but may also occur in blunt force and gunshot injuries. The location of the breached artery will dictate whether excessive internal bleeding or external bloodshed will occur. Severe internal bleeding within the abdominal cavity occurs with damage to the aorta, and external bleeding may not be significant. Damaged arteries that are closer to the surface of the body release blood under pressure to the external environment resulting in gushing or spurting. In either case, the blood will continue to surge from the damaged artery as long as the heart continues to pump and there is available blood. Medical intervention is critical for survival.

Some common arteries close to the surface are as follows:

Temporal	Temples
Carotid	Neck
Subclavian	Under clavicles
Axillary	Axilla
Brachial	Arms
Radial	Wrists
Femoral	Upper thighs
Popliteal	Knees
Tibial	Ankles

Blood Volume

Blood pressure must be kept above 100/40 mm Hg to maintain blood supply to the brain and the other tissues of the body. The brain is the most sensitive organ of the body to the reduction in blood supply. Circulation of blood must be continuous to provide oxygen and nutrients to the tissues and to take away carbon dioxide produced by the metabolism of the tissue. To maintain blood pressure and thus circulation, the blood volume must be maintained at about 7% of the mass of the body. Thus for a person weighing 150 lb, he or she will have about 10.5 lb of blood. (0.07 × 150 = 10.5). Likewise, a 100-lb person has 7 lb of blood and a 200-lb person has 14 lb of blood. Blood is generally measured in metric units. Thus if a person weighs 150 lb that is equivalent to 68 kg (2.2 kg/lb as the conversion factor) the blood volume will be 4.8 kg (7% of 68 kg). Blood has a volume of approximately 1 L to 1 kg of mass. Thus, 4.8 kg of blood is approximately 4.8 L of blood. Blood for transfusions is donated and used in 500 mL (0.5 L) amounts; these are called units. Thus a person weighing 150 lb has 9.6 units of blood. A person weighing 200 lb has 12.7 units. A person weighing 100 lb has about 7 units of blood. This drop in volume of blood with weight is why, generally, a person weighing less than 100 lb is not allowed to donate blood.

The control of blood volume loss is maintained by thirst. Water replaces the volume or serves as a volume expander. The other materials such as protein, salts, sugar, and cellular elements take longer to replace, but blood pressure is maintained if the volume is maintained. Excessive volume is eliminated by the kidneys with the production of increased amounts of urine.

Blood consists primarily of water, which explains why it has a weight of approximately 1 kg/L. However, it has a large amount of protein dissolved in the plasma calculated at approximately 7% (7 g of protein per 100 mL of blood). Other chemicals are found in lesser amounts. Suspended, not dissolved, in blood are red cells, white cells, and platelets. Figure 2.2 shows the separation of the plasma and cellular components of the blood in a collection tube. Red cells are composed primarily of hemoglobin and are red as a result of oxygenated hemoglobin. Hemoglobin is an iron-containing protein that carries oxygen from the lungs to the rest of the tissues of the body and assists in the return of carbon dioxide. Hemoglobin must be packaged within red cells because hemoglobin and especially its contained iron are poisonous to the body. Red cells are very specialized cells that, when mature, do not have nuclei and thus do not contain DNA. Normally there are about 4.5 million red cells in 1 mL of blood or a total of about 43 billion red cells in a person weighing 150 lb. The amount of hemoglobin contained in 100 mL of blood is 14 g; this is ordinarily all contained within red cells. Thus a 150-lb person has 672 g of hemoglobin in the blood, or about 1.5 lb. The concentration of suspended red cells is reduced in anemia as is the concentration of hemoglobin.

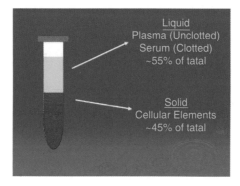

Figure 2.2 Schematic representation of the proportion of plasma/serum and cellular elements in human blood.

Birds, reptiles, and mammals all have high-pressure circulatory systems. These circulatory systems only work if there is blood under pressure in the arteries. Injury or disease that disrupts an artery or arteriole will result in bleeding and thus loss of blood volume from the vascular system. It makes no difference whether the bleeding is in the body, such as the chest or abdominal cavity, or whether the loss is outside the body. The importance is the loss of volume from the vascular system. If about 30% of blood volume is lost, the person will be in shock and will be able to function very little. Lying down is about all that can be done. For a 150-lb person, 30% of blood volume is 1.5 L or 3 units. If about 50% of blood volume is lost within an hour (i.e., 5 units or 2.5 L in a 150-lb person), the blood pressure in most young people will drop to levels so low that life cannot be maintained. With loss of blood, as the blood pressure drops there will be a slowing of the rate of loss because blood pressure drives the loss of blood.

Causes of Blood Volume Loss

In the broadest of senses, blood loss can be caused by disease or injury. This discussion shall be primarily directed to the external loss of blood, blood that can be seen at the scene of an injury or illness. However, internal bleeding such as hemothorax (bleeding into the chest cavity) or hemoperitoneum (bleeding into the abdominal cavity) can both result in exsanguination (bleeding to death) without any external loss of blood.

External bleeding provides patterns that allow the investigator to determine important issues of what, where, and how a bloodshed event occurred. To begin this analysis of patterns requires an understanding of the causes of bleeding.

First, not all bleeding is from injury, although most is. Thus, the first issue to address in an investigation is to make an effort to determine whether the bleeding is from injury or disease. In cases of death, the best way to make this determination is by autopsy examination of the body. However, autopsies are generally done tomorrow and answers are needed today. Decisions as to whether to call out the crime laboratory, specialized investigative units, and supervisory personnel need to be made based on the determination of whether the person was injured or suffered some natural disease. These decisions must be made before the autopsy is done and generally before accurate information can be obtained from the hospital if the person survived.

Blood Loss Resulting from Medical Conditions

Bleeding from disease generally does not require extensive investigation, and calling out the crime lab, specialized units, and supervisors for a disease process will generally be looked on as a waste of valuable resources. The following guidelines will assist in identifying bleeding caused by disease.

Bleeding from the Skin

Bleeding from the skin is from an injury unless it is from one of the following:

1. Ruptured varicose vein in the lower extremities. This condition is associated with venous insufficiency syndrome. Chronic venous insufficiency syndrome is a medical condition where blood pools in the veins of the lower legs and affects 2 to 5% of the population in Western countries. It is often seen in women older than age 60 years and in people who are alcoholics. The disease creates varicose veins and may cause

the feet and lower legs to become swollen, often accompanied by a dull ache that can worsen with prolonged standing and walking. As this disease progresses, the skin can darken in color and ulcers may occur including indurations of skin and even skin breakdown with overt ulceration as shown in Figure 2.3a.

Simply stated, veins return blood to the heart subsequent to the distribution of oxygenated blood. Veins possess one-way valves that enable the blood to flow upward through the venous system and prevent the downward backflow, as previously discussed. Varicosities develop when these valves become damaged, allowing blood to flow backward and pool, creating abnormally high pressure in the veins. As the pressure increases, capillaries rupture, creating a reddish-brown discoloration of the skin. Veins deteriorate and become engorged with blood, and the outer venous wall becomes thin and protrudes just under the skin. The blood-engorged, thin-walled vein is easily ruptured as the result of motion or slight contact such as a bump or scratch. Fatalities have been reported in medical literature. There will be only one break in the skin, it will be oblong or circular, no more than $1/4$ in. (0.5 cm) in diameter, and located on the lower leg or foot (Figure 2.3b).

2. Postmortem injury. Injury that occurs after death may result in significant blood loss from postmortem lacerations. Rarely, a person dying a sudden cardiac death will collapse and strike their head. Falling in this way can produce lacerations if the head strikes concrete, asphalt, or the edge of a wooden or plastic surface. If the deceased's laceration is below the level of any part of the rest of the body, external bleeding from postmortem laceration can occur, driven by the force of gravity with blood draining from the injury. Of course, autopsy examination should confirm the scene interpretations. If there is a small amount of blood present (less than 1 cup, or 250 mL) or there is no scalp hematoma or the deceased has a history of heart disease or is older than 50 if male or 55 if female, then it may be presumed that the bleeding is postmortem and therefore not from injury in life.

With the exception of these two phenomena, all bleeding from the skin should be considered to be from injury and thus require possible criminal investigation.

Bleeding from the Mouth and Nose

The cause of the bleeding may be from injury or from disease. The common diseases producing bleeding from the nose and mouth are as follows:

Cancer of the lung eroding into the pulmonary artery, the aorta, or a major pulmonary vein
Tuberculosis, when a tuberculoma (tumor produced by tuberculosis) erodes into the pulmonary artery or a major pulmonary vein
Aortic aneurysm that erodes into the lung
An arteriovenous malformation (birthmark) of the lung that ruptures
Ruptured varicose vein in the esophagus (esophageal varices)
Nasal cancer or nasal infection eroding into the veins of the nose or sinuses
Peptic ulcer

A history of these diseases in a person who dies with blood from their nose and mouth points toward disease-caused bleeding. It is worth noting, however, that a gunshot wound from a small-caliber weapon fired into the mouth will present as bleeding from the mouth

Medical and Anatomical Aspects of Bloodshed 17

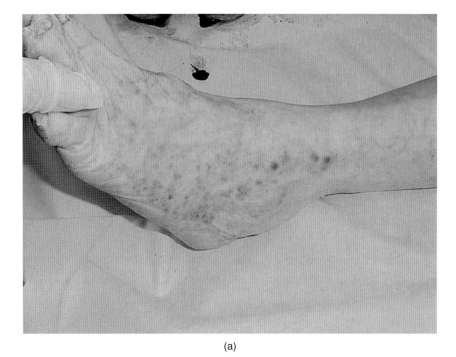
(a)

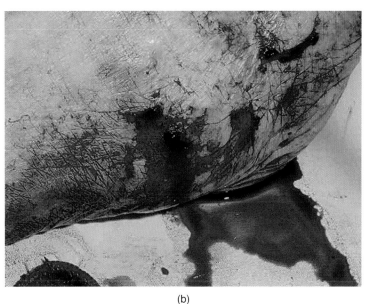
(b)

Figure 2.3 (a) Area of varicose veins in lower leg of elderly female. (b) Ulcerated varicose vein in lower leg of elderly female that caused fatal hemorrhage.

and nose and may be mistaken for disease-caused bleeding. As with all cases, autopsy confirmation should be obtained. Varicose veins in the esophagus referred to as esophageal varices only occur in severe cirrhosis of the liver, almost always from alcoholism. Cancer or tuberculosis is generally known to the deceased and thus to the witnesses to the death (Figures 2.4a and b). Bleeding from the stomach from cancer or ulcer or severe inflammation

(a)

(b)

Figure 2.4 (a) Victim, a known terminal lung cancer home hospice care patient, found outside his residence with a large quantity of blood on the ground and around his mouth. (b) The victim hemorrhaged extensively from his lungs into his toilet prior to exiting his residence. Note the presence of air bubbles within the blood.

Medical and Anatomical Aspects of Bloodshed 19

of the stomach lining (gastritis) almost never looks like blood. Blood that has been in the stomach for even a short period is partially digested and looks like coffee grounds in water. Indeed, the description for this is "coffee ground emesis." Thus, coffee ground emesis will not generally be confused with bleeding from injury (Figures 2.5a, b, and c).

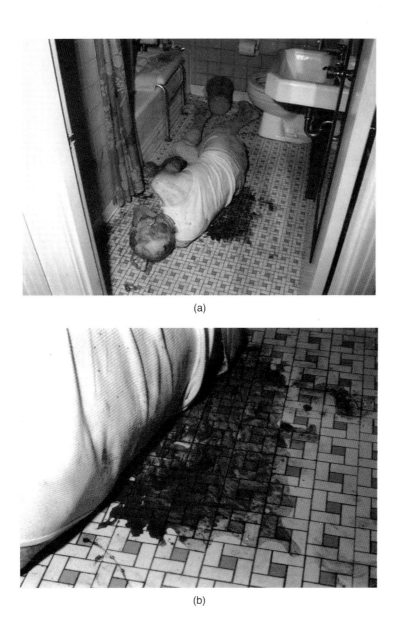

Figure 2.5 (a) Victim of hemorrhaging peptic ulcer found on the floor in his bathroom. Note the brown color of the blood on the floor. (b) Closer view of blood on bathroom floor. (c) The victim vomited blood into his toilet prior to his death. Note the dark brown clumps of denatured blood referred to as "coffee ground emesis."

(c)

Figure 2.5 (Continued).

Bleeding from the Respiratory System

The respiratory system of the body provides oxygen to the circulatory system for distribution to the tissues and organs of the body and the elimination of carbon dioxide through rapid gas exchange. Air enters the respiratory system through the nose and mouth during the inspiration or inhalation and travels though the pharynx and larynx and into the trachea. The trachea divides bilaterally into the bronchi and further subdivides into bronchioles, allowing the passage of air into the alveoli of the lungs where the exchange of oxygen and carbon dioxide occurs. (See Figure 2.6.) The lung capacity of a normal adult is 4 to 6 L of air. The maximum air expelled or exhaled after maximum inspiration or inhalation is 3 to 5 L.

Physical trauma or damage to the respiratory system including the nose, mouth, airway, and lungs may produce an accumulation of blood in the airway passages. Trauma to the head or face may cause blood to accumulate in the nose, mouth, or sinus areas. The presence of blood in one or more of these areas provides an opportunity for blood to be forcibly projected or expelled from the nose, mouth, or open wound of the airway or lungs by exhalation or compression.

If the bleeding started in the lung or the airways of the lung or if the blood is from the nose, throat, or esophagus and is inhaled into the lung, the blood will be mixed with air. Thus it will be frothy in appearance. Coughing and sneezing of blood may occur when there is blood present in the nose, mouth, throat, lung, or esophagus and produce fine droplets of blood.

Bleeding from the Gastrointestinal Tract

Bleeding with red-colored blood from the rectum can be from disease and frequently result from engorged hemorrhoids or cancer of the colon or rectum. Bleeding from higher in

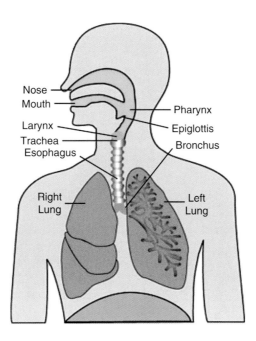

Figure 2.6 Schematic representation of the human respiratory system. (Courtesy of Jim Wallace, University of Tennessee, Memphis.)

the gastrointestinal tract (esophagus, stomach, small and large intestine) will come out black or tar colored, rarely with a red tinge, as shown in Figures 2.5a and b.

Blood Loss Resulting from Injuries

Any disruption of the skin by injury may produce bleeding. The amount and pattern of bleeding is determined by the object causing the injury, the location of the injury, the characteristics of the injury, and certain characteristics of the injured person.

Abrasions

Abrasions are caused by scraping of the skin by friction relative to the surface of an object with a rough surface, which causes removal of the superficial layers of the skin. The epidermal surface is rubbed away, exposing the deeper structures comprising the skin, as shown in Figures 2.7a and b. Abrasions may bleed, but the bleeding is primarily capillary in origin and will be of a small amount. A forceful impact may produce more bleeding. A tangential contact produces an abrasion that exhibits directionality, whereas direct contact that crushes the skin produces an imprint abrasion. Figure 2.8 shows a large area of abrasion referred to as "road rash."

Lacerations

Lacerations are caused when the skin or internal organs are torn by a shearing or crushing force as a result of impact with blunt objects. They tend to be irregular with abraded contused

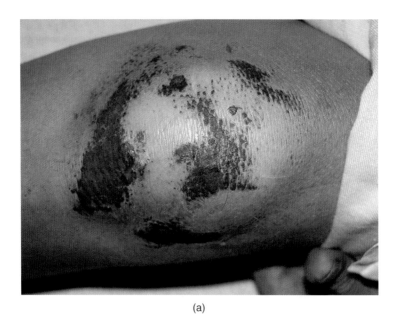

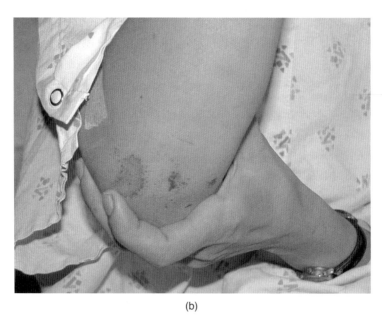

Figure 2.7 (a) Abrasion on knee caused by fall. Note the small amount of serum near the edge of the abraded area. (b) Abrasion on elbow caused by fall. Note the friction scraping of the epidermis and lack of blood.

margins and occur most commonly over bony prominences where the skin is more easily stretched and torn (Figure 2.9). There are usually tissue bridges within the depth of the wound, as shown in Figure 2.10. Underlying skull fractures may be present. Lacerations may bleed significantly, especially those on the head, but may not on other parts of the body because blood vessels stretch more than other tissue and thus remain intact.

Medical and Anatomical Aspects of Bloodshed

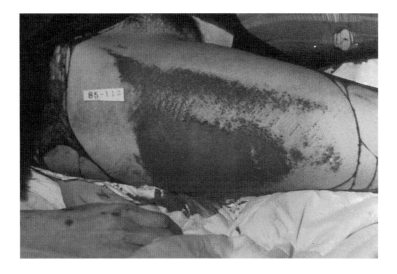

Figure 2.8 Victim of vehicular/pedestrian collision vaulted onto roadway incurred wide abraded area on upper right thigh referred to as "road rash" in addition to fatal head injuries.

The head is unique in two ways. First, there is bone very close to the surface, making lacerations relatively easily made compared with the rest of the body (Figure 2.11). Secondly, the blood supply to the head is significantly greater than in the rest of the body. The combination makes lacerations to the scalp quite bloody. The shape and dimensions of a laceration or underlying fracture may correspond to the object that produced it. A good example would be circular or crescent-shaped lacerations and fractures of the head caused by a round hammer head, as shown in Figure 2.12.

Figure 2.9 Multiple linear lacerations on top and left rear of head of victim.

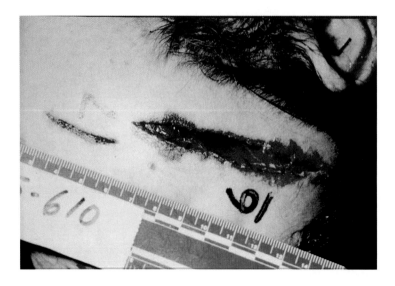

Figure 2.10 View of laceration exhibiting irregular edges and tissue bridges.

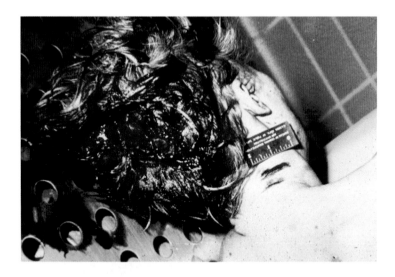

Figure 2.11 Gaping laceration on rear of head caused by repeated beating against floor.

Incised Wounds

Incised wounds are cutting injuries produced by sharp objects such as knives or razors. Unfortunately, most physicians not trained in forensic science call all breaks in the skin lacerations or "cuts" whether they are actually lacerations (caused by blunt objects) or incised wounds (caused by sharp objects). Incised wounds often result in significant cuts in blood vessels and thus copious bleeding. The edges of the wounds are usually sharply defined as

Medical and Anatomical Aspects of Bloodshed

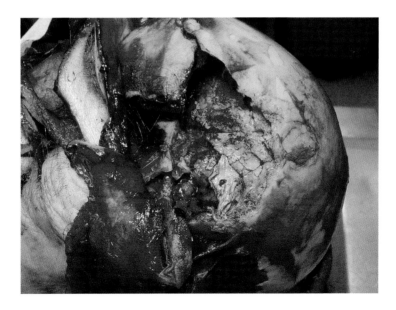

Figure 2.12 Multiple lacerations and associated skull fractures caused by blows with hammer.

compared to lacerations (Figure 2.13). Indeed, incised wounds can often be fatal if large arteries are cut, such as in the groin (femoral artery) or the arm (brachial artery) or neck (carotid artery), as shown in Figures 2.14 and 2.15. If arteries are missed in these vulnerable spots, but a vein cut, the vein injury may not result in exsanguination.

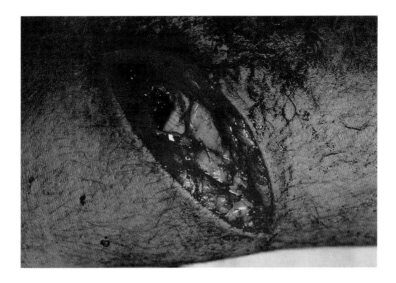

Figure 2.13 Incised wound with sharp, well-defined edges. Compare to laceration shown in Figure 2.10. (Courtesy of Todd Thorne, Kenosha, Wisconsin Police Department.)

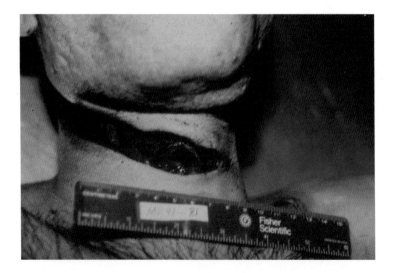

Figure 2.14 Incised wound of right neck that severed right carotid artery and trachea that resulted in death.

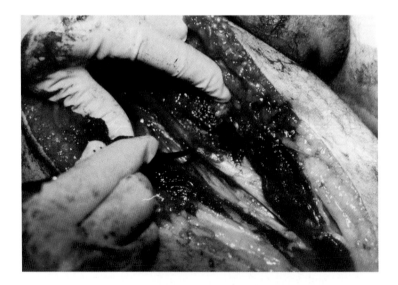

Figure 2.15 Incised wound of right thigh that partially severed the femoral artery and resulted in death.

Incised wounds of the wrist are often made by people attempting suicide. These efforts are very rarely successful because of the difficulty in cutting the radial artery (the one in which you can feel a pulse) because it is protected from injury by the radial bone (Figure 2.16). Generally, only the veins of the wrist are cut, and the bleeding tract will clot before significant blood can be lost.

Stab Wounds

Stab wounds are caused by sharp objects and are distinguished from incised wounds by being deeper than they measure on the surface (Figure 2.17). Generally, stab wounds have

Medical and Anatomical Aspects of Bloodshed

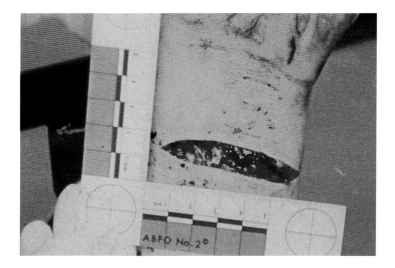

Figure 2.16 Nonfatal incised wound of wrist that did not involve artery.

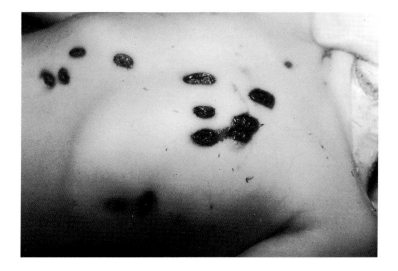

Figure 2.17 Multiple stab wounds in chest of victim.

little external bleeding. An exception to this is a stab wound of the chest where a major vessel or the heart is penetrated. This will cause massive internal bleeding, and the wound may drain the blood in the chest by gravity because of the postmortem position of the victim. This will result in extensive external blood loss. This is dependent on the posture of the victim and the location of the stab wound. Abdominal stab wounds, even fatal ones, rarely have significant external bleeding. In many domestic violence cases, fatal incised and/or stab wounds are inflicted with knives from the wooden kitchen knife block (Figure 2.18).

Gunshot Wounds

In the United States, gunshot wounds are the most common cause of inflicted fatal injury (injury intentionally produced, whether suicidal or homicidal). Accidental gunshot wounds

Figure 2.18 The wooden knife block in the kitchen—"weapons of opportunity."

are far less common. A penetrating gunshot wound exhibits an entrance wound with no exit, and a perforating gunshot wound exhibits both an entrance and exit wound. Because of the velocity of the projectile, the tissue in front of the projectile is compressed, then stretched, and then lacerates as the projectile moves through the tissue. This produces a massive blunt trauma injury that has a diameter of destruction that is larger or much larger than the diameter of the projectile. The amount of damage is primarily a function of the velocity of the projectile and indeed goes up as the square of the increase of the velocity. High-velocity projectiles, those traveling faster than 2000 ft/sec (600 m/sec), usually produce near amputation of an arm or leg. Such wounds obviously bleed copiously.

In civilian practice most gunshot wounds are of low or moderate velocity (500 to 2000 ft/sec). External bleeding of such wounds is often quite modest. The exceptions are the same as those discussed in stab wounds, where a chest wound may have extensive postmortem drainage. Another exception is the head. Head wounds generally bleed more than anywhere else on the body for the same reason that lacerations to the head bleed more. Another distinguishing fact of head gunshot wounds is the devastating pressure wave associated with the passage of a projectile in the head. The brain is basically incompressible, and thus the shockwave from the projectile is transmitted throughout the skull, if the bullet enters the skull. This causes fracturing of the bones making up the frontal base of the skull (the orbital plates, the ethmoid and sphenoid bones). The orbital plates are the thinnest bones of the skull, and they frequently fracture. This results in bilateral black eyes if blood pressure continues for a brief time after the shooting. The ethmoid and sphenoid bones make up the top of the nose internally. There are huge veins that run along these bones, which lacerate when the bones shatter. This results in massive bleeding in the upper portions of the nose, which results in near instantaneous external bleeding from the mouth and nose. The actual mechanism of death in many people who are shot in the head is drowning in their own blood.

Gunshot wounds of entrance are classified based on their characteristics that relate to the distance the firearm was discharged from the target.

Medical and Anatomical Aspects of Bloodshed

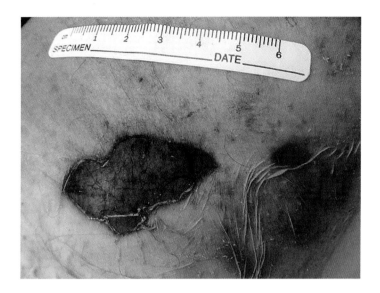

Figure 2.19 Contact gunshot wound of the chin. Note the presence of smoke, small laceration, and red carboxyhemoglobin.

Contact Wounds

Contact or near-contact wounds demonstrate the blackening effect of the gases and soot that accompany the projectile. Blackening around the edges of the wound from the soot and cherry-red discoloration of the blood as a result of the formation of carboxyhemoglobin are visible. There is usually a thin band of contusion in the margins of the wound (Figure 2.19). A muzzle imprint may be present on the skin. Contact wounds in areas where there is bone immediately beneath the skin frequently produce varying dimensions of lacerations because of the expanding gases between the bone and skin. Often stellate in appearance, they are frequently seen in contact wounds of the head (Figure 2.20). These

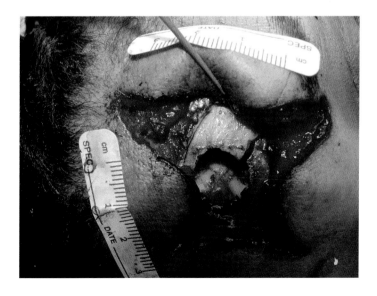

Figure 2.20 Stellate-shaped contact wound of the forehead.

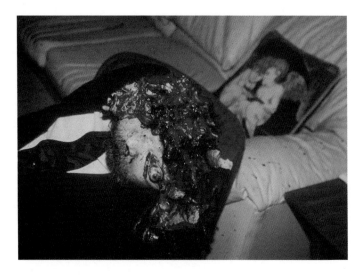

Figure 2.21 Contact wound of head with massive destruction of the cranium and evacuation of the brain.

types of gunshot wounds are capable of producing significant quantities of back spatter of blood and occasionally tissue. Explosion of the head and evacuation of the brain are not uncommon, as shown in Figure 2.21.

Intermediate-Range Gunshot Wounds

As the distance from the barrel of the firearm to the skin increases, the effects of the gases diminish and only the unburned or partially burned powder particles and the projectile are capable of penetrating the skin around the entrance wound. The powder particles produce stippling or tattooing around the entrance wound (Figure 2.22). This effect is generally

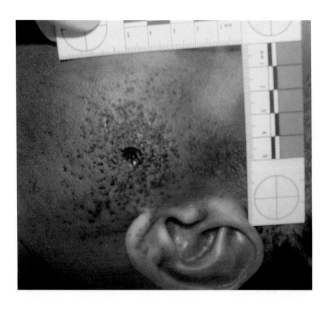

Figure 2.22 Intermediate-range gunshot wound exhibiting circular area of stippling or tattooing of gunshot powder particles.

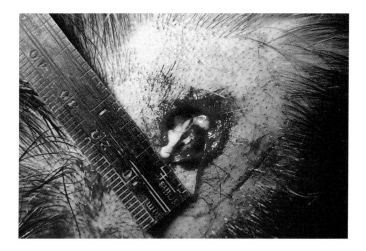

Figure 2.23 Distant (approximately 15 ft) gunshot wound with absence of stippling. (Courtesy of Todd Thorne, Kenosha, Wisconsin Police Department.)

observed up to distances from near contact to up to 3 and occasionally 4 feet muzzle-to-target distance, depending on the weapon and type of ammunition.

Distant or Long-Range Gunshot Wounds

Gunshot wounds of entrance classified as distant or long-range occur when the muzzle of the weapon is sufficiently far enough from the body so that there is no deposition of soot or stippling (Figure 2.23).

Gunshot Exit Wounds

Gunshot wounds of exit are frequently lacerated, as shown in Figure 2.24, and often produce significant forward spatters of blood that travel in the direction of the projectile.

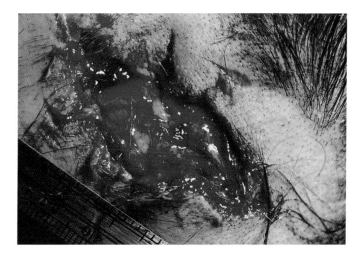

Figure 2.24 Large lacerated gunshot wound of exit of victim in Figure 2.23. (Courtesy of Todd Thorne, Kenosha, Wisconsin Police Department.)

Although the conventional wisdom is that gunshot wounds of exit are larger than gunshot wounds of entrance, this is not always the case. This is often seen with contact wounds of the head that are often much larger than the corresponding exit wound.

Postmortem Aspects of Blood in the Body

After a person dies, the blood ceases to circulate. This results in color changes during the agonal or perimortem period, which is the period of time while a person dies. The color change during this period may provide clues to the cause of death. Obviously, if someone dies from exsanguination, then there is a remarkable pallor or absence of the color, which is ordinarily imparted to the skin and gums of a person (Figure 2.25). This absence of color is the result of shock, where blood is shunted away from the skin as the blood pressure drops from loss of blood. This is much more noticeable on light-skinned people than in darker skinned individuals. Additionally, if there has been interference with oxygenation of the blood, then there will be a darkening or bluish discoloration of the skin. This occurs because blood combined with oxygen is a rather bright red, whereas blood that is not combined with oxygen is a darker red. When viewed through the skin, the darker blood has a blue color to it. Interference with oxygenation of the blood can be caused by airway obstruction, by café coronary or manual strangulation, or by problems in the lung such as pneumonia.

After death has occurred and the blood stops circulating, the red cells in the blood settle because of the effects of gravity. This is referred to as lividity. Blood is composed of plasma, which contains a large amount of protein and red blood cells. The red blood cells make up from 42 to 50% of the volume of blood. This value is known as the hematocrit. The red blood cells are denser than blood plasma and sink. The rate of sinking is called the sedimentation rate, and the rate of sinking increases in a number of inflammatory diseases.

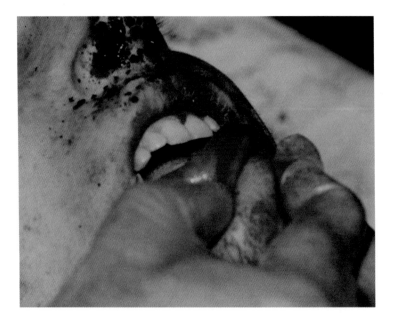

Figure 2.25 Pallor of gums resulting from absence of blood caused by exsanguination.

Medical and Anatomical Aspects of Bloodshed   33

In dead bodies, in as little as 20 min in light-skinned people, the dependent or lower portions of the body show evidence of livor mortis or lividity, a red or red–blue discoloration resulting from the settling of blood. When death has occurred as the result of carbon monoxide poisoning, lividity is cherry red because of the replacement of oxyhemoglobin with carboxyhemoglobin (Figures 2.26a and b). Cyanide poisoning also imparts a cherry-red color to the blood. During the first 24 h after death, lividity is nonfixed. This means that finger pressure on the discolored area will blanch the lividity. This provides a good way to tell whether discoloration of the skin is a bruise or lividity, but it only works during the first 24 h after death. After 24 h the lividity becomes fixed. This fixation is caused by the rupture of the red blood cells as they die. When they rupture they release the hemoglobin contained

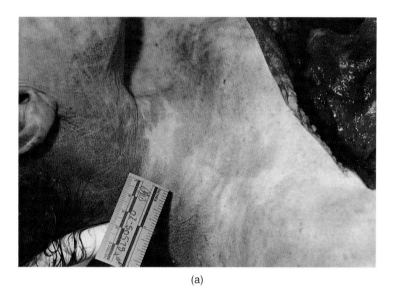

(a)

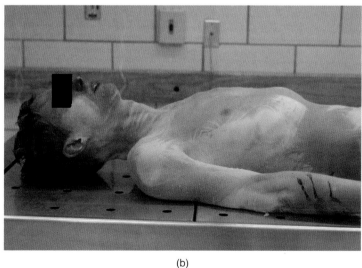

(b)

Figure 2.26 (a) Postmortem lividity of the neck as a result of settling of blood in that area. (b) Victim of carbon monoxide poisoning. Note the presence of the characteristic "cherry-red" lividity. (Courtesy of Cynthia Gardner, MD, Hamilton County Coroner's Office, Cincinnati, OH.)

within the cells, and this stains the tissue. Thus after 24 h (variable if there is anything that accelerates red blood cell death), lividity cannot be eliminated by finger pressure. This gives some help in determining the time of death within a 24-h period.

Lividity may not occur or may be very difficult to discern in people who have bled to death, particularly if the bleeding was slow.

Bloodstain Patterns within the Body

The documentation of the physical characteristics of bloodstain patterns within the body is an important part of the forensic autopsy protocol. These patterns are the result of several mechanisms.

Patterns Associated with Postmortem Lividity

Patterns of lividity that do not match the position of the body at the time of observation indicate that the body has been moved postmortem. This is permanent if the body is moved after 24 h because of lividity fixation. This nonmatching lividity can provide important investigative leads in death investigation.

Finally, although lividity is dependent, meaning on the lower parts of the body, the actual weightbearing portions of the body will not have lividity because the blood cannot settle in blood vessels that are compressed by the weight of the body. This phenomena is called pressure blanching of lividity. The blanched or void areas may depict a pattern or shape of the surface or object that the portion of the body was in contact with, as shown in Figure 2.27.

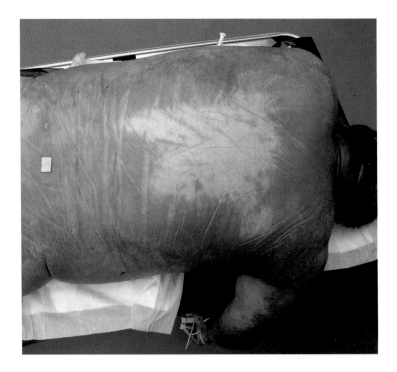

Figure 2.27 Postmortem lividity on back of victim. Note blanched area as a result of pressure on back in that area. Because lividity is gravity dependent, it will be absent in areas of the body where the weight produces pressure and occludes the capillaries.

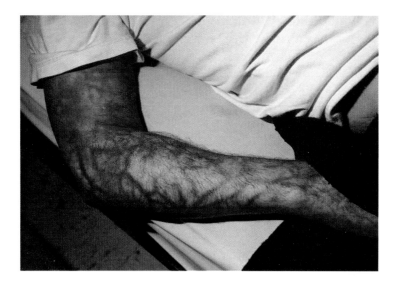

Figure 2.28 An example of marbling of the blood vessels beneath the skin on right arm of deceased. (Courtesy of Todd Thorne, Kenosha, Wisconsin Police Department.)

Patterns Associated with Postmortem Marbling of the Skin

Marbling is a characteristic of postmortem decomposition associated with bacterial gas formation and swelling of the body. It is produced by hemolysis of blood in vessels with the reaction of hemoglobin and hydrogen sulfide and the development of greenish-black discoloration of the vessels, as shown in Figure 2.28.

Petechial Hemorrhaging (Tardieu Spots)

Petechiae are pinpoint hemorrhages produced by the rupture of small capillaries close to the skin as a result of increases in intravascular pressure. Larger petechiae are referred to as ecchymoses. They may be seen within the skin, conjunctiva of the eye, or mouth and may be associated with asphyxial deaths such as strangulation, as shown in Figure 2.29.

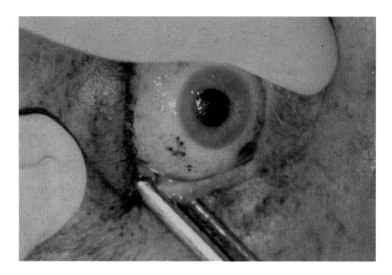

Figure 2.29 Petechial hemorrhages in eye of victim of strangulation.

Patterns Associated with Bruises

Bruises, or contusions, are areas of hemorrhage into soft tissue resulting from rupture of blood vessels from blunt trauma. Hematomas are contusions that are engorged with blood. Contusions and/or hematomas are often present on the skin as well as internal tissues and organs (Figures 2.30a and b). Bruises may exhibit a recognizable pattern of the offending object, such as seen in Figure 2.31 and Figure 2.32.

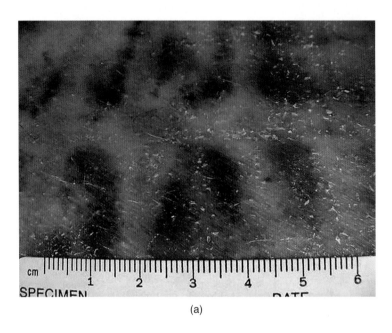

(a)

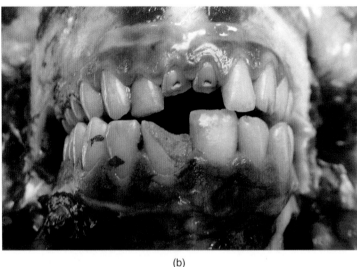

(b)

Figure 2.30 (a) An example of multiple bruises. (b) Bruised gums associated with blunt trauma to the mouth and teeth.

Medical and Anatomical Aspects of Bloodshed

Figure 2.31 Patterned bruises on back of victim associated with whipping with a dog leash.

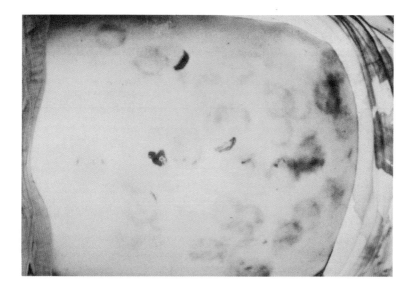

Figure 2.32 Patterned bruises on back of victim associated with the round head of a hammer.

Patterns Associated with Abrasions

Abrasions, as previously described, involve friction removal of the superficial layers of the skin. A forceful impact may produce more bleeding. Bite marks often result in patterned abrasions with extensive underlying hemorrhage or contusions. Patterned abrasions may also be associated with a particular object. Examples of identifiable patterned abrasions are shown in Figure 2.33, Figure 2.34, and Figure 2.35.

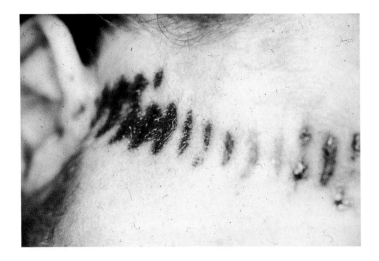

Figure 2.33 Patterned abrasion on head of victim associated with a rebar.

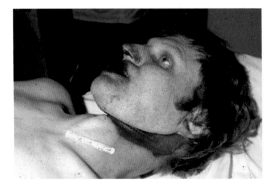

Figure 2.34 Patterned abrasion on neck associated with autoerotic hanging with belt.

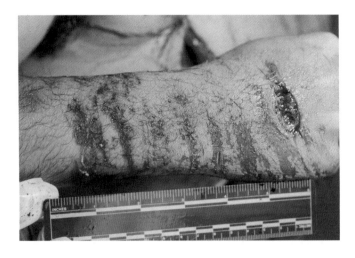

Figure 2.35 Patterned abrasions associated with impact to left arm of pedestrian victim with grill from vehicle. (Courtesy of Todd Thorne, Kenosha, Wisconsin Police Department.)

Conclusion

The high-pressure circulatory system, which all vertebrate animals have as a necessity for life, causes the loss of blood to occur from disease or injury. External loss of blood provides physical evidence that can be analyzed to determine where, when, and what caused the bleeding. The understanding of bloodstain pattern analysis requires knowledge of the physiology of circulation and the diseases and injuries that result in loss of blood.

Biological and Physical Properties of Human Blood 3

Biological Functions of Human Blood

Human blood is a liquid connective tissue that serves two primary functions: transport and defense. The role of blood in carrying oxygen to the cells and carbon dioxide to the lungs for disposal through respiration is probably the most well-known function. Blood is also responsible for transporting food in the form of glucose, lipids, and amino acids from the alimentary canal to the cells. It also serves as a carrying medium for hormones that are secreted directly into the blood by the endocrine glands and electrolytes such as sodium (Na^+), calcium (Ca^{++}), and bicarbonate (HCO_3^-) that serve a vital role in the maintenance of the proper acid-base balance needed to sustain life and neuroconduction.

Blood serves a supply function, and it also plays an integral part in waste removal. Metabolism leads to the production of waste such as carbon dioxide, urea, creatinine, lactic acid, and uric acid at the cellular level. This waste is stockpiled into the tissue fluid, but it must be removed or the levels will become toxic and cause death. Blood completes this path from the cellular level to the organs of removal, usually the liver or the kidneys. In addition to delivering the basic metabolic needs, drugs and other foreign chemicals — both beneficial and detrimental — make their way from the route of ingestion to the target organ via the bloodstream. Residual levels are then removed via the bloodstream.

Bringing essential substances to the body and then aiding in the subsequent waste removal are only parts of the role blood serves to keep the organism alive. To thrive, an animal must also be able to adequately regulate body temperature. Extremes to either end of the spectrum could result in death. Obviously, more active regions of the body will generate more heat, but the body must be able to equalize this heat production. The bloodstream plays an important role in the regulation of body temperature by distributing the excessive heat throughout the body, thereby allowing it to dissipate.

Another important function of blood is the role it plays in defense of the body. White blood cells (WBC), or leukocytes, are produced in the bone marrow and the lymphatic tissue. The body stores reserve supplies at the site of formation until they are needed as a replacement for aging white cells or as emergency responders in case of overwhelming infection. Once WBCs are called into action, they circulate throughout the body via the bloodstream. In this regard, WBCs can be considered the body's mobile readiness system of defense.

Another important bodily function provided by blood is clotting. The prevention of excessive blood loss is no less crucial to survival than the delivery of food, removal of waste, and protection from infection. Without a functional coagulation system, even the smallest childhood injury could easily lead to death from exsanguination. While the ability to clot is important, the control mechanism to prevent excessive clotting is equally important. One can only marvel at the intricacies of the human clotting mechanism upon realizing that this system begins clot formation the instant an injury to the circulatory system occurs while simultaneously leaving the circulating blood in a liquid state. Platelets, fibrinogen, prothrombin, calcium, and clotting factors — the necessary components for functional clotting — are all components of blood.

Blood certainly serves many important functions, all of which are possible because it is given access to all areas of the body by a series of arteries, veins, and capillaries. Oxygenated blood leaves the left side of the heart through the aorta, which may measure as much as 2.5 cm in diameter. The arteries become progressively smaller and culminate into capillaries that may measure as small as 5 to 10 μm in diameter (1/10 the diameter of a human hair). As small as the capillaries are, they are the sites where oxygen is released and waste products are collected. Veins, ranging in size from 1 mm to 1.5 cm in diameter, serve to return the blood from the capillaries to the heart and lungs.

Composition of Blood

Whole blood consists of cells and cell fragments suspended in a liquid medium called plasma. If a whole blood specimen were separated by centrifugation, the red blood cells (RBCs) would go to the bottom of the container. Proportionately, the red cell fraction comprises 55% of the total volume. Lying on top of the red cell layer would be a relatively small fraction consisting of the WBCs followed by an even smaller layer of platelets. This layer is referred to as the buffy coat. Above the buffy coat is the liquid portion of blood. This liquid fraction is called plasma if clotting has not occurred, but the term is changed to serum once clotting takes place. Since the buffy coat makes up such a small portion of the specimen, the plasma is said to comprise 45% of the total (Figure 3.1).

Volume and pH

Whole blood has a normal pH of 7.4 and comprises approximately 8 to 9% of the total body mass. Adults have approximately 60 to 66 mL of blood per kilogram of body weight. Converting into the English system, this would be approximately 2 oz or 4 tbsp of blood for every 2.2 lb of body weight. The average adult has 4.5 to 5.0 L, or roughly 1 gal, of blood. (Adult males average 5.7 L or 1.5 gal, and adult females average 4.3 L or 1.125 gal.) Corresponding to their lower body weight, children and infants have smaller blood volumes. An 80-lb child normally has 2.4 to 2.5 L of blood. An 8-lb infant would be expected to have about 8.5 oz (~0.25 L) of blood.

Plasma

Normal plasma is a pale yellow, transparent fluid. Individuals with liver problems typically have markedly yellow plasma that is described as icteric. Milky, whitish plasma is indicative

Biological and Physical Properties of Human Blood

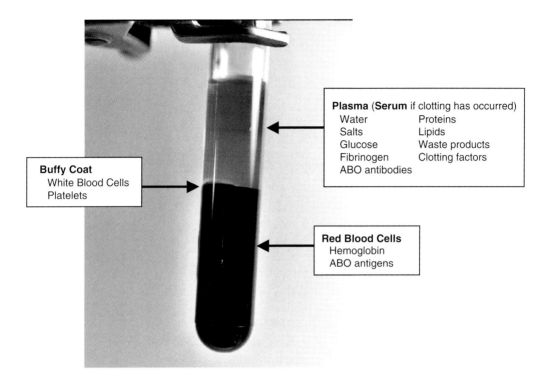

Figure 3.1 Components of whole human blood.

of an elevated concentration of lipids and is referred to as lipemic. Plasma is 90% water and 10% solutes such as proteins, salts, lipids, and glucose. The blood plasma is the primary transport medium for nutrients, electrolytes, and hormones taken to the cells and waste products removed from the cells. Without plasma, the cellular elements in blood would be effectively stranded and unable to move throughout the body to perform their duties. Table 3.1 describes the components and composition of human plasma.

Table 3.1 Components and Composition of Plasma

Water 90%	
Solutes 10%	
Plasma proteins ~7%	Albumin
	Globulins
	Fibrinogen
Nutrients ~2%	Lipids
	Glucose
	Vitamins
	Steroid hormones
	Amino acids
Inorganic salts ~1%	Sodium
	Potassium
	Calcium salt

White Blood Cells (Leucocytes)

White blood cells (WBCs), or leucocytes, are responsible for protecting the organism from pathogens. They are divided into two major categories: granulocytes and nongranulocytes, according to the presence or absence of visible granules in their cytoplasm. Granulocytes include neutrophils, eosinophils, and basophils and are produced in the bone marrow. Nongranulocytes consist of lymphocytes and monocytes and are produced by the lymph nodes such as tonsils, spleen, thymus, and intestinal mucosa. The density, or normal count, of WBCs in the blood is 5000 to 7000/mm^3, making them much less plentiful than RBCs (approximately 1 WBC per 600 RBCs). The nucleated WBC is an important source of DNA for forensic testing.

The intracellular protection afforded by the WBCs is accomplished by destruction of the offending bacteria. Bacterial infection causes the production of chemicals by two mechanisms. One is the inherent by-product of the metabolism of the bacteria itself. The other is a direct result of the body's reaction to bacterial exposure. These chemicals act much like a homing signal to direct the WBCs to the bacteria. Once the white cells locate the bacteria, they use the process of phagocytosis to engulf and quarantine the bacteria. The white cells then release enzymes that will destroy the bacteria by digestion and liquefaction.

Neutrophils are the most abundant WBCs and are largely responsible for preventing the overpopulation of commensal bacteria. Eosinophils are in relatively low numbers in the bloodstream but can increase greatly in conditions such as parasitic infection. Basophils leave the bloodstream and concentrate at the site of infection or inflammation. The granules in basophils release mediators that increase blood flow to the site and otherwise aid the inflammatory response. Any allergy sufferer is all too familiar with the effects of the histamine released by basophils.

Lymphocytes are the second most abundant white cell and include B cells and T cells. The body will recognize foreign protein such as bacteria, molds, parasites, and viruses as an antigen. In response to this exposure, the B cells produce antibodies, giving the organism a line of extracellular defense. T cells are responsible for killing cells that are virally infected. They also enhance antibody production by B cells and recruit other leukocytes to the site of infection or tissue damage.

While the WBC is floating along in the bloodstream in search of infection, they also perform other light housekeeping duties such as removing small blood clots, depleted RBCs, and other small particulate waste. Left unchecked, these minute fragments will accumulate and lead to damage or even demise. WBCs remove this cellular debris using the same engulf-and-liquefy mechanism used against bacteria. This is the primary function of monocytes. Table 3.2 shows the proportion of leucocytes in normal blood.

Table 3.2 Proportion of White Blood Cell Types

Type	Percentage
Neutrophil	50–70
Lymphocyte	20–40
Monocyte	3–8
Eosinophil	2–4
Basophil	0.5–1

Red Blood Cells (Erythrocytes)

Red blood cells (RBCs) are manufactured in the bone marrow and are the most abundant cellular component in blood. Normal RBC counts are 4.8 million to 5.4 million red blood cells per µl of blood, a level that equates to 1 billion RBCs in only 2 to 3 drops of blood. The immature red cell has a nucleus that is lost on differentiation into a mature cell. Of more importance to the red cell is the fact that having no nucleus means having no mitochondria which, in turn, contributes to their short life span of 120 days. With such a short life span, 3 million red cells die every second and must be removed by the liver and spleen.

RBCs are not only plentiful but also extremely well designed to be as efficient as possible in their transport function. Their very small size (7 to 8 µm) enables them to make their way through the smallest of capillaries — some so small that the RBCs must actually pass single file. Although the routes red cells must use to perform their duties are small, they must also present as much surface area as possible to maximize oxygen and carbon dioxide transport. The geometric shape of the human red cell allows it to satisfy these seemingly conflicting requirements. Human RBCs have a biconcave shape that maximizes the surface-to-volume ratio while simultaneously allowing the cell to be small enough to pass through the capillaries. Figure 3.2 shows the various cell types in the blood of normal individuals.

Hemoglobin

Hemoglobin resides in the cytoplasm of the RBCs and accounts for 90% of their dry weight. It is the hemoglobin that gives red cells the ability to transport O_2 and CO_2. A hemoglobin molecule consists of four polypeptides, each of which has a heme group attached. One atom of iron is at the center of each heme and is capable of binding to one molecule of O_2. As a result, each hemoglobin molecule has the capacity to carry up to four molecules of O_2. The red color of blood is directly attributable to the hemoglobin contained in the cytoplasm of the RBCs. Arterial blood contains oxygenated hemoglobin, called oxyhemoglobin, which has a characteristic bright red color. The blood returning through the venous system has already released its cargo of oxygen and is carrying carbon dioxide to the lungs instead. Hemoglobin bound to carbon dioxide is a darker, blue-red color.

In the capillaries of the lungs, hemoglobin is exposed to cool temperatures, higher pH, and increased oxygen pressure. These conditions are favorable for the binding of oxygen, and hemoglobin is converted to oxyhemoglobin $[Hb(O_2)_4]$. Once the oxyhemoglobin reaches the tissues, it must release the oxygen and then attach to the carbon dioxide waste. At the level of exchange, the red cell is exposed to an acidic environment with warmth and lower oxygen pressure — all factors that facilitate the release of the oxygen needed by the cells. Carbon dioxide is then able to bind to a different part of the hemoglobin molecule and be transported back to the lungs for disposition. This is shown in Figure 3.3.

Platelets (Thrombocytes) and Clotting Factors

Platelets are irregularly shaped cell fragments produced by megakaryocytes. They are the smallest component of human blood with a diameter of only 2 to 4 µm. Platelets are essential in protecting the body from excessive blood loss resulting from overt trauma or even normal daily activities that could result in minor blood vessel damage. Without platelets, even minor damage would produce irreversible leakage. No animal could survive long with such a handicap.

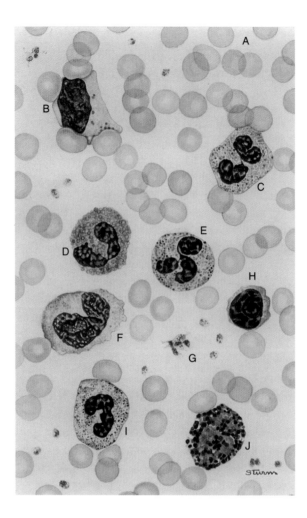

Figure 3.2 Cell types found on peripheral blood smears from normal individuals. A, Erythrocytes; B, large lymphocyte with purplish red (azurophil) granules and deeply indented by adjacent erythrocytes; C, neutrophilic segmented; D, eosinophilic segmented; E, neutrophilic segmented; F, monocyte with gray-blue cytoplasm, coarse linear chromatin, blunt pseudopods; G, platelets (thrombocytes); H, lymphocyte; I, neutrophilic band; J, basophil. The arrangement is arbitrary, and the number of leukocytes in relation to erythrocytes and thrombocytes is greater than would occur in an actual microscopic field. (From Sallah, Sabah, and Bell, Ann, *The Morphology of Human Blood Cells*, 6th edition, Abbott Diagnostics, 2003, p. 42. Reproduction of "Morphology of Human Blood Cells" has been granted with approval of Abbott Laboratories. All rights reserved by Abbott Laboratories.)

The physical disruption of a blood vessel allows the platelets to be exposed to the collagen underneath the endothelial lining of the vessel. The exposure to collagen causes the platelet to begin to change shape, forming numerous spiny projections protruding from their membranes that adhere to the ends of the damaged vessel. At the same time platelets are changing shape, they also become "sticky" and begin to coalesce and form the physical barrier that stems minor bleeding. This initial barrier is effective and necessary, but it is only a stopgap measure. Platelets then further decrease blood loss by releasing

Biological and Physical Properties of Human Blood

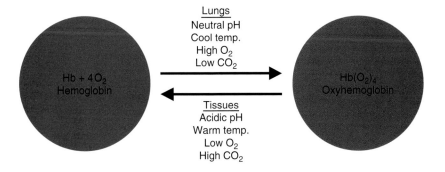

Figure 3.3 Environmental conditions during oxygen transfer by hemoglobin.

substances that cause vasoconstriction at the site of injury. Finally, the platelets work in consort with the tissue fluids from the site of injury to initiate the formation of a meshwork of insoluble fibrin threads. This mesh of fibrin threads traps blood cells and change the flowing blood into a gelatinous mass. To stabilize the clot, the web of fibrin constricts, squeezing out the liquid portion — now called serum. This conversion of fibrinogen to insoluble fibrin is the end-product of the clotting cascade. The clotting cascade provides for conversion of fibrinogen into fibrin, but it also provides a feedback control mechanism to prevent rampant clotting.

Tissue factor (TF) is released as a result of tissue damage. TF combines with Factor VII, creating a protease that acts on Factor X. In turn, Factor X binds to and activates Factor V, creating prothrombinase. Prothrombinase converts prothrombin (or Factor II) into thrombin. Thrombin cleaves fibrinogen and also activates Factor XII, which acts on the fibrinogen to create an insoluble meshwork. The clotting cascade is demonstrated in Figure 3.4. An image of a clot taken with a scanning electron microscope is demonstrated in Figure 3.5.

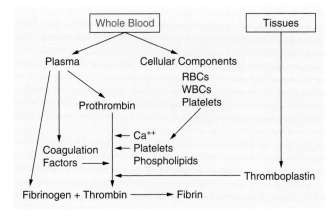

Figure 3.4 Diagrammatic representation of clotting cascade.

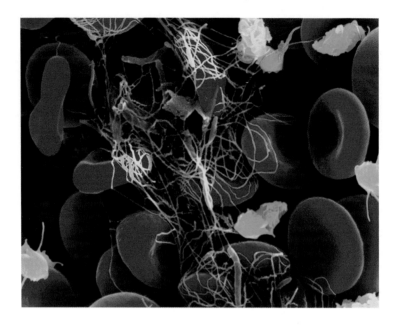

Figure 3.5 Formation of blood clot taken with a Hitachi Field Emission Scanning Electron Microscope (FESEM). The red concave discs are red blood cells; the blue irregular cellular elements are activated platelets intermixed with yellow strands of fibrin. The magnification was 1445× for an image 24 mm in the narrow dimension. (Copyright Dennis Kunkel Microscopy, Inc., Kailua, Hawaii.)

Physical Properties of Blood

Blood is a fluid, or more precisely a liquid. In considering the physical properties of blood, it is important to understand the characteristics of a liquid. A liquid has the capacity to flow and also the ability to assume the shape of its container. The capacity to flow is made possible by the fact that the molecules in a liquid are able to move around freely. Although the molecules in a liquid can move freely, the spacing between the molecules remains fixed unless there is a change in temperature or pressure.

The flight characteristics and the resultant stain characteristics of blood are a function of the physical properties of blood. Three physical properties are especially important to the study of bloodstains:

1. Viscosity
2. Surface tension
3. Relative density (or specific gravity)

Viscosity

Viscosity of a liquid is a measure of its resistance to change shape or flow. In simpler terms, it is the "thinness" or "thickness" of a liquid. Water is thinner, and therefore less viscous, than corn syrup. Blood is approximately four times more viscous than water, although its specific gravity is only slightly higher than that of water.

Viscosity is a direct result of the mutual attractions between the molecules making up a liquid. The membranes of the RBCs have a high concentration of sialic acid, which produces a large electronegative charge on the surface of the RBC. It is this large negative charge that gives blood its viscosity.

Pressure changes do not significantly alter the viscosity of a fluid, but temperature changes do have an effect. As a general rule, the viscosity of a liquid decreases when temperature is increased. A good example of the effect of heat on viscosity is the heating of corn syrup. As any cook knows, corn syrup becomes thinner or less viscous when it is heated. Once the heat is removed and corn syrup cools, it will thicken, representing a return to its original, more viscous state.

Blood — A Non-Newtonian Fluid

To better understand the definition of a non-Newtonian fluid, perhaps it is best to first define a Newtonian fluid such as water and gasoline. The viscosity of a Newtonian fluid is constant (neglecting the effects of temperature and pressure). By doubling the speed at which layers of water slide past each other, the resisting force is doubled. A classic example is a water pistol. To get water to squirt out of a water pistol, one must pull the trigger. If the trigger is pulled harder, the water will squirt out faster.

A non-Newtonian fluid, such as blood, *does change viscosity*, and that change in viscosity is not dependent on the rate of shear produced by layers flowing over each other (velocity). The viscosity of non-Newtonian fluids such as blood, ketchup, pie fillings, and gravy is completely independent of the rate of shear. Doubling the speed at which the layers of non-Newtonian fluids slide past each other does not result in a doubling of the resisting force. With some non-Newtonian fluids, such as ketchup, the resisting force may less than double, whereas others exhibit resisting force that more than doubles their viscosity. This is the reason pie fillings thicken when stirred and struggling in quicksand makes escape even harder. Blood is a pseudoplastic, non-Newtonian fluid, which means that the viscosity of blood decreases with increases in velocity gradient. Pseudoplastic, non-Newtonian fluids are also called shear thinning fluids. At higher shear rates, such as would be encountered in the tiny capillaries, blood becomes less viscous, thus facilitating the rate of flow. Figure 3.6 shows a view of the capillaries beneath the epidermal layer of the skin. The dark color of the capillaries is the result of the postmortem marbling during decomposition of the body.

Poiseuille's Equation

If a fluid is flowing through a pipe, the velocity of the flow produces a resistive force. The fluid closest to the walls of the pipe is moving more slowly than the fluid near the center of the stream. Basically, friction is created between the areas within the fluid that are traveling at different velocities.

The simplest model for a fluid flowing through a pipe is streamline or laminar flow through a uniform straight pipe. This model is governed by Poiseuille's equation:

$$\text{Volume rate of flow (Q)} = \acute{\eta} r^4 (P_1 - P_2)/(8\eta L)$$

where

Q = volume rate of flow or volume per unit time
$P_1 - P_2$ = the pressure difference between the ends of the pipe
L = the length of the pipe
r = the radius of the pipe
$\acute{\eta}$ = the coefficient of viscosity

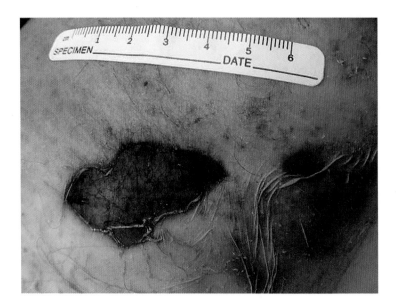

Figure 3.6 View of capillaries beneath epidermal layer of skin.

Poiseuille's equation actually *does not strictly apply to blood flow*—blood vessels are certainly not uniform, straight pipes, and blood flow is pulsatile, especially in the arterial portion of the circulatory system. Discussing Poiseuille's equation in relation to the circulatory system is an oversimplification at best. It is useful, however, because of how dramatically the equation shows the effect of the radius of the pipe on the resulting flow rate. For the circulatory system, this equates to the effect of the radius of the blood vessel in the circulatory system on the resulting flow rate. The volume per unit of time is proportional to r^4; therefore a relatively small reduction in the radius will produce a drastic reduction in the velocity of the fluid. Some of the capillaries in the human body are 1/10 the diameter of a human hair. If such a small number were entered into Poiseuille's equation, the rate of flow would be drastically slow. Because blood is non-Newtonian, its viscosity is reduced when it is moving through the very small capillaries, thus facilitating its flow (Table 3.3).

Surface Tension

Although liquids are not rigid and the molecules are moving freely, the spacing between the molecules is fixed under controlled conditions. The electrical attractions between the molecules in this fixed spatial arrangement create cohesive forces. The strength of these cohesive forces is exemplified by the amount of energy required to break these attractions. A prime example of this is the amount of energy, in the form of heat, required to break the cohesive forces and result in the boiling of water.

In a container of liquid, the molecules at the interior of the container are attracted equally from all surrounding planes. Because there are no molecules above those lying at the surface, there is an inequity in intermolecular attractions resulting in increased cohesive

Table 3.3 Viscosity of Common Fluids

	1 Centipoise = 0.01 Poise		
	Dynes/sec² or Poise at 20°C	at 37°C	Centipoise at 70°F
Water	0.01		1
Whole blood		0.04	4
Plasma		0.015	1.5
Ethanol	0.012		1.2
Glycerine	14.9		1,490
Mercury	0.016		1.6
Milk	0.03		3.0
SAE 10 Motor Oil	0.85–1.4		85–140
SAE 30 Motor Oil	4.2–6.5		420–650
SAE 40 Motor Oil	6.5–9.0		650–900
Corn syrup	50		5,000
Honey	100		10,000
Ketchup	500		50,000
Mustard	700		70,000
Peanut butter	2,500		250,000

forces between the surface molecules (Figure 3.7). The molecules are attempting to achieve the most stable, low-energy configuration possible by minimizing the exposed surface area. The increased cohesive forces at the surface of a liquid result in what is often referred to as a skin or membrane of surface tension. Surface tension is expressed in terms of force per unit length (dynes per centimeter; a dyne = 10^{-5} Newton) and represents stored energy. The surface tension of blood is 50 dynes/cm at 20°C.

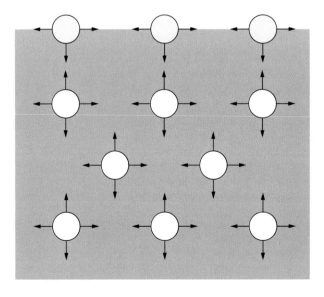

Figure 3.7 Representation of cohesion and surface tension.

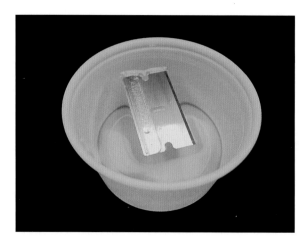

Figure 3.8 Razor blade of steel floating on water.

Surface tension is demonstrated in many daily occurrences, such as a water bug walking across a pond. The water bug is able to do this because the skin created by the surface tension of the water supports the weight of the bug. The surface tension of pure water is 72.8 dynes/cm at 20°C; that is not only enough to support a water bug but enough to support a razor blade composed of steel, a substance 7.8 times heavier than water. The trick is to lower the razor blade onto the surface of the water without penetrating the skin of surface tension (Figure 3.8). If any part of the blade pierces the surface, the skin of surface tension will be broken and the razor blade will sink. The same physical principle is used in the procedures we follow to make clothes washing more efficient. The surface tension of the water tends to bridge the gap in the weave of the fabric instead of penetrating the fabric and removing as much debris as possible. When detergent, a substance with a very low surface tension, is added to the laundry, the cumulative effect is to lower the surface tension of water from 72.8 dynes/cm to approximately 3.0 dynes/cm, hence the adage that "soap makes water wetter." Warm water cleans better for the same reason. The heating has broken the cohesive forces between the water molecules, leading to a reduction in surface tension. With the reduction of the surface tension, the tendency of the water to lie across the weave of the fabric is reduced and better cleaning results. Table 3.4 shows the surface tension values of common substances.

Table 3.4 Surface Tension of Common Fluids

Fluid	dyne/cm at 20°C
Ethanol	22.3
Soap	25
Olive oil	32
Blood	50
Glycerine	63.1
Water	72.5
Mercury	465.0

Biological and Physical Properties of Human Blood

Adhesion, Cohesion, and Capillarity

Adhesion, cohesion, and capillarity are terms often used in the description of the physical characteristics of a liquid, and it would be a good idea to explain their meanings at this point. Adhesion is the attractive forces between *unlike* molecules. A drop of blood is affixed to a wall or a knife by adhesive forces.

Cohesion differs from adhesion in that it is the attractive forces between *like* molecules. In a container of blood or in a single drop of blood, it is the cohesive forces resulting from the electrical attractions that are holding the molecules together. As described earlier, surface tension is the direct result of the especially strong cohesive forces at the surface of a liquid (Figure 3.9).

Capillarity or capillary action describes the phenomena in which surface tension causes a liquid to be drawn upward in a container in such a manner as to be opposing gravity. By merely touching the tip of a small capillary tube to the surface of a blood sample, blood will be drawn a great distance "up" the capillary tube. This happens because the adhesive attraction between the blood and the walls of the glass tube exceed the cohesive forces between the molecules of blood (Figure 3.10). When blood, or any other liquid, is drawn into a porous material, it is doing so by capillary action. Conversely, if the cohesive attraction between the molecules of the liquid exceeds the adhesive attraction between the liquid and the container, the meniscus is convex and capillary action works in reverse (Figure 3.11).

Relative Density

The density of a substance is the measure of its weight per unit of volume (d = m/v) and is expressed in terms of grams per cubic centimeter (g/cm^3). Iron has a density of 7.8 g/cm^3, meaning that a cube of iron measuring 1 cm × 1 cm × 1 cm would weigh 7.8 g.

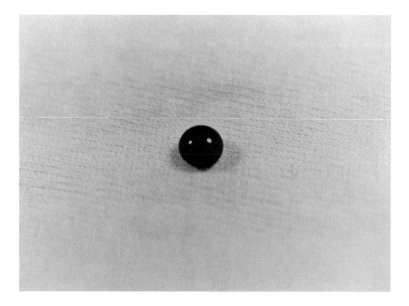

Figure 3.9 Blood drop carefully placed on cotton cloth maintains its shape as a result of cohesion and surface tension.

Figure 3.10 Concave meniscus (water in glass pipette).

The term relative density has replaced the term specific gravity. As the term suggests, relative density is a comparison of the ratio of the density of any given substance to the density of water. The density of water is 1 g/cm^3 (1 cm^3 equals 1 mL). Therefore, the relative density of iron would be (7.8 g/cm^3)/(1.0 g/cm^3), which equals 7.8. Relative density is a

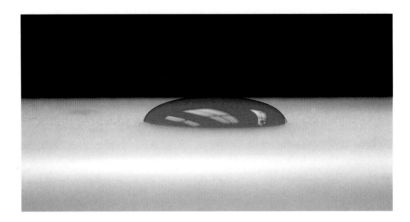

Figure 3.11 Convex meniscus.

Table 3.5 Densities of Common Fluids

Substance	Density (gm.cm^3)
Water	1.0
Air	0.0013
Blood	1.06
Bone	1.8
Iron	7.8
Mercury	13.6
Silicon	2.4
Earth	5.6
Jupiter	1.3

unitless quantity because it is a ratio of two numbers, each with the same unit of measure and the units cancel each other. Neglecting buoyancy, any substance with a density lower than the density of water, (i.e., lower than 1.0 g/cm^3) will float in water. Substances with a density greater than 1.0 g/cm^3 will sink in water. Blood has a relative density of 1.060. Table 3.5 shows the densities of common fluids.

Effects of Drugs and Alcohol on Bloodstains

Research conducted by Herbert Leon MacDonell has shown that the bloodstain patterns produced from the blood of an intoxicated individual will not be significantly altered as a result of the blood alcohol level. The alcohol concentration of the samples tested covered a range of concentrations up to 1.0%, which is two times the recognized lethal level. The physical properties of the altered samples were not appreciably changed by the addition of the alcohol, and therefore the patterns produced will remain essentially the same.

Although the alcohol content of the blood itself does not alter the patterns, the long-term effects of excessive alcohol consumption may have a deleterious effect on the individual's clotting function. Alcoholism often leads to cirrhosis, which, in turn, affects the role of the liver in clotting factor production and the function of factor produced. In addition to the liver damage, chronic alcoholism often leads to poor nutrition as well. Vitamin deficiencies, resulting from nutritional inadequacies and/or by the liver's diminished ability to use vitamins ingested, will lead to increased clotting times and potentially excessive bleeding. A contributing factor, no doubt, is the effect alcohol plays in determining how a person reacts to an incidence of bleeding. The death of actor William Holden is a good example. Holden was drinking alone and apparently hit his head on a table, causing a 3-in. laceration to his forehead. Although there was a working telephone nearby, Holden did not seek help and death resulted from hemorrhage.

Prolonged clotting times can also result from anticoagulation drug therapy. Clinically, anticoagulants are designed to inhibit clotting as a desired treatment for certain disease states. The physician must closely monitor the effect of prescribed anticoagulants. Today there is widespread use of low-dosage (81 mg) aspirin as a preventive measure against heart disease. Aspirin affects the production of platelets, both in number and function.

Elective surgery is typically preceded by the cessation of aspirin ingestion for a minimum of 14 days as a precaution against excessive bleeding.

Blood Use in Experiments

In some situations it will be helpful, or even necessary, to devise experiments or controlled reproductions to answer questions presented by certain cases. Access to sufficient amounts of blood to approximate case circumstances can be problematic. Blood banks may be willing to supply outdated units of human blood for experimental use, and this is certainly an option one might explore. MacDonell compared the physical properties of freshly drawn blood and samples with added anticoagulants and preservatives such as EDTA (ethylenediamine tetraacetic acid), citrate, or oxalate and found no measurable difference in the way they acted.

Be aware, however, that units of whole blood — which most closely approximates the blood one would find at a crime scene — are in limited supply. Blood banks routinely harvest components when a unit is nearing expiration, and, as a result, seldom have outdated whole blood on hand. It is advisable to establish a working relationship with a local blood bank and to notify them in advance if the need for significant volumes of whole blood is anticipated. The blood bank may impose a fee to recover their processing costs. The unit will have been tested for several potential sources of infection, and a certificate or statement will be available should you so desire.

Use of animal blood is an acceptable alternative to human blood and is available in large quantities from slaughterhouses. M.A. Raymond compared porcine blood, up to 2 weeks of age, to human blood and found significant similarities in the physical properties of surface tension, viscosity, and specific gravity (relative density). Because it is the physical properties that influence the pattern, porcine blood is an acceptable alternative to human blood for bloodstain experiments for duplication of patterns. The authors as well as other bloodstain analysts have used bovine (cow), canine (dog), and equine (horse) blood for teaching and experiments and noted no significant differences in the patterns produced.

Unless an anticoagulant is used immediately, the animal blood will clot extensively as a reaction to its exposure to a foreign surface, namely the collection container. Over time,

Table 3.6 Summary of Facts about Blood

Viscosity	0.04 dynes/sec^2 at 37°C
Surface tension	50 dynes/cm
Relative density (sp gr)	1.060
pH	7.4
Volume	Adults: 60–66 ml/kg of body weight
	Males ~ 5.7 L or 1.5 gallons
	Females ~ 4.3 L or 1 1/8 gallons
	Children: 2.4–2.5 L for 80-lb child
	Infant: 0.25 L for 8-lb infant
% by volume	40%–45% red blood cells
	55%–60% plasma
Red cell count	4.8–5.4 million/μl of whole blood
White cell count	5000–7000/mm^3

some of the clot will lyse of its own accord and liberate cells, but the resultant specimen will usually be quite dilute in this case. If absolutely necessary, agitation will help to liberate some of the cells caught in the clot, producing a more concentrated specimen. Agitation should be used carefully because it can lead to physical destruction of cells. Adding an anticoagulant, such as EDTA, to the collection container will not only prevent most of the clotting that occurs but will also act as a preservative if the blood is to be stored for a while. Ideally, the anticoagulant should be added to the collection container prior to use and collection should be allowed to proceed with gentle agitation.

Physical Properties of Bloodstain Formation

Drop Formation and Travel

Consider a drop of blood forming on a person's fingertips. Blood will continue to accumulate on the most dependent portion of the fingertips until a drop is able to break free and fall. According to Newton's First Law of Motion, in the absence of the influence of a net force, a body at rest remains at rest and a body in motion remains in motion at a constant speed in a straight line or constant velocity. This law gives us the definition of a force: a force is any influence that can change the velocity of a body.

When a drop is forming in a dependent location, gravity is exerting a downward force on the drop. Simultaneously, the cohesive forces of the surface tension of the blood are trying to reduce the amount of exposed surface area and are exerting an upward force in opposition to gravity. As long as these forces remain equal, there is no net force applied to the drop, and therefore the forces — gravity and surface tension — cancel out each other. In the absence of a net force, there is no acceleration or no change in the velocity and, in accordance with Newton's First Law of Motion, a body at rest remains at rest. In the example of a newly forming drop of blood, the formation is arrested and no drop will break free.

The resistance of any body to a change in velocity is inertia. The inertia of a body is related to its mass. The more mass a body has, the more resistance it presents to any force attempting to change its state of rest or of constant velocity. For example, if a baseball bat is swung at a constant velocity and strikes first a baseball, and on the second swing, it strikes a bowling ball, there will be less acceleration of the bowling ball than of the baseball although the bat was swung at the same velocity (net force) both times. A bowling ball has more mass than a baseball; therefore, it has more inertia.

Newton's Second Law of Motion defines the relationship between inertia and mass, the applied force, and the net effect. This law states that the net force acting on a body equals the product of the mass and the acceleration of the body and links the cause (which is the force) and the effect (which is acceleration).

$$\text{Net force} = (\text{mass})(\text{acceleration}) \text{ or}$$

$$F = ma$$

(Note: ΣF may substitute for F to represent the sum of the forces or net force.)

In our example of a freely forming drop of blood, the upward force of surface tension is in opposition to the downward force of gravity. As more and more blood accumulates in the newly forming drop, the weight of the drop increases, as does its mass. To achieve a net force (F), it is necessary to break the stalemate between surface tension and gravity. The equation $F = ma$ in Newton's Second Law of Motion tells us the factors that must be changed to result in a net force—namely, mass (m) and acceleration (a).

As the weight of the drop increases in our example, there is a corresponding increase in mass. Mass, however, is a scalar quantity—meaning there is no direction associated with mass. It is not exerting an upward force with surface tension, and it is not exerting a downward force with gravity. Mass merely measures the response of a body to an applied force. That leaves the accumulating weight of the drop to create the overall net force, but weight is not mentioned in Newton's Second Law of Motion. How does the accumulating weight of the drop achieve a net downward force and allow the drop to break free of its source?

Weight is the gravitational force with which the earth attracts a body. As such, weight is the force that causes a body to be accelerated downward with the acceleration of gravity (g). Consider Newton's Second Law of Motion in terms of what we know about weight (w), the net force (F) equals the weight of a body (w), and the acceleration of the body (a) equals the downward acceleration of gravity (g).

$$\text{Newton's Second Law of Motion } F = ma$$

If F = weight of a body (w) and a = the downward pull of gravity (g), then $F = ma$ becomes $w = mg$, which may also be expressed as weight = (mass)(acceleration of gravity). In a more conventional form,

$$\text{If } F = w \text{ (weight of a body) and}$$

$$a = g \text{ (the downward pull of gravity)}$$

Then substituting into the equation for Newton's Second Law

$$F = ma \text{ becomes } w = mg$$

In simple English, this says that the weight of a body is equal to the product of the mass of the body multiplied by the downward pull of gravity.

$$\text{weight} = (\text{mass})(\text{acceleration})$$

Whereas the weight of a body can certainly vary between earth and outer space, that constitutes a very large change in distance from the earth's surface. For most situations, and certainly for most situations of concern in bloodstain pattern analysis, gravity can be considered a constant. That being the case, the weight of a body would be proportional to its mass—the body of larger mass would weigh more than the body of smaller mass.

Referring back to the equation $w = mg$, if g is a constant, then $w = m$. The accumulating weight of the drop represents a net downward force, and the drop is able to break free and fall away from the source.

The Shape of a Falling Drop

When the drop initially breaks free, it is very slightly elongated. As it continues to fall, the effects of air resistance will cause the sphere of blood to flatten slightly. A drop of blood has the same physical properties as a container of blood. In a drop of blood, the cohesive forces exerted by the molecules on the surface of the sphere are greater than the forces being exerted by the molecules in the interior of the drop. The attractive forces of surface tension cause the drop to assume the shape that exposes the least amount of surface area possible; that shape is a ball or sphere (Figures 4.1 and 4.2), not the frequent media representation of a "teardrop" shape of a falling raindrop or other liquid, as seen in Figure 4.3.

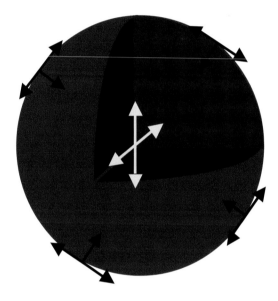

Figure 4.1 Spherical drop of blood in cross-section showing the forces of surface tension.

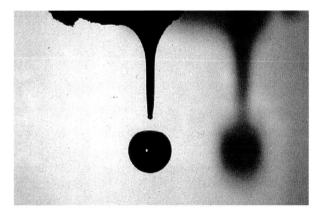

Figure 4.2 Spheroid shape of a blood drop resulting from the effect of surface tension as it falls through air after separating from a blood-soaked cloth.

Figure 4.3 Artist conception of a falling drop as "teardrop" shaped.

The spherical shape of a drop of blood results from the cohesive forces of surface tension, but the viscosity of blood is responsible for the retention of this spherical shape. A great deal of study has been done regarding the configuration of falling raindrops. A drop of water may actually oscillate so much in flight that the length or vertical axis is 97% of the drop's horizontal axis, making the drop rounded at the top and slightly flattened on the bottom. Because air resistance and the movement of the molecules within the drop influence the oscillations, larger drops of water will exhibit a greater degree of oscillation in flight. Conversely, smaller drops do not oscillate as much. Studies of raindrops by Ivars Peterson have shown that droplets 0.5 mm in diameter exhibit little if any oscillations or change in shape as they fall, whereas large rain drops, 5 mm in diameter, displayed distinctive changes in shape. The same holds true with drops of blood. Large drops of blood, such as a drop produced from a dropper or a nonimpact event, also exhibit distinct changes in shape while in flight. This has been confirmed by several studies, such as Pizzola using drops with a diameter of 3.7 mm; MacDonell with drops up to 4.6 mm in diameter; and Bevel and Gardner citing drops as large as 5.3 mm. Impact events result in drops of blood that are much smaller in volume and, thus, diameter. Drops created as the result of impact will demonstrate less oscillation during flight as a result of their smaller size.

If one were to consider a drop of blood and a drop of water of exactly the same diameter, the length of time over which the drop of blood would exhibit fluctuations in shape would be much shorter than that of the drop of water because of the damping time of blood. Damping time is a measure of the length of time any fluid will be distorted prior to reassuming the spherical shape caused by surface tension and is inversely proportional to the viscosity of the fluid. Because blood is a relatively viscous fluid, the oscillations or changes in shape that occur during flight are quickly restrained. In comparison to water, because blood is four times more viscous, damping occurs four times more quickly (Figure 4.4).

A study by M.A. Raymond showed that a passively formed drop of blood with a diameter of 2 mm in flight (4- to 5-mm resultant stain) would distort down to 1% of its original amplitude. This distortion was damped in the time required for the drop to free-fall approximately 7 cm. A drop of blood with the initial velocity of 8 m/sec would oscillate for only the first meter of travel. Ross Gardner studied the distortion demonstrated by smaller drops in the range of those resulting from impact. His studies showed that the majority of

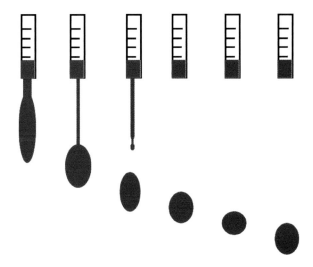

Figure 4.4 Drop shape as a function of time. Each frame represents a lapse in time.

oscillations were damped within 0.05 sec. Within reasonable limits, we can state that drops of blood are spherical in shape.

Volume of a Drop of Blood

The accepted standard volume of a drop in a clinical setting is that 1 mL equals 20 drops. This means than a single drop would contain 50 μL of liquid. Herbert Leon MacDonell examined this measure in an experiment that included blood samples from males and females ranging in age from 2 to 89 years old. Specimens included freshly drawn blood, as well as blood that had been anticoagulated using ethylenediamine tetraacetic acid (EDTA), citrate, and oxalate. His findings showed that the average volume of a drop was 0.0505 mL or ~50 μL. The factor MacDonell did identify, however, that would affect the drop volume was rapid bleeding. In situations where there is rapid bleeding, such as the complete exposure of one of the larger vessels, the drops that result will be slightly larger than the typical 50-μL drop. Conversely, slower bleeds do not create drops with a volume smaller than 50 μL.

The foregoing discussion is applicable only to freely forming drop volumes originating from situations designed to duplicate venous flow as closely as possible. MacDonell was the brunt of a great deal of criticism for his use of the wording "normal" drop and has subsequently changed the wording to "typical" drop. Bevel and Gardner cite that, although they were able to create stable drops with a volume of as much as 60 μL (5.5-mm diameter), MacDonell's original 50-μL (4.6-mm diameter) "typical" drop more closely represents what one might expect when drops form freely during venous flow.

Drop Volume as a Function of Source

Drop volume is very much dependent on the source object from which the drop originates. As discussed earlier in this chapter, the amount of blood required to break surface tension dictates the volume of the drop. If a drop is forming on a larger surface, the increased surface area presents an increased surface tension. As a result, drops that originate from a large surface such as a baseball bat will be larger in volume than drops originating from a small surface such as a knife. A single instrument such as a hammer exhibits a large and small surface area from which drops of blood may originate, as shown in Figures 4.5a and b.

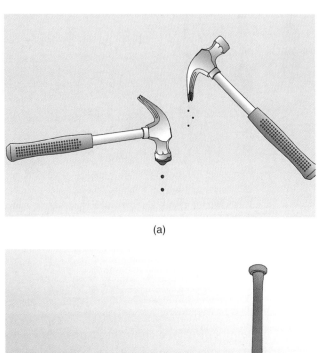

(a)

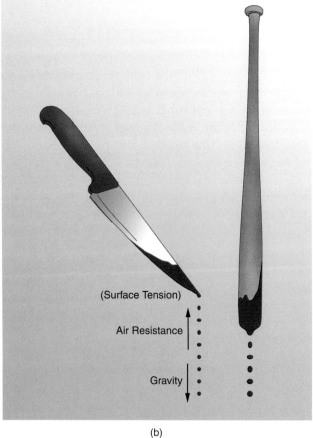

(b)

Figure 4.5 Effect of increased surface area of the source object and resultant drop volume. (Courtesy of Alexei Pace, Malta.)

Table 4.1 Variation in Drop Volume Caused by the Source Object

Object	Size and Position	Stain Diameter (in millimeters)
Glass tube	1.3-mm inner diameter, vertical position	14.5
Glass tube	4.0-mm inner diameter, vertical position	24.8
Steak knife	12.5-cm blade length, 1.3-cm blade width	15.8
Knife	15.3-cm blade length, 2.6-cm blade width	17.7
Screwdriver	Phillips #2, 6-mm diameter shaft	17.5
Crescent wrench	30-cm length (drop from fixed side of jaw)	23.1
Hammer head	2.6-cm diameter head, horizontal position	26.9
Hammer claw	1.3-cm claw width, horizontal position	25.0
Heavy cotton fabric	Vertical position	21.0
Index finger	Vertical position	19.6
Thumb	Vertical position	21.1

Experiments conducted by Terry Laber have demonstrated the drastic variation in drop volume caused by the source object. Laber found drops as small as 13 µl falling from a single human hair compared with drops of 160 µL originating from a terry cloth sock. MacDonell published data in 1990 that demonstrated the fact that stain diameter would be affected by drop volume. His experiments held the height of the dropping distance constant, which dramatized the effect the drop volume had on the stain diameter. R. Bruce White measured passive falling blood drops from various sources and the diameter of the resultant stains. He used smooth white cardboard as a target surface and a falling distance of 1 meter. Table 4.1 is a summary of his findings.

Distance Fallen

All free-falling bodies near the earth's surface have the same downward acceleration because of gravity. Gravitational acceleration (g) is 9.8 m/sec or 32 ft/se. In a vacuum, velocity (v) is represented by the following:

$$v = (\text{gravitational acceleration})(\text{time}) \text{ or } v = gt$$

Therefore, a freely falling body would accelerate 32 ft/sec for every second of fall. At the end of 1 second, the velocity would be 32 ft/sec; at the end of second #2, the velocity would be 64 ft/second; etc.

The real world is not a vacuum. The effects of air resistance cannot be ignored. Buoyancy of the falling object must also be considered when comparing the rate of fall for two different objects. Starting from a resting position, a drop of blood continually accelerates and increases its velocity over time. As the drop is falling, however, air resistance or friction is also accumulating. Eventually the downward force (gravitational acceleration) is equaled by the friction of air resistance and the drop will cease to accelerate. The drop will then maintain a constant velocity, referred to as maximum terminal velocity.

The maximum terminal velocity of a typical 50-µL drop of blood is 25.1 ft/sec ± 0.5 ft/second. This value is an absolute maximum for falling drops in air. Maximum terminal velocity is achieved in the length of time required for a typical 50-µL drop of blood to fall a distance of 14 to 18 ft. From a practical standpoint, there is very little appreciable change in drop size seen after a falling distance of 4 feet. Because gravitational acceleration depends

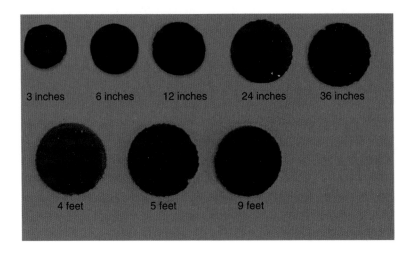

Figure 4.6 Influence of dropping height on stain diameter. Target surface of glass; all drops were 50 µL in volume.

on weight, smaller drops will achieve maximum terminal velocity more quickly but they will be traveling at a slower velocity. Upon achieving maximum terminal velocity, the drop will cease to accelerate and the resultant stain diameter ceases to increase significantly (Figure 4.6). The increase in stain diameter relative to distance a typical 50-µL drop of blood has fallen is represented graphically in Figure 4.7.

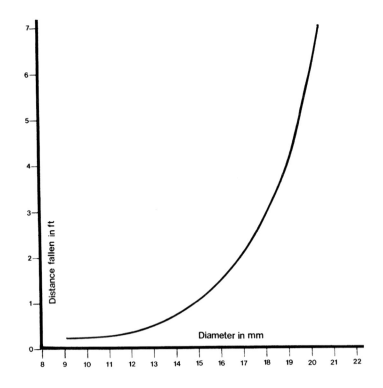

Figure 4.7 Graphic representation of the distance fallen in feet by single drops of blood dripped from fingertips versus diameter in millimeters.

The estimation of the distance a drop of blood has fallen as merely a function of the diameter of the resultant stain can result in gross error in interpretation and is not recommended. The importance of the following statement by MacDonell in *Bloodstain Pattern Interpretation* cannot be overemphasized:

> Any estimate of the distance a drop of blood fell before impact as a function of the diameter of its resulting stain is essentially an exercise in futility.

The volume of the drop creating the stain must be known, and the effect of the target surface must also be taken into consideration. Bloodshed at a crime scene may produce many drops of different volumes that create stains on different surfaces. Reproducing the target surface is easy enough. Determination of the original drop volume is possible, but it is likely to be extremely difficult. One must determine how the drop was produced during violent activity. Did the drop that produced the stain originate from the skin surrounding a bloody head wound or from a single hair when the head was moved? Did the drop fall from a weapon or other object? In actual practice, these variables would be difficult to ascertain with any degree of certainty.

Classification of Bloodstain Patterns

An understanding of how the bloodstain pattern analysis terminology has been created and is evolving is necessary. A review of the literature reveals that many of the early researchers used terms based on the physical appearance of the stain patterns. Often they would associate the appearance of the patterns with items they had previously observed in nature or the laboratory. Terminology is necessary to convey research as well as scene observations in a clear and concise manner. The terminology of this discipline is continually being refined in an effort to create a clear and concise language in which scientists can convey information.

When we examine bloodstain pattern analysis terminology, two trends stand out in regard to how terms were and are derived:

1. From the physical APPEARANCE of the stain pattern
2. From the MECHANISM in which the pattern was created

Bloodstain patterns are initially visually examined to evaluate the physical characteristics of the stains, including size, shape, distribution, location, and concentration. Our analysis of the physical appearance of a stain pattern would have minimal significance without prior knowledge of how blood will behave under known conditions. In evaluating the physical characteristics or *appearance* of a pattern, we cannot help but to be assessing the potential *mechanism(s)* by which the stain was created.

Bloodstain pattern analysts are trained to correlate the physical characteristics of a static stain pattern with the dynamic mechanism(s) by which they may be created. Analysts acquire this basic knowledge of the discipline through experimentation where stain patterns are created by a variety of known mechanisms, and then they assess the physical characteristics, *appearance*, of the resulting stain patterns.

Analysts must have knowledge of the mechanism(s) in which bloodstain patterns are created prior to initiating an actual case evaluation. The *appearance* and *mechanism* do

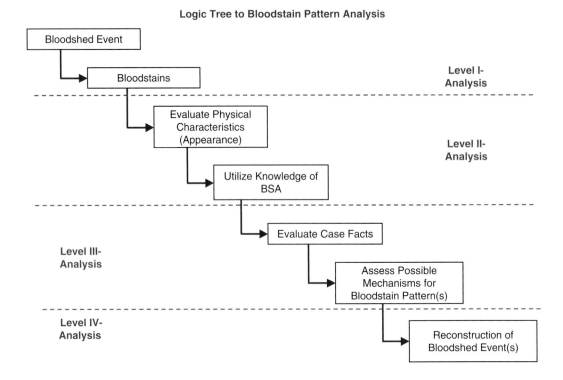

Figure 4.8 Logic tree to bloodstain pattern analysis.

interrelate, but during our initial evaluation of the patterns we should be more concerned with categorizing the bloodstain patterns based upon their physical *appearance*. The physical *appearance* of the stain patterns in conjunction with case facts and prior knowledge of how blood behaves under known conditions is what allows us to establish the *mechanism(s)* by which a pattern was created.

As you will come to understand when we use bloodstain pattern analysis to reconstruct past events, we need to evaluate more than the *appearance* of the stain patterns. The analysis of bloodstain patterns may be divided into four specific levels of analysis, as shown in Figure 4.8. The four *levels* of analysis correlate directly with our actual bloodstain *pattern* taxonomy. After recognizing the substance as blood, we evaluate the physical characteristics, or *appearance*, of the stain patterns during our initial examination of the bloodstain patterns. Based on their physical characteristics, the bloodstain patterns may be classified under one of three primary bloodstain pattern categories, as seen in Figure 4.9. These categories will be expounded upon in later chapters.

Passive

Passive bloodstains are patterns whose physical features indicate they were created without any significant outside force other than gravity and friction.

Spatter

Spatter bloodstains exhibit directionality, vary in size, and are associated with a source of blood being subjected to an external force(s), in addition to gravity and friction.

Physical Properties of Bloodstain Formation

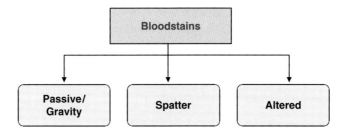

Figure 4.9 Primary bloodstain categories.

Altered

Altered bloodstains are patterns whose appearance indicates the blood and/or pattern has undergone a physical and/or physiologic alteration.

In Figure 4.10, the subcategories or *mechanism*-based terms relative to bloodstain pattern analysis are listed below their corresponding *appearance*-based categories. Additional information such as types of injuries, stain location, history of the evidence, and interventions by police and emergency medical technicians on the scene will assist the analyst in further categorizing the stains as to the mechanism that created them. How far we are able to go in regard to establishing the *specific* mechanism(s) that created a pattern is often dependent on factors beyond the general appearance of the bloodstain patterns.

This taxonomy will assist you in your analysis of bloodstain patterns by establishing levels your analysis may attain dependent on available facts. As you proceed further down

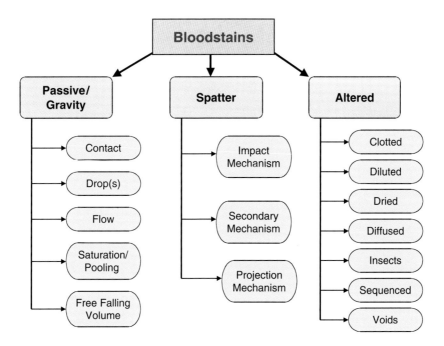

Figure 4.10 Primary bloodstain categories with subcategories.

the taxonomy you will notice there is a certain degree of unavoidable overlap. A classic example would be clotted spatter, which would fit under two categories: spatter and altered. This would be referred to as a complex pattern in that more than one mechanism applies to the stains, a spatter-producing mechanism as well as the physiology of blood clotting. Facts such as limited staining, highly irregular appearance, surface texture issue, and so on may limit your analysis of a stain pattern to inconclusive.

Passive Bloodstains

5

Introduction

The term passive refers to bloodstains and patterns that have been produced from free-falling drops or a volume of blood that has been subjected to the external forces of gravity, air resistance, and the nature of the surface on which the stains have formed. Passive bloodstains are distinguished from the spatter groups in that the blood has not been subjected to a significant external force or impact at its source. They comprise a diverse group of bloodstains and patterns that are commonly encountered at bloodshed events (Figure 5.1).

Free-Falling Drops on Horizontal Surfaces

As discussed in Chapter 4, a drop or drops of blood falling through air will retain their spherical shape and will not break up into smaller droplets unless the surface tension is disrupted by an external force such as an impact. Blood may drip from an exposed wound or from blood-soaked clothing, hair, weapons, or any object that contains a sufficient volume of blood to permit the formation of a drop that breaks away from the primary source of blood.

Theoretically, bloodstains that result from spherical free-falling drops of blood onto a horizontal surface are circular in shape. Their resulting diameters are a function of the volume of the drop, the distance fallen, and the surface texture on which it makes impact. On impact with smooth, hard, nonporous surfaces, the surface tension of blood drops will resist rupture and uniformly circular stains will be produced independent of the falling distance. Conversely, rough-textured or porous surfaces will disrupt the surface tension of blood drops and cause them to rupture on impact. The resultant bloodstains will exhibit distortion, irregular shapes, and spiny edges. Spines are the pointed-edge characteristics of a bloodstain that radiate away from the central area of the stain. In addition to spiny edges, the bloodstain may exhibit some secondary peripheral satellite spatter. It is important to understand that the degree of distortion and spattering of bloodstains resulting from free-falling drops is a function of the surface texture of the target rather than the distance fallen. Concrete, unpolished wood and fabric material are considered to be rough surfaces. Newspaper and tissue paper, although not considered rough surfaces, often exhibit very irregular and distorted bloodstains, whereas glass, porcelain, and tile, as well as hard smooth cardboard, exhibit the least distortion with respect to bloodstains on their surfaces. Figure 5.2a–j show examples of passive stains on various surfaces.

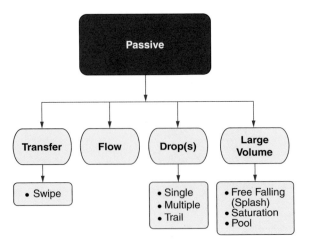

Figure 5.1 Taxonomy of passive category of bloodstains.

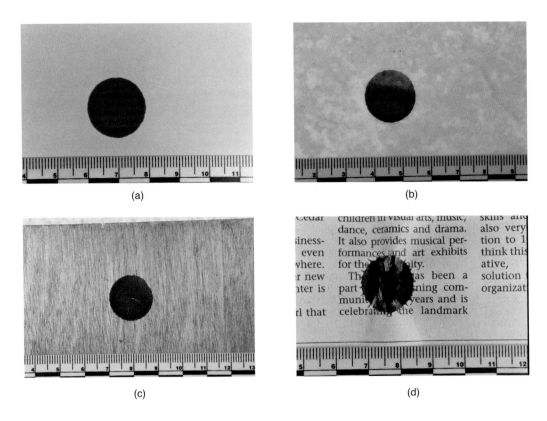

Figure 5.2 Effects of target surface textures on bloodstain characteristics and degree of spatter produced from single drops of blood that fell 36 in. Note the production of spines and small spatters around the periphery of the central bloodstain as a result of contact with rough surfaces. a) Glass; b) smooth tile; c) finished pine board; d) newspaper; e) linoleum; f) pressure-treated decking; g) corrugated cardboard; h) denim; i) 100% cotton T-shirt; j) carpet.

Passive Bloodstains

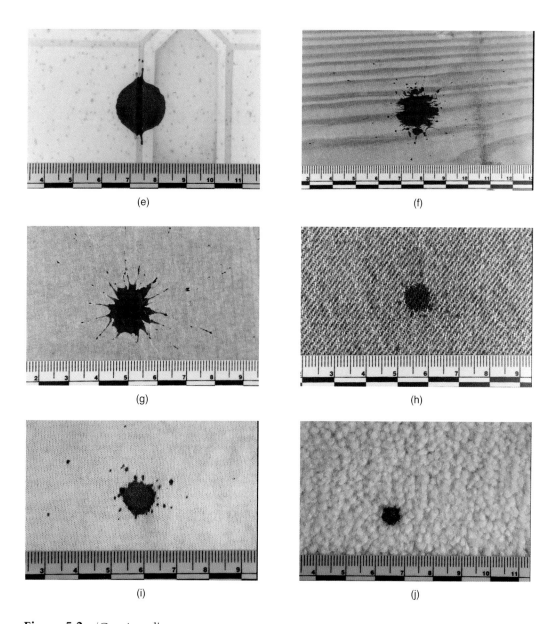

Figure 5.2 (Continued).

Free-Falling Drops on Angular Surfaces

When spherical, free-falling drops of blood traveling through the air strike a nonhorizontal surface, the resultant bloodstains are more oval or elliptical and elongated in shape rather than circular in shape. The elongation is produced by the blood drop contacting and wiping or skidding against the target surface. The more acute the angle of impact, the greater the elongation of the bloodstain as the width decreases and the length increases. This is more evident in stains that have resulted from droplets that have fallen at angles of impacts of 60° or less. Figure 5.3a–i shows the increase in elongation of stains as the angle of impact decreases from 90° to 10° in 10 degree increments. The narrowest end of elongated bloodstains produced

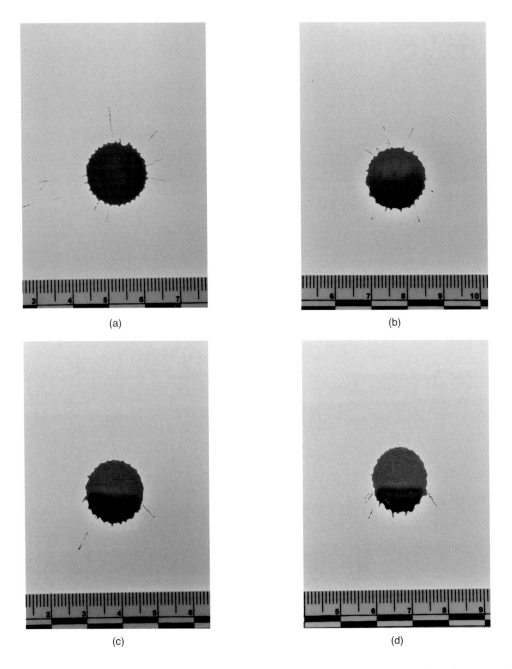

Figure 5.3 Increasing elongation of bloodstains relative to decreasing angles of impact of single drops that fell 36 in. onto smooth cardboard. a) 90°; b) 80°; c) 70°; d) 60°; e) 50°; f) 40°; g) 30°; h) 20°; i) 10°.

from free-falling drops striking a nonhorizontal surface points in the direction of travel. Depending on the type of surface impacted, the characteristics at the terminal edge of the bloodstain may exhibit tapering, spiny projections; a small dotlike spatter; or a tadpole-shaped wave cast-off which has flipped from the parent stain. Whereas the tail of the parent drop

Passive Bloodstains

Figure 5.3 (Continued).

points in its direction of travel, the tail of the wave cast-off points back to the parent bloodstain (Figure 5.4a and b). In 1986, Pizzolla described the event as follows:

> The distortion of the drop is limited to its lower area in contact with the impacted surface. As the drop continues its travel, it gradually collapses downward with respect to the target surface accompanied with little change in the shape

(i)

Figure 5.3 (Continued).

of the upper hemisphere. The top of the upper hemisphere falls further while the fluid displaced during the drop collapse is forced out radially, forming a ring at the circumference. The surface tension counters the lateral spreading of the drop.

Shortly after the collapse, the center region is significantly depressed. Following the formation of this depression or involution, the fluid forced to the rim retracts, coalesces, and progresses forward into a somewhat prolate form at the leading edge. The lower portion of the prolate area adheres to the impact surface while the upper portion grows or forms a droplet as it rises away from the impact surface.

If the surface is at a sufficient angle and the drop has sufficient energy, the kinetic energy of the moving blood initially overcomes the surface tension and pulls away from the main body of blood to form a droplet, which becomes approximately spherical, drawing out a fine filament in the process. If the velocity is sufficiently high, the droplet formed can separate from the filament. Angular stains on rough-textured surfaces will exhibit irregular and distorted shapes, and accurate measurements are difficult to establish. Figure 5.5a–c shows the effect of passive stains on a denim surface produced from falling drops of blood at 60°, 30°, and 10° respectively.

Drip Patterns

When there are multiple free-falling drops of blood produced from a stationary source onto a horizontal surface, drip patterns will result from blood drops falling into previously deposited wet bloodstains or a small pool of blood. These drip patterns will be large and irregular in shape with small circular to oval satellite spatters around the periphery of the central stain

Passive Bloodstains 77

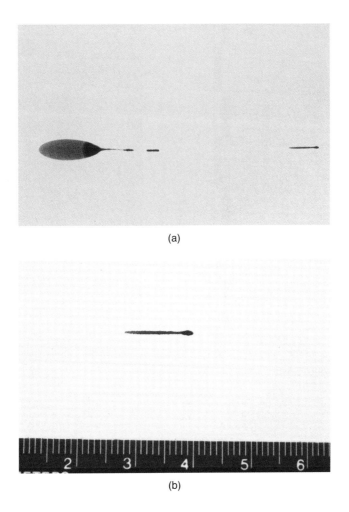

Figure 5.4 (a) View of elongated bloodstain with elongated terminal edge pointing in the direction of travel from left to right. The wave cast-off is at the right and points back to the parent stain. (b) Closer view of wave cast-off stain.

on the horizontal as well as nearby vertical surfaces. Satellite spatters are the result of smaller droplets of blood that have detached from the main blood volume at the moment of impact (Figure 5.6, Figure 5.7, and Figure 5.8). If blood drips into itself from a height of less than 12 in. on a smooth, hard surface, *no* satellite spatters will occur. It is possible to have a pool of blood created by a drip mechanism and create no satellite spatters. Chapter 6 discusses secondary or satellite spatter in more detail.

From a practical point of view it is important that the investigator be able to recognize the types of bloodstains and patterns resulting from free-falling drops based on their size, shape, and distribution and to document their locations. The stains should be related to the possible sources and movement of these sources through the recognition of drip patterns and drip trails.

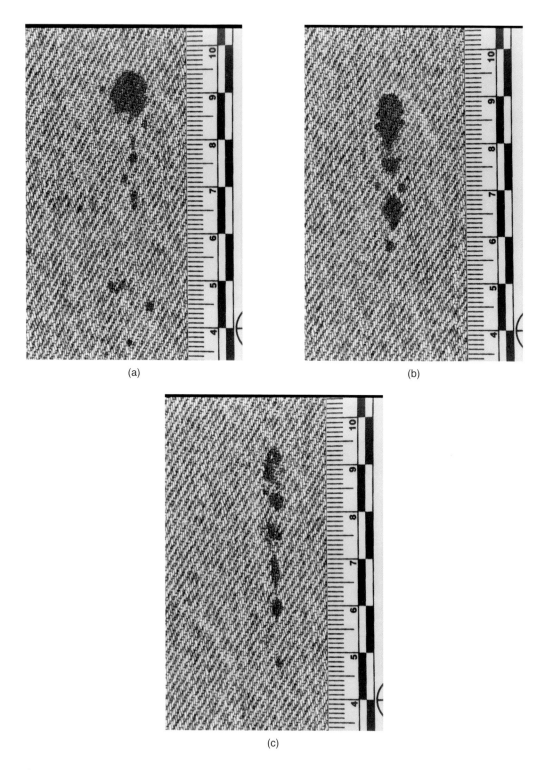

Figure 5.5 Increasing elongation of bloodstains relative to decreasing angles of impact of single drops that fell 36 in. onto denim. a) 60°; b) 30°; c) 10°.

Passive Bloodstains

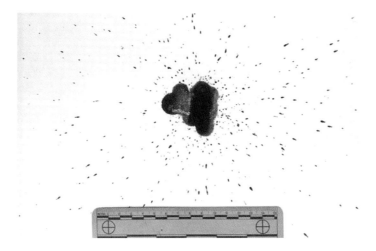

Figure 5.6 Drip pattern created by free-falling drops of blood that fell into each other onto smooth cardboard from a height of 36 in. Note the satellite spatter around the periphery of the parent stain.

Figure 5.7 Pool of blood with a large amount of satellite spatter on the nearby horizontal and vertical surfaces created by blood dripping from the landing above the area where victim was found. (Courtesy of Martin Eversdijk, The Netherlands.)

Drip Trails

When free-falling blood drops are subjected to a force producing horizontal motion as well as the downward pull of gravity, they may have an angular impact on vertical or horizontal surfaces. Directionality may be established by the stain shape and edge characteristics. A person bleeding while in a fast walk or run would provide the source of blood with sufficient horizontal motion to allow the free-falling drops to impact a horizontal surface such as a floor at an impact angle of less than 90°. A series of bloodstains resulting from free-falling

Figure 5.8 Drip pattern of blood with satellite spatter on entryway steps where victim was reported to be seen upright and bleeding prior to his collapse on a bed inside the residence.

blood drops traveling in a particular direction with sufficient horizontal motion may exhibit scalloped edges, indicating the direction of travel of the source (Figures 5.9a and b). A person bleeding while in a slow walk may produce nearly circular bloodstains, and directionality based on edge characteristics of the bloodstains may not be obvious (Figure 5.10). If a drop with horizontal motion falls a substantial distance before striking a surface, it will be more circular because gravity will affect the horizontal motion. Vertical surfaces such as walls may receive an angular impact when the surface is in proximity to the free-falling drops.

If the source of blood is not an active bleeding site, the quantity of blood available for dripping may be limited and the distance of the trail pattern may be short. This applies to bloody weapons or objects carried around or from a scene by an assailant. The edge characteristics of the parent drop may show the direction of travel. When blood is dripping from a moving active source of bleeding, the trail may be quite long, depending on the amount of blood available (Figure 5.11).

The longest blood trail that came to the attention of the authors was a case in Appleton, Wisconsin, where police followed an interrupted blood trail for more than 1.5 miles. Apparently a man was locked out of his house without his keys and had cut himself after shattering the window to gain entry. He wandered around bleeding profusely for more

Passive Bloodstains

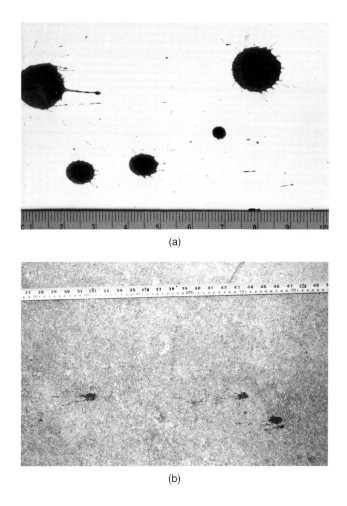

Figure 5.9 (a) Drip trail exhibiting scalloped edges indicating direction of travel from left to right. (b) Drip trail exhibiting scalloped edges and elongated spines on cement surface indicating direction of travel from right to left.

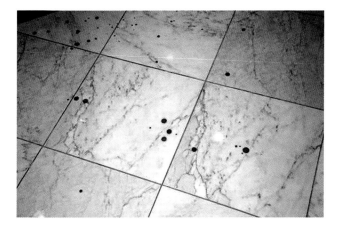

Figure 5.10 Drip trail consisting of nearly circular bloodstains. Edges do not indicate direction of travel.

Figure 5.11 Extensive drip trail on plank deck as a result of active bleeding from a source. Photo courtesy of Inv. Marlin Hohnstein, Lincoln Nebraska Police Department.

than 3 h. He eventually walked home, where he was let in by his wife who had arrived home in the interim. The man was unsure where he had been and had no explanation for his wandering. No crime had been committed.

Splashed Bloodstain Patterns

When a quantity of blood in excess of 1.0 mL is subjected to a minor impact (such as stepping into a pool of existing blood) or is allowed to free-fall at least 4 inches to a surface affected only by the force of gravity, a splashed bloodstain pattern will be produced. Splashed bloodstain patterns are usually characterized by a large central stain exhibiting minimal distortion with surrounding, long, narrow spatters. There will be very little satellite spatter present. The periphery of the central stain often exhibits minimal distortion or spine formation as opposed to a projected bloodstain pattern. This is variable depending on the impact velocity and the surface texture. The surrounding spatters are long and narrow because they have struck the surface at a very acute angle close to the horizontal. H.L. MacDonell has demonstrated that 90% of spatters created by splashed blood exit the original pool at angles of impact of less than 40 degrees.

These long, narrow spatters that surround the central stain in a splashed bloodstain pattern have a tendency to exhibit reverse directionality. Instead of the tails of these spatters pointing in the direction of travel, they are often reversed, with the tails pointing back toward the central stain. Some find the tendency for these spatters to reverse directionality to be of great concern. This should not be the case because the factors that cause the reversed directionality are well understood. They should be considered simply a characteristic that can be used to determine the type of force that led to the creation of the pattern.

The tendency of these spatters to reverse directionality is attributable to two causes:

1. Very acute angle of impact
2. High incidence of colliding with each other

Passive Bloodstains

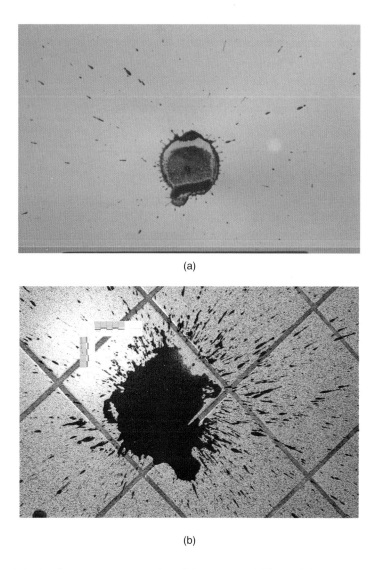

(a)

(b)

Figure 5.12 (a) Bloodstain pattern produced by 1 mL of blood falling 36 in. onto smooth cardboard. (b) Bloodstain pattern produced by 5 mL of blood falling 36 in. onto tile floor.

Spatters may impact a target surface at a very acute angle of impact either because of their path of travel or the position of the target relative to the trajectory of the droplets. Reversed directionality is also attributable to the relatively low amount of energy being exerted on the source of blood coupled with the fact that more than 90% of the spatters are exiting 40° or less from the horizontal. These factors cause a very high probability that the spatters exiting the pool will collide with each other in flight and reverse their direction of travel.

Secondary blood splashing or ricocheting may occur as a result of the deflection of large volumes of blood after impact from one surface to another. When sufficient bleeding has occurred, splashed patterns may be produced by the movement of the victim or assailant. Larger quantities of splashed blood will create more spatters (Figures 5.12a and b). Stepping into a pool of blood can cause spatters on the inner aspects of footwear but not on the outer aspect as blood is splashed from one shoe to the other.

Flow Patterns

A flow pattern is an accumulation of a volume of blood with generally regular margins that has moved across a surface from one point to another as a result of the influence of gravity and the contour of the surface. Depending on the initial volume of blood, there may be single or multiple flows on a surface (Figures 5.13a and b). Flow patterns may be observed on the body or clothing of the victim as well as the surface on which the victim is found (floor, bed, chair, etc.) (Figure 5.14). The alteration of the direction of flow patterns is a common observation. The directional change in flow or other alteration may result from obstructions or changes on the surface or a change in position of the person or object that contains the wet blood (Figures 5.15 and 5.16). The altered flow pattern shown in Figures 5.17a and b was produced on painted drywall. The closer view of this stain shows a textured detail within the stain that resembles a transfer pattern but is actually the pattern of the painted drywall. Flow patterns of blood are important to recognize because they may yield important information such as movement of a victim and/or assailant during bloodshed as well as postmortem movement or disturbance of the body and alteration of the scene of the death.

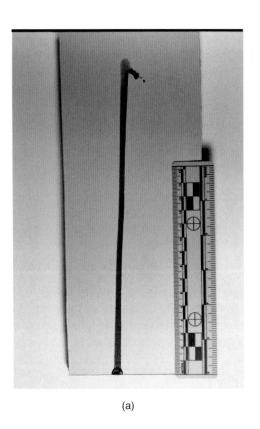
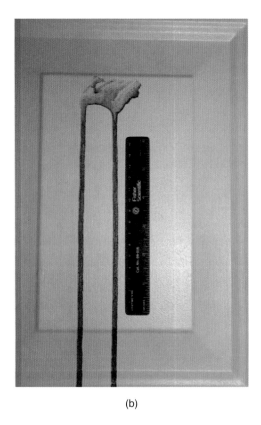

(a) (b)

Figure 5.13 (a) Single flow pattern on wall indicating downward directionality. (b) Multiple flow patterns on wall indicating downward directionality. Note that the transfer stain from the source of the blood is larger than that in Figure 5.13a.

Passive Bloodstains

Figure 5.14 Multiple flow patterns on body, indicating movement of victim.

Figure 5.15 Flow patterns on wall indicating directional change in flow resulting from surface texture of the wall. (Courtesy of L. Allyn DiMeo, Arcana Forensics, La Mesa, California.)

Figure 5.16 Multiple flow patterns on bathroom walls and adjacent area that resulted from massive head injury resulting from self-inflicted gunshot.

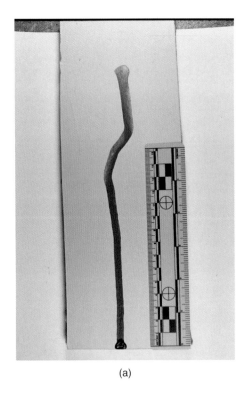
(a)

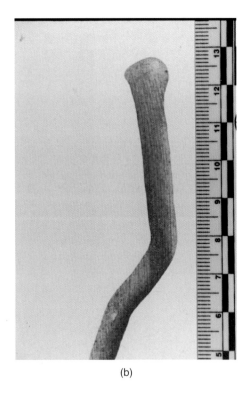
(b)

Figure 5.17 (a) Altered flow pattern on painted drywall showing texture detail of wall. (b) Closer view of altered flow pattern on painted drywall showing texture detail of wall.

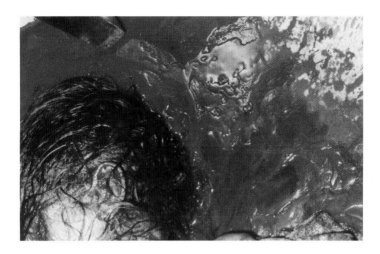

Figure 5.18 Large pool of blood accumulated around victim who received blunt force injury to head.

Blood Pools

A blood pool is an accumulation of a volume of blood on a surface whose shape is not specific but it conforms to the surface contour. A pool of blood is often associated with extensive bleeding of a victim in a position that allows the accumulation to occur (Figure 5.18). It may also be associated with flow patterns resulting from the effect of gravity on a sloped surface with the accumulation of blood on a surface. The blood is not totally absorbed into the surface and will eventually dry or clot, and there may be evidence of serum separation and formation of heavy crusts of blood, which will be discussed in a later chapter. The quantity of blood within flow patterns and pooling of blood or lack thereof should be observed with respect to the injuries sustained by the victim to help determine whether the victim received the injuries at one area or another location after injury or death. Figure 5.19 is a view of a bathroom where passive bleeding and pooling of blood occurred in the area of the sink. However, the bathroom did not show signs of a struggle. The altercation had occurred in the kitchen, and the victim was found on his bed. The blood in the bathroom was that of the victim.

Saturation

A saturation stain is an accumulation of a volume of blood without specific shape that has absorbed into the surface rather than pooled on top of the surface (Figure 5.20). When there has been extensive bleeding by a victim, large saturation bloodstains may occur on carpets and bedding as well as the clothing of the victim and can obliterate other stains of interest that occurred prior to the saturation. Large volumes of blood can saturate through carpet and carpet padding to the floor below as well as through several layers of bedding. A wicking effect or diffusion of blood may extend the size of the original saturation stain and is recognizable by its generally lighter color and irregular edges at the periphery of the diffused areas. There may not be evidence of clot formation if most of the blood is absorbed into the material.

Figure 5.19 View of a bathroom with pooling of blood on edge of sink and downward flow patterns on front of sink cabinet. Note that the water was running and created a void in the sink as well as a large amount of satellite spatter on the rear of the sink, splash guard, and lower portion of bathroom mirror.

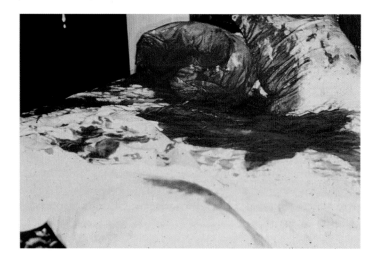

Figure 5.20 Large blood saturation stains on bed linens and pillows.

Transfer Bloodstains

The subclassification of passive bloodstains in the category of transfer includes bloodstains produced by a variety of events. In general, the term transfer refers to stains produced as the result of an object wet with blood contacting a nonbloody surface and transferring a quantity of blood onto that surface. The term transfer also applies to the physical alteration of an existing wet bloodstain caused by a nonbloody object touching or moving through it. Transfer stains require contact between two surfaces. The recognition of transfer bloodstains is important to establish movement and activities of the victim and the assailant both during

Passive Bloodstains

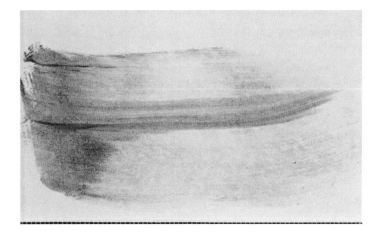

Figure 5.21 A swipe exhibiting a feathered edge indicating motion from left to right.

and after the assault. It is important to examine the scene, the body of the victim, and the clothing of the victim and that of the suspected assailant for bloodstains of this type.

Blood smear is a term that is often used to describe a nondescript transfer stain on a surface that has been produced by contact with a bloody object. The stain contains no recognizable features to indicate the object that produced it. A *swipe* is a transfer stain resulting from a moving object wet with blood contacting a nonbloody surface that may exhibit directionality (Figure 5.21). Commonly observed swipes are often recognizable as having been produced by a specific object such as bloody hair and fabrics (Figure 5.22). Ross M. Gardner described research concerning the determination of the directionality of swipes. He concluded that the determination of direction of motion is possible in some swipe patterns, using the orientation of five identified characteristics. The recurring pattern evident across all surfaces is the presence of the irregular demarcation on the contact side,

Figure 5.22 Bloody hair swipes on linoleum floor. (Courtesy of Detective Jan Ewanow, Binghamton, New York Police Department.)

accompanied by the presence of one or more of the four other characteristics (feathered demarcation, striations, diminished volume, welling) on the departure side of the stain. Of the four departure side characteristics, striations in the stain are the least valid characteristic to consider, whereas diminished volume appears to be the most valid characteristic.

As with all bloodstain patterns, determination of direction of motion is not always possible. Situations in which both departure and contact edges share common characteristics (e.g., both have an irregular demarcation) are common. In such instances, it would be inappropriate to attempt to identify the direction of motion.

A *wipe* pattern is characterized as a bloodstain created when an object moves through a preexisting wet bloodstain, thereby removing blood from the original stain and altering its appearance. The nature of the object producing swipes or wipe patterns may sometimes be determined by careful observation as having been produced by a hand, finger, fabric, etc. Figure 5.23, Figure 5.24, and Figure 5.25 relate to a case where a woman was beaten on her bed and subsequently moved to the floor. Narrow, linear wipes within the saturated blood on the bed were consistent with having been produced by a broom.

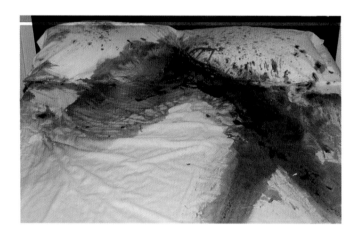

Figure 5.23 Blood saturation area on bed where a victim was beaten with a hammer prior to their final position on the floor.

Figure 5.24 Closer view of area of saturation on bed linens. Parallel linear wipes indicate attempted clean-up effort with a broom.

Passive Bloodstains

Figure 5.25 Bloodstained broom found at scene.

A *drag* pattern is a type of wipe created when an existing pool or source of blood is altered by the movement of a victim or other object on a surface. The feathering of the leading edge of the pattern may indicate the direction of the movement (Figures 5.26a–c). Wipe patterns may also be classified as altered bloodstain patterns because at least two mechanisms are involved in their production.

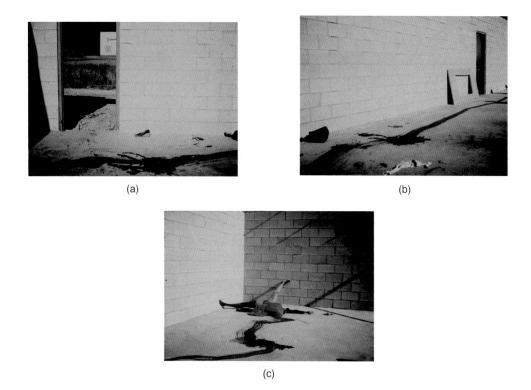

Figure 5.26 (a) Initial point of attack on victim indicating pooling of blood and start of drag pattern. (b) Continuation of drag pattern with a second pool of blood indicating a pause in the forward motion. (c) Terminal point of drag pattern and final location of victim at the scene.

Recognizable Blood Transfer Stains

Transfer bloodstain patterns with recognizable features that have resulted from contact with a bloody object are commonly observed at bloodshed events. A recognizable mirror image of the original surface or at least a recognizable portion of the original surface may be transferred to the second surface. Recognizable features may be present in direct transfers as well as swipe and wipe patterns. Common examples of recognizable transfer bloodstain patterns on surfaces are palm, finger, feet, or shoe impressions and fabric and weapon impressions. In some cases, class and or individual characteristics may be determined from the bloody transfer impressions and are frequently subjected to blood enhancement techniques, as will be discussed in Chapter 16. Figures 5.27 through 5.34 show various examples of recognizable blood transfer stains or impressions.

Smudges, wipes, and transfer patterns of victim's blood are frequently produced on and within an assailant's vehicle while leaving the scene. Door handles, steering wheel, gearshift apparatus, console, and floor pedals should be examined carefully for these types of bloodstains. Chapter 13 will describe examination of vehicles for bloodstains in greater detail.

Conclusion

Passive stains as a group are usually recognized without difficulty. The subcategory of transfer stains is of special significance to the analyst because they often capture a recognizable mirror

Figure 5.27 Bloody finger transfers on door casing.

Passive Bloodstains

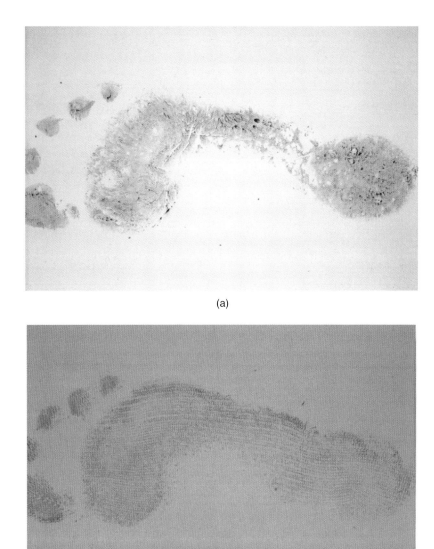

(a)

(b)

Figure 5.28 (a) Transfer pattern in blood of right foot exhibiting sole and toe detail. (b) Transfer pattern in blood of the same right foot clad in a sock showing distinct weave pattern of fabric.

image that can be associated with specific items such as weapons and impressions of footwear, handprints, and fingerprints. When attempting to determine whether an object could have produced a particular transfer pattern, it is usually necessary to conduct a series of experiments using items similar to those in question because it is never a good practice to add blood to an evidentiary item. It is also important that test patterns be reproduced on a similar surface on which the original pattern was produced.

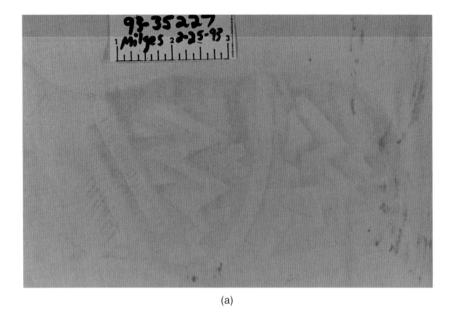

(a)

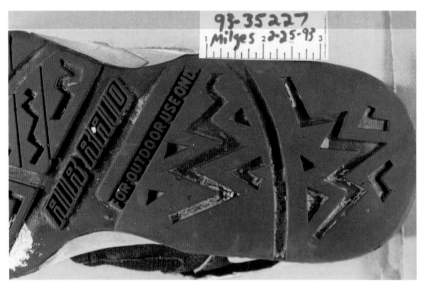

(b)

Figure 5.29 (a) Bloody transfer impression of shoe sole with triangular pattern. (b) Shoe with sole characteristics similar to transfer impression depicted in Figure 5.29a.

The differentiation between a small transfer pattern and an impact spatter pattern may determine whether he or she could have been a perpetrator or merely someone who came into contact with the source of blood. The determination of whether the bloodstains are the result of spatter or transfer is not always easy and often requires microscopic examination of the stained area and experimentation. Transfer stains on fabric usually coat the top of the fabric weave, as shown in Figure 5.35, whereas spatters are usually driven down between the fibers.

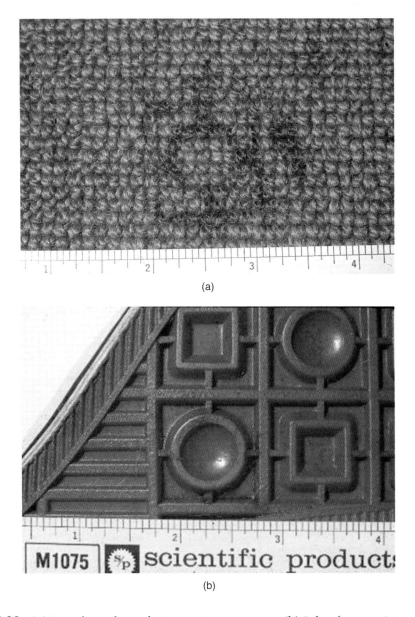

Figure 5.30 (a) Partial sneaker sole impression on carpet. (b) Sole of suspect's sneaker containing victim's blood.

Finally, the estimation of blood volume in a pool or saturated bloodstain is often attempted in an effort to determine whether there was sufficient blood loss to indicate that death has occurred in a certain location. This issue arises frequently where there is a scene with substantial quantities of blood present in the absence of a victim. There may be a question as to whether the location of a body represents a primary or secondary scene. MacDonell developed an experiment that used volumes of blood in 50-mL increments added to a surface similar to that containing the unknown quantity of blood. Area measurements were made of the known and unknown stained areas with a grid

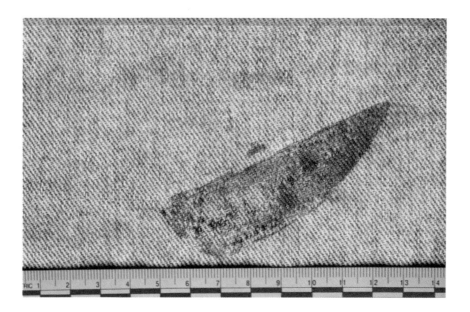

Figure 5.31 Partial knife impression in blood.

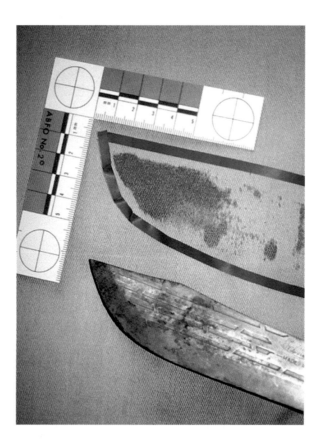

Figure 5.32 Bloody knife impression compared to suspect's knife.

Passive Bloodstains

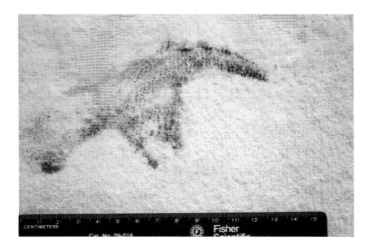

Figure 5.33 Transfer pattern of claw-type hammer on towel.

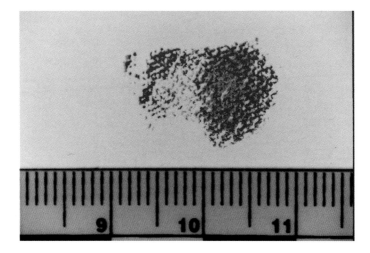

Figure 5.34 Fabric impression on wall.

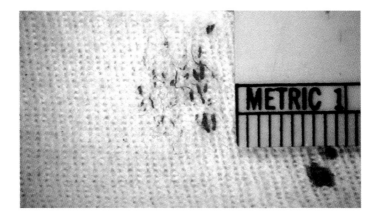

Figure 5.35 Transfer stain on left indicating bloodstains on top of fibers with spatter on right that has penetrated the fibers.

device. Another method for estimation of blood volume on absorbent surfaces involved weighing the bloodstained surface and subtracting the weight of an unstained area of that surface similar in size. The difference in weight was used to estimate the volume of blood deposited on the surface. Any estimation using these techniques should be conservative.

Formation of Spatter and Spatter Associated with a Secondary Mechanism

6

Introduction

Spatter is defined as a dispersion of blood spots of varying size, created when a source of fluid blood is subjected to an external force(s). The individual spots within the dispersion typically exhibit directionality characteristics that indicate the location from which they originated. Spatter is one of the three major bloodstain pattern categories within our bloodstain pattern taxonomy (Figure 6.1). Spatter is created by a wide variety of mechanisms and is commonly observed at scenes where bloodshed has occurred. The minimum requirements for the production of spatter are a fluid source of blood and an external force great enough to overcome the physical properties of the blood. On many surface textures, the shape and distribution of the individual spatters will allow you to determine the origin or the location of the source of the spatter.

Physical Properties of Blood Relative to Spatter Formation

The physical properties of blood including specific gravity, viscosity, and surface tension tend to maintain the stability of exposed blood and/or blood drops, causing them to be resistant to alteration or breaking up. When a source of exposed blood is subjected to a sufficient external force, energy is transferred to the blood, causing it to break up into droplets smaller than the original parent volume of blood. These droplets become airborne and move away from the parent source of blood as individual spheroids. They will follow a straight/flat or a parabolic flight path. A straight trajectory flight path will deposit droplets at approximately 90° or upward or downward at a lesser angle of impact. A parabolic arc flight path will deposit droplets with a downward trajectory when the surface contacted by the blood droplet is close to the downward side of the arc. While these droplets are traveling through the air, they are being acted on by air resistance and gravity. The droplets will travel through air until they strike a surface or until air resistance and gravity overcome their motion. On striking a surface, these three-dimensional droplets are now referred to as spatters. The resulting spatters will often exhibit directionality and have observable differences in their sizes and exhibit distributions relative to the mechanism(s) that created them. The size of the resulting spatter is a function of its original droplet volume.

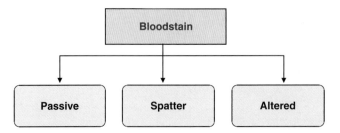

Figure 6.1 Major bloodstain pattern analysis taxonomy categories.

Weaker external forces acting on a fluid source of blood often result in larger droplets, whereas stronger external forces result in smaller droplets. When spatter is produced, there will be a size range. The larger droplets resulting from the ratio between their mass and surface area will travel farther from the source than the smaller ones, which are more affected by air resistance. In the laboratory setting this phenomenon can be demonstrated by depositing blood onto a moving fan blade that has been modified to work as a miniature paddle. The entire apparatus is enclosed, and only a small door is made for the resulting spatters to exit. The resulting spatters are depicted in Figures 6.2a and b.

Physical Characteristics of Spatter Patterns

When examining any type of bloodstain pattern, we initially examine the stains' physical characteristics or appearance. These physical characteristics would include the size, shape, and distribution. In addition, you should consider the target surface texture, where the blood has been deposited, the quantity or concentration of spatters, and the location of the spatters on the target surface.

Size Range and Quantity of Spatters

The size range and quantity of individual spatters produced by a single mechanism can vary significantly depending on the volume of the parent source of blood and the amount and/or type of force being exerted on the fluid blood. The effect the parent blood volume has on the size range and quantity of spatter produced may be observed in Figures 6.3a and b. The spatters will range in size even when produced by a single spatter mechanism. The size range and quantity can assist in ascertaining the mechanism by which a spatter pattern was produced. The quantity or number of individual spatters relates to the volume of available blood and the mechanism by which it is being disrupted.

As can be observed in Figures 6.3a and b, the quantity of blood being impacted can influence the size range as well as the quantity of spatters produced. In the laboratory environment, the amount of force applied to a blood source, the manner in which the force is applied, and the volume of blood being used is easily controlled. However, in actual casework we have only limited knowledge as to the amount of blood being acted on or the amount of force acting on the blood. Thus, an understanding of how both blood volume and the amount of force being applied to a blood source will affect the resulting spatter pattern is essential and can only be ascertained through firsthand experimentation.

Formation of Spatter and Spatter Associated with a Secondary Mechanism

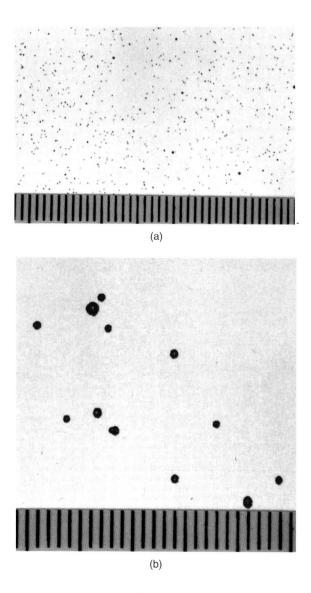

Figure 6.2 Both a and b were produced when blood was deposited onto a modified fan blade while the blade/paddle was moving, causing the blood to be projected out across a series of cardboard targets. (a) Representative sample of the size of the spatters at the opening of the apparatus. (b) Representative sample of the spatters at approximately 100 in. from the opening of the apparatus.

Because of the overlap of spatter size ranges between various spatter-producing mechanisms, it is not practical to differentiate between spatter-producing mechanisms based solely on the size range of the spatters alone. This overlap in spatter size can be easily observed in Figure 6.4. Therefore, the differentiation between specific spatter-producing mechanisms generally requires more information than the size of the resulting spatters. Occasionally, even with additional information you will not be able to reduce the spatter-producing mechanism down to one specific mechanism.

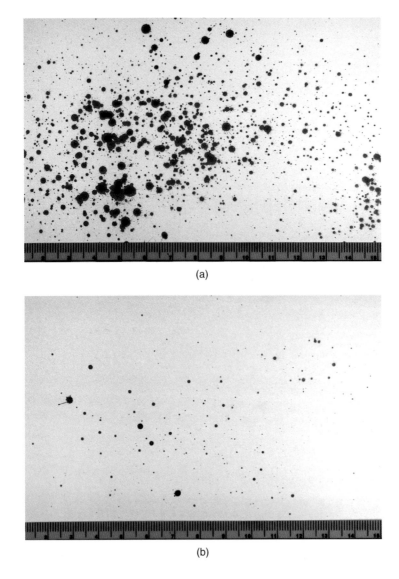

Figure 6.3 Effects of altering the parent blood volume while maintaining a constant amount of force and target distance. (a) Parent blood volume of 1 mL. (b) Parent blood volume of 0.1 mL.

Distribution and Location of Spatter

Any stain created with blood as the medium may be referred to as a bloodstain pattern. The term "pattern" takes on a different degree of significance when referring to a *spatter pattern*. In this context, the term pattern refers to a *distribution* of individual spatters. It is not possible to have a distribution with a single spatter (spot). Hence, a single blood spatter does not constitute a "spatter pattern."

On examination, the analyst must identify a pattern as a spatter pattern prior to attempting to ascertain the specific mechanism by which it was created. Determining the specific mechanism(s) that created a spatter pattern often requires more information than merely the pattern. Therefore, it is advisable to refer to these bloodstain patterns as simply "spatter patterns" until all of the available information has been reviewed.

Formation of Spatter and Spatter Associated with a Secondary Mechanism

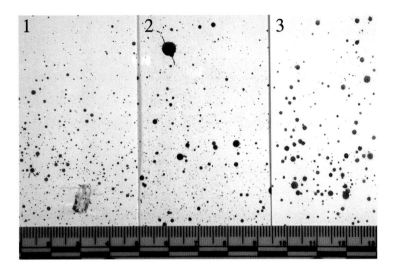

Figure 6.4 Comparison of the size range of spatter produced by (1) gunshot, (2) expiration, and (3) beating.

The distribution of the spatters in a spatter pattern can vary significantly. For the analyst to refer to the distribution as a spatter pattern, they must be able to identify a sufficient number of individual spatters to constitute a pattern. The individual spatters that make up a spatter distribution usually will exhibit similar directionalities that can be traced back to a common area of convergence. The distribution of the spatters can assist in determining the mechanism that created the spatter pattern. Figure 6.5 is a case example of a pattern with a linear distribution that is associated with a cast-off mechanism.

The location of the individual spatter(s) can assist in determining the mechanism that created the spatters. An example is a beating case in which spatters are identified on the ceiling of the scene as well as on the back of the accused perpetrator's clothing; both locations are common sights to observe spatters associated with a cast-off mechanism. In a case reviewed by the author, the accused claimed they never had their shoes off at the crime scene, but on examination blood spatters were located within the shoes. The accused version of the events was immediately in question based solely on the location of the spatters within their shoes.

Shape of Spatters

The shape of the individual spatters within a pattern will be dependent on the following:

- Flight path of the spatter prior to striking the target surface
- The texture of the target surface: absorbent versus nonabsorbent, smooth versus coarse
- The shape and/or angle of the surface being struck by the spatters

The shape of the spatters can be used to establish a common area of convergence and origin of a group of spatters. This will be discussed in detail in Chapter 10. The directionality and angle of impact of the spatters is established by their shape. Typically, spatters that have struck nonporous and nonabsorbent surfaces such as walls, countertops, cabinets, etc., will have round to elliptical shapes that represent the angle(s) in which they struck the surfaces.

Figure 6.5 Example of a cast-off pattern located on a closet door of a homicide scene.

Figure 6.6 depicts spatters with a left-to-right directionality on the lower portion of a door at a stabbing scene; however, eg spatters having only very minimal volumes may appear only as circular stains on striking a surface, regardless of what angle they strike the surface. Measurements of width and length of the stains will be difficult to establish.

Figure 6.6 Impact spatter pattern on the lower portion of a door at a homicide scene involving a stabbing.

When evaluating the shape of spatters on a surface, consideration should be given to the target surface's horizontal and vertical position at the time the spatter was deposited onto the target surface. The concern is whether the elliptical shape of the spatters is caused entirely by the angle in which the spatter struck the surface or is it because of angle or curvature of the target surface. An example is where spatters strike a curved target surface at 90 degrees but the resulting spatters have an elliptical appearance. Hence, the shape of the spatters in this situation do not completely represent the angle in which they struck the curved object, but undoubtedly they represent both the angle of the curved target surface as well as the angle of impact (Figure 6.7 and Figure 6.8).

Another issue to consider when evaluating the shape of spatters on a surface is spatter deposited on walls, which are at a significant vertical angle (Figure 6.9). The interiors and the exteriors of motor vehicles have a vast number of curved, angular, and irregular surfaces, which makes them all the more complex to analyze. This issue often arises with spatter distribution on footwear because it is not always possible to determine the orientation of the shoe at the time the spatters were deposited onto the shoe, not to mention the fact that

Figure 6.7 Impact spatters located on a hot water tank in a homicide scene involving a beating. (Courtesy of David Williams, Rochester Police Department, Rochester, New York.)

Figure 6.8 Closer view of the spatters on the hot water tank depicted in Figure 6.7. (Courtesy of David Williams, Rochester Police Department, Rochester, New York.)

Figure 6.9 An angular wall within a death scene.

shoes in general have numerous curves and irregularities. The authors have reviewed numerous cases where analysts have attached unwarranted specificity to angle of impact of spatters identified on footwear.

Effect of Target Surface Texture on Spatter Appearance

As with all bloodstain patterns, the target surface texture must be carefully evaluated prior to rendering opinions in regard to the stain patterns. Clothing fabrics, floor coverings, and upholstery are some of the more common surface textures in which blood is spattered onto

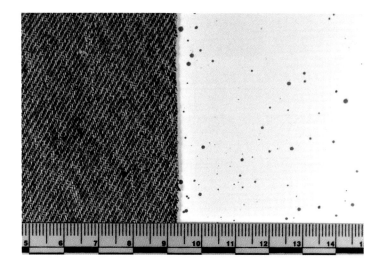

Figure 6.10 The substrate the blood strikes can affect the overall appearance of spatter pattern. The denim to the left and the smooth index stock to the right were stained at the same time.

during a bloodshedding event. These substrates can affect the size of the spatters as well as the shape of the spatters. When droplets with diameters less than 2 mm in size impact absorbent fabrics, their volumes usually will be too small to actually show any directionality. The texture and color of the substrate can mask the appearance of the stains, necessitating the use of high-intensity lighting and/or a microscope during evidence examinations (Figure 6.10). The size and shape of the individual spatters can be affected by the target surfaces they strike. Spatters may appear larger on an absorbent fabric as opposed to painted drywall, and their shape can be influenced by the weave and absorbance of the fabric with which they are coming into contact.

Mechanisms Associated with the Creation of Spatter Patterns

The mechanisms in which spatter is created are divided into three categories:

1. Secondary mechanism(s)
2. Impact mechanism(s)
3. Projection mechanism(s)

Spatter created by a secondary mechanism(s) are created when blood impacts a surface and/or itself on a surface, causing the physical properties of the blood to be overcome by the secondary impact. Spatter created by an impact mechanism(s) refers to a spatter pattern resulting from an object directly impacting (a primary impact) a source of fluid blood. Spatter created by a projection mechanism refers to a spatter pattern produced by the disruption of the physical properties of the blood without an impact. Figure 6.11 depicts our taxonomy of specific spatter-producing mechanisms. The specifics regarding the spatter-producing mechanism(s) will be discussed in greater detail in subsequent chapters.

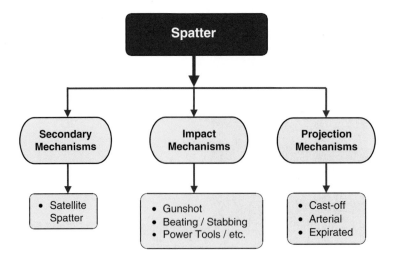

Figure 6.11 Bloodstain pattern analysis "spatter" taxonomy diagram.

Significance of Spatter Identification

After identifying a blood distribution as a spatter pattern and then gathering pertinent scene-, medical-, and case-related facts, the analyst may then be able to establish the specific mechanism(s) by which the pattern was created. Spatter patterns can assist the analyst in reconstructing past events.

The identification and analysis of spattered blood may do the following:

1. Allow for the determination of an area or location of the origin(s) of the blood source. Where did the spatter events occur?
2. Link the accused's clothing to the location of a spatter-producing event.
3. Link the accused to the homicide.
4. Allow for the determination of the mechanism by which the spatter pattern was created.
5. Be used to corroborate or refute an accused's account of how the blood was deposited on their clothing.
6. Link an item of evidence to a spatter-producing event.

Spatter Associated with a Secondary Mechanism

Spatters associated with a secondary mechanism refer to spatters produced other than by the actual bloodletting event (e.g., shooting, beating, stabbing). These spatters are produced when the physical properties of blood are overcome by a "secondary impact." This secondary impact occurs when the fluid droplets strike another object or blood itself, resulting in a distribution of spatters around the periphery of the parent stain (Figures 6.12a and b). Spatter

Formation of Spatter and Spatter Associated with a Secondary Mechanism

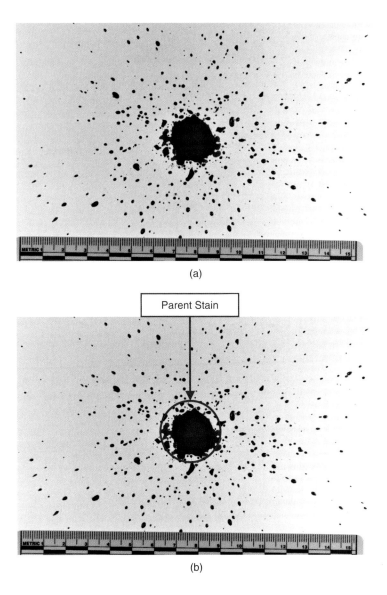

Figure 6.12 (a) A drip pattern with satellite spatter surrounding the parent stain. All of the surrounding satellite spatters were created when passive drops impacted the parent stain in the center of the pattern. (b) Same pattern as A with the parent stain labeled.

associated with a secondary mechanism is referring directly to satellite spatter (Figure 6.13). Some analysts refer to all spatters created by this secondary-type mechanism as secondary spatter rather than satellite spatter.

The size, shape, distribution, quantity, and location of satellite spatter correlates to the manner in which they were produced. Satellite spatters have a horizontal and a vertical perspective. The vertical perspective accounts for their shape and distributions around the parent stain.

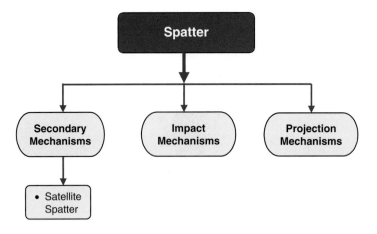

Figure 6.13 Secondary spatter taxonomy diagram.

Satellite Spatters on Horizontal Surfaces

The shape of satellite spatters relates to the mechanism by which they were created and the angle in which they struck surrounding objects. Satellite spatters associated with drip patterns are typically round to slightly elliptical, whereas the satellites associated with larger arterial spurts are more elongated and even spinelike in appearance. On a horizontal surface the satellite spatter tends to become more elliptical or elongated or spinelike as the distance from the initial impact point increases. While on a horizontal surface, their impact angle is proportionate to the distance they travel away from the impact site. Figure 6.14 depicts a portion of a horizontal target from a drip pattern. Depending on the dropping height and the surface texture of the target surface, a single drop of blood may create satellite spatters (Figure 6.15). In cases where a volume of blood is projected with force or when an object falls into an existing pool, the resulting satellite spatters will be very elongated and spinelike. Spinelike satellite spatters are commonly observed within arterial patterns and splash patterns (Figures 6.16a and b).

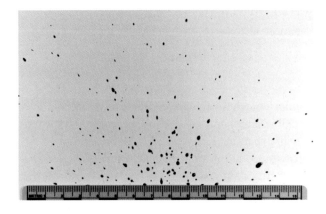

Figure 6.14 Depicts only the satellite spatter portion of a drip pattern.

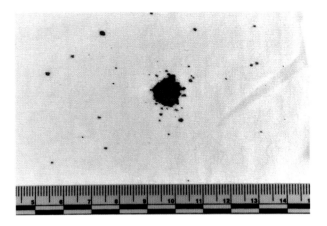

Figure 6.15 A single passive drop of blood fell 48 inches prior to striking a piece of cotton cloth. The surface texture of the cloth overcame the drop's surface tension, resulting in a large parent stain surrounded by satellite spatters.

The overall size range of satellite spatters tend to be the most uniform of the entire spatter-producing mechanisms. When satellite spatters result from blood dripping into blood or from blood passively striking a coarse surface, the resulting satellites have a fairly constant size of less than 1 mm to 4 mm in diameter. The exception is the spinelike spatters observed within arterial and splash patterns. The size range of satellite spatters overlaps the size range of the spatters produced by several other spatter-producing mechanisms and should always be considered as a possible spatter mechanism when evaluating clothing.

The distribution of satellite spatters is constant in that these spatters will be deposited around the periphery of a parent stain. This may include any item around the periphery of the parent stain (e.g., walls, shoes, pant legs) and are often observed on the feet and ankle region of an injured person who has ambulated around while bleeding. The number of satellite spatters produced can vary depending on a number of factors. Satellite spatters are commonly located on items near to the ground or floor such as flooring surfaces, base molding, feet, shoes, table legs, and pant legs. In the cases where blood struck the top of a table, countertop, stair tread, or other elevated surface, the resulting satellite spatters will have a commiserate distribution.

On a horizontal surface it is normally not difficult to establish that they are in fact satellite spatters because the parent stain is alongside the spatters. The difficulty arises when evaluating potential satellite spatters on a vertical surface, such as a shoe, where the item with the spatter has been moved away from the parent staining.

A number of factors have an influence on the appearance of satellite spatters:

1. Target surface texture
2. Volume of the parent stain
3. Mechanism by which the parent stain is produced (passive dripping single drops, free-falling large volume, a projected large volume, etc.)
4. Distance traveled by blood prior to striking the surface

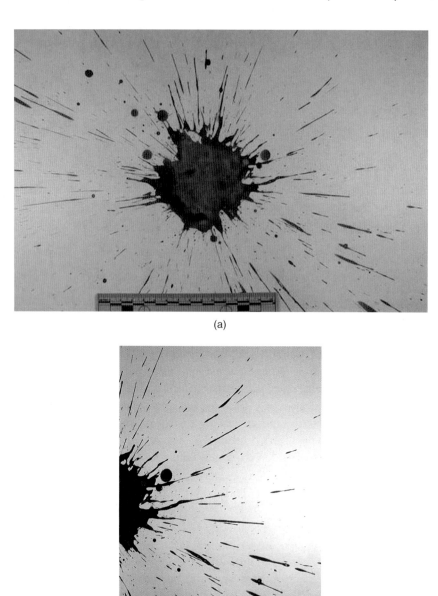

Figure 6.16 (a) Pattern created by a volume of blood projected onto a cardboard target. (b) Spinelike satellite spatters created by increasing the force beyond that of gravity.

Satellite Spatters on Vertical Surfaces

In 1996 Paul Kish at the Annual Meeting of the American Academy of Forensic Sciences Meeting presented research on the topic of satellite spatters. His research dealt primarily with the vertical perspective of satellite spatters from single drops of blood. Prior to this, little

Formation of Spatter and Spatter Associated with a Secondary Mechanism

Figure 6.17 Diagrammatic representation of blood dripping into blood and the resultant production of satellite spatter.

mention of the vertical perspective of satellite spatters appeared in the literature. When blood strikes a surface and satellite spatters are created, there must be a vertical perspective of the individual satellite spatters in order for them to be distributed around the periphery of the parent stain. This is illustrated in Figure 6.17. The fact that the shape of a resulting spatter depends on the angle in which it struck a surface indicates the round to elliptical satellite spatters must ascend vertically prior to striking the surrounding horizontal surfaces.

The vertical height which these satellite spatters will achieve depends on the surface texture being impacted, volume of blood impacting the surface, dropping height of the blood prior to striking the surface texture, and the distance from the impact site the vertical surface is located. Each satellite spatter will follow its own parabolic flight path away from the parent stain/impact site. The spatters will continue until they either strike a vertical surface or fall off as a result of gravity to a nearby horizontal surface. When the impact site and vertical surface are on the same plane, the maximum height the satellite spatters will attain will always be less than the height from which the blood fell (Figure 6.18, Figure 6.19, and Figure 6.20).

Satellite spatters on shoes and pant cuffs can be misinterpreted as being impact spatters from a beating, impact spatters from a shooting, and/or expired bloodstain patterns. When evaluating the significance of spatters, you must evaluate scene facts to determine whether any of the spatters on the accused's shoes and lower pant legs may be associated with satellite spatters (Figure 6.21, Figure 6.22, and Figure 6.23).

Issues to consider when evaluating whether the spatters identified on the accused could in fact be from satellite spatter are as follows:

1. What is the maximum height of spatters on the accused person?
2. Is the bloodstaining limited to the shoes and/or pant cuffs?
3. Has the accused rendered assistance to the bleeding victim?

Figure 6.18 Satellite spatters captured on a vertical cardboard target that was positioned next to the impact site of passive drops falling into one another. Zero on the scale represents the same plane as where the blood drops were impacting.

4. Are there locations within the scene where blood has been deposited on a flooring surface conducive with the production of satellite spatters (e.g., passive drops on coarse surfaces, drip patterns, splash patterns, arterial patterns, etc.)?

It is especially critical to consider this potential explanation for spatters on an accused's shoes when there is evidence to support that the accused was assisting the bleeding victim and there is evidence of dripped blood on the ground and/or flooring surfaces. Figure 6.24 depicts the toe region of the left shoes, which was worn by the accused in a shooting case. Stains consistent with being the result of passive drops with accompanying satellite spatters

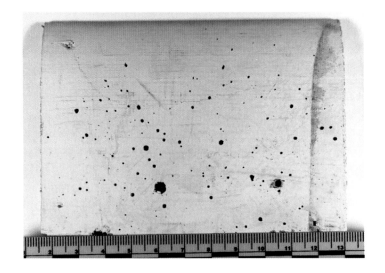

Figure 6.19 Satellite spatters on a piece of base molding.

Formation of Spatter and Spatter Associated with a Secondary Mechanism   115

Figure 6.20 The appearance of satellite spatters that were deposited onto a wall next to a drip pattern.

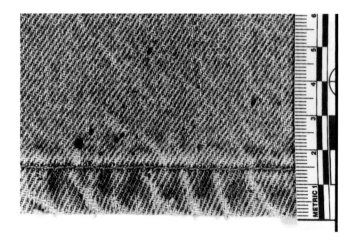

Figure 6.21 Satellite spatters on the lower cuff edge of a pair of blue jeans.

Figure 6.22 Satellite spatters on the insole region of a sneaker.

Figure 6.23 Satellite spatter on the sole of a work boot that was being worn by a victim who received a single gunshot wound to the head. He was seated in a chair bleeding from his wound onto the floor prior to coming to rest on the floor. The location of satellite spatters on footwear is partially dependent on the position of the footwear at the time of staining.

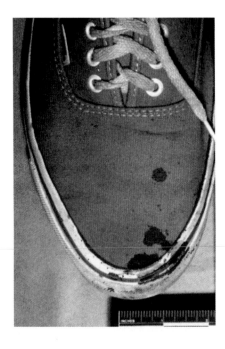

Figure 6.24 Passive drops of blood and satellite spatters located on top of the toe region of the shoe of the accused.

Formation of Spatter and Spatter Associated with a Secondary Mechanism 117

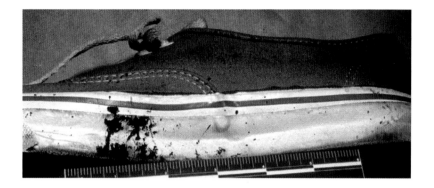

Figure 6.25 Satellite spatters on insole region of the shoe of the accused.

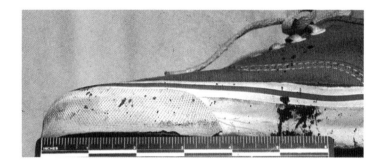

Figure 6.26 Satellite spatters on insole region of the shoe of the accused.

were present on the toe region of this shoe. Additional satellite spatters were located on the insole region of the shoe (Figure 6.25 and Figure 6.26). The victim had a single gunshot wound to the head. The accused carried the victim out of their dwelling, placing her into his vehicle and leaving a trail of blood drops on various floor and ground surfaces. These spatters were misinterpreted as being a direct result of spatter from the gunshot rather than satellite spatters created as the accused carried the victim from the scene.

Satellite spatters will achieve a maximum vertical height that is dependent on surface texture being impacted and the blood volume. Usually, satellite spatters will be located on the shoes or feet as well as the lower pant legs of the person who was in close proximity of blood that has impacted a horizontal surface. If the vertical height of the spatters is in question, the analyst would have to conduct a series of experiments that take into consideration the case factors such as the surface texture the blood struck, blood volume striking the surface, potential distance between the blood impact site and the vertical surface, and the potential blood-dropping heights.

The location and distribution of satellite spatters along with case factors will often allow the analyst to either exclude or include the spatters on the accused's clothing as coming from a secondary mechanism as compared to an impact or projection mechanism (e.g., beating, shooting, expired blood). Because of the similarity in the size range of the spatters produced by satellite spatters, spatter associated with gunshot, beating, and expired blood, there will be situations where the analyst cannot exclude satellite spatter as one of the possible mechanisms that may have produced the spatter identified on the accused person.

Impact Spatter Mechanisms 7

Introduction

When a bloodstain pattern has been identified as spatter based on the size, shape, or distribution of the stains, it is frequently possible to further classify the pattern as impact spatter (Figure 7.1). Impact spatter results from an object directly striking a source of exposed blood. The resultant stains may range in size from 0.01 mm to 3 to 4 mm or larger in diameter. Further analysis is required to associate the spatter pattern with beating, stabbing, gunshot, power tool, or explosive events. The identification of impact spatter is case driven in that the size of the stains is not the deciding factor in forming that conclusion. This determination requires assimilation of specific case information pertaining to the assessment of the scene and location of the stains as well as knowledge of the nature of the injuries sustained by the victim(s). If the victim was removed from the scene to another location by the assailant, further caution is advised in making specific determinations. It is important to consider all other mechanisms that are capable of producing bloodstains in a similar size range. There is considerable overlap of stain sizes between impact spatter and mechanisms such as satellite spatter and some projected mechanisms. Some of these mechanisms are nonviolent in nature and not directly associated with the actual infliction of injuries to the victim. Blows administered to a victim with blunt instruments (fists, club, hammer, rock, golf club), as well as with sharp objects such as knives and axes, may produce impact spatter that overlaps the size range of spatters observed in many gunshot cases. When spattered blood is present on clothing and footwear, it is important to obtain the history of this evidence as to when and where the items were collected from a victim or suspect. It is often claimed by suspects that they found the victim, shook them, or attempted cardiopulmonary resuscitation (CPR) or some other activity that may produce small spatters on their clothing.

Impact Spatter Associated with Beating and Stabbing Events

Impact spatter similar to that produced in actual beating and stabbing events is easily reproduced in the laboratory with the use of a modified rat trap apparatus (Figure 7.2). This is one of the experiments that students participate in during basic courses in bloodstain pattern analysis. Varying amounts of fluid blood are subjected to controlled forceful impacts with targets consisting of cardboard and other surfaces such as cloth (Figures 7.3a, b, and c). The amount and direction of force and the resultant effect on the size range and distribution

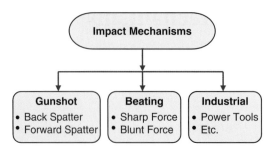

Figure 7.1 Taxonomy of impact mechanisms.

of impact spatter on these surfaces has been studied extensively. Figures 7.4a and b demonstrate the comparison of a smaller and larger quantity of blood struck with equal force. However, the amount of force applied to the amount of available fluid blood at the impact site is always an unknown variable in actual casework. The lack of bloodstains on an accused person should be viewed as an issue of further inquiry. Often the lack of bloodstains on an accused person or his or her clothing will relate to the types and locations of the deceased's injuries or the actions taken by the accused during or subsequent to the violent act. Therefore, the analyst should exercise caution as to what the size of the stains or quantity of spatter is expected to be present at a scene or on a suspect's clothing. Radial patterns of blood spatter produced by impact may be seen as distributed in a fashion not unlike the spokes of a wheel (Figure 7.5). The direction of force and resultant location of impact spatter was studied by Dr. Eduard Piotrowski and published in 1895. Figures 7.6a, b, and c demonstrate some of his experiments.

Figure 7.2 Impact spatter apparatus designed to create a reproducible impact to known quantities of blood. A spring-loaded door hinge is used as the source of energy. This causes the upper jaw to be forced down onto the blood located on the base of the apparatus. The amount of energy can be controlled by adjusting the tension on the hinge spring or by releasing the upper jaw from different distances away from the lower base.

Impact Spatter Mechanisms

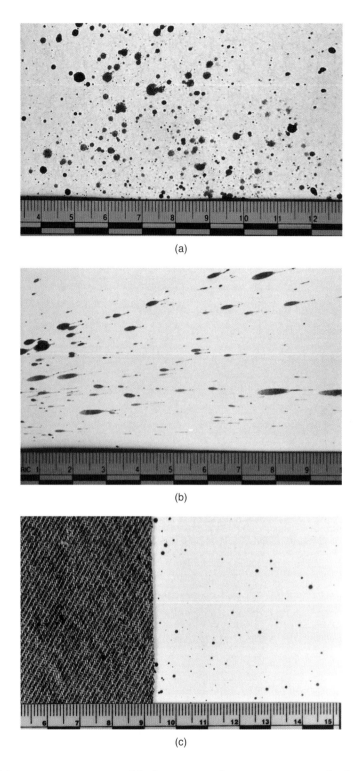

Figure 7.3 (a) Impact spatter created by beating mechanism on a vertical white cardboard at a distance of 16 in. (b) Impact spatter created by beating mechanism on a horizontal white cardboard at a distance of 16 in. (c) Comparison of impact spatter created by beating mechanism produced on a vertical white cardboard and blue denim at a distance of 16 in.

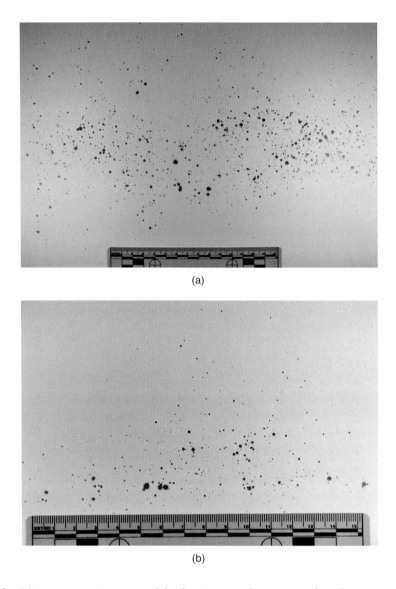

Figure 7.4 (a) Impact spatter created by beating mechanism produced on a vertical white cardboard at a distance of 6 in. using 0.5 mL of blood. (b) Impact spatter created by beating mechanism produced on a vertical white cardboard at a distance of 6 in. using <0.25 mL of blood. The amount of force was the same as utilized in Figure 7.4a.

It is important to remember that in actual beating cases that a single blow usually is not sufficient to produce significant blood spatter except perhaps in the case of a massive crushing injury. The source of blood must be exposed when receiving the impact(s) in order to create impact spatter. The determination of the directionality and angle of impact enables the analyst to establish the area(s) of convergence and origin of these bloodstains in specific locations. Frequently, the spatter distribution and location reflects the position of the victim when blows were struck without having to resort to a stringing technique. This is seen in Figure 7.7, in which the victim received blunt injury while on the floor of a hardware store close to a snow shovel. Figure 7.8 demonstrates the distribution of impact spatter on two sides of a

Impact Spatter Mechanisms

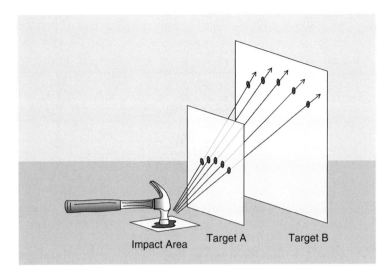

Figure 7.5 Representation of the radial distribution of impact spatter. The further away from the impact, the greater the separation between the bloodstains. (Courtesy of Alexei Pace, Malta.)

filing cabinet exhibiting upward directionality. The area of origin is easily seen as close to the floor below where there is a heavy saturation of blood. The victim in this case was a night watchman at a department store who was beaten with a heavy credit card imprint device. If the victim is on a floor or other surface during the course of a beating, the spatter may be observed radiating away from the area of impact, producing streaking patterns on the victim and floor and impacting at a low level on nearby walls or other objects within range (Figure 7.9). Sometimes a sector or portion of the radial pattern will appear free of blood. This void area may represent interception of impact spatter by the assailant or other intervening object(s). Figure 7.10 shows that the location of impact spatter indicated that the door was open when a victim was assaulted.

Multiple blows to the head with a blunt instrument often create considerable quantities of impact blood spatter, sometimes accompanied with tissue and skull fragments. Figure 7.11 shows a husband who was attacked while in bed by his wife and stepson. He was struck multiple times with a sledge hammer. The size and quantity of spatter produced as the result of beating and stabbing mechanisms vary considerably from case to case. In terms of spatter production, stabbing events are considered a beating with a sharp object. Once blood has been exposed from a knife wound, the impact of the knife handle or the hand of the assailant wielding the knife on the exposed blood is the mechanism for spatter production. However, multiple stabbing events generally do not create the quantity of spatter observed with multiple blows from a blunt object. Figure 7.12 demonstrates impact spatter produced from a multiple stabbing event that occurred close to the rear of a brick fireplace. Additional factors that must be considered with respect to size, quantity, and distribution of impact spatter include the volume of blood struck by an object and the number and force of the blows administered to the victim. In the laboratory the volume of blood exposed to impact and the force of the impact can be controlled. Additionally, there are occasions where the victim's head is covered with thick hair, bedding, towels, or other materials prior to or during a beating, which may decrease the quantity of blood spatter produced.

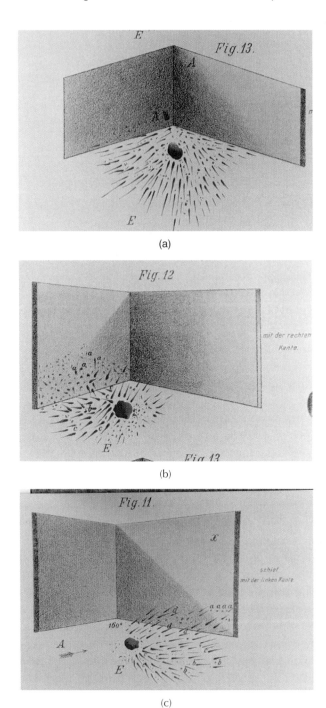

Figure 7.6 (a) Dr. Eduard Piotrowski's experiment that demonstrated that the impact spatters were directed toward him as he struck the blood source with the front edge of the hammer from above. (b) Dr. Eduard Piotrowski's experiment that demonstrated that the impact spatters were directed to his left away from him as he struck the blood source with the hammer from right to left. (c) Dr. Eduard Piotrowski's experiment that demonstrated that the impact spatters were directed to his right away from him as he struck the blood source with the hammer from left to right. (Courtesy of Herbert Leon MacDonell, Corning, New York.)

Impact Spatter Mechanisms

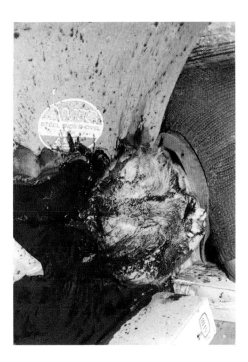

Figure 7.7 View of victim beaten with wrench on floor of hardware store. Note the upward directionality of the impact spatter on the snow shovel.

Figure 7.8 Impact spatter exhibiting upward directionality on both sides of file cabinet, indicating that the night watchman was beaten while on the floor.

Figure 7.9 Radial distribution of impact spatter on floor and lower wall of laundry room. The victim's head was repeatedly struck against the floor.

Figure 7.10 Impact spatter on door casing and inner edge of door indicating that the door was open when the spatter was produced.

Impact Spatter Mechanisms

Figure 7.11 Distribution of impact spatter and tissue on headboard of bed. The victim was struck repeatedly with a sledge hammer.

In many cases impact spatter is present on the same or nearby surfaces with other patterns of blood associated with the event, and these should also be documented and evaluated. Figure 7.13 shows impact spatter on a wall that resulted from the victim being beaten with a hammer while on his bed. There are also transfer stains and flow patterns on the wall surface with heavy saturation of blood and clots on the bed. The analyst should be aware that excessive pooling of blood on a floor or saturation on absorbent surfaces can obliterate spatters and light transfer patterns.

During beating and stabbing events the assailant may receive significant blood spatter on shoes, trousers, shirt, and other garments. Often there is little or none present. The quantity

Figure 7.12 Rear wall of fireplace in dining room close to location where victim was repeatedly stabbed.

Figure 7.13 Impact spatter and transfer stains on wall close to bed where victim was beaten with a hammer.

and location of blood spatter deposited on an assailant depends on the relative position of the assailant and the victim, the angle of the force, and the number and location of blows struck. For example, an assailant delivering blows with overhead swings to a prone victim in many cases would receive blood spatter on their lower legs as well as the hand and arm that wielded the weapon (Figure 7.14). However, when blows are struck to a victim at an angle with the direction of force away from the assailant such as side swings, little if any spatter may impact on the assailant. An assailant may also wear protective garments, may change clothes prior to apprehension, or may not be wearing any clothing during the attack on the victim. It should be recognized that the absence of blood spatter on the clothing of a suspect does not necessarily exonerate the accused. The issue of the significance of the lack

Figure 7.14 Impact spatter on denim jeans of son accused of beating his mother to death at home and burning down the house to eliminate evidence.

Impact Spatter Mechanisms

of blood on a suspect was effectively addressed by Paul Kish and Herbert MacDonell of the Laboratory of Forensic Science in Corning, New York. They cowrote an excellent guest editorial in the *Journal of Forensic Identification* entitled "Absence of Evidence Is Not Evidence of Absence" in 1996. The authors stressed that conclusions should be drawn based on the presence of bloodstains, whereas the absence of bloodstains on clothing should neither implicate nor exonerate the accused. The following is a summary of the variables affecting the size, shape, and distribution of impact spatter associated with a beating mechanism:

- Shape of weapon
- Weight and length of weapon
- Number of impacts
- Amount of force applied
- Direction of force applied
- Location of wounds
- Movement of victim and assailant during attack
- Amount of blood available for a given impact

Experience has shown that other objects worn by an assailant may contain blood spatter that may not be apparent to a person attempting to wash bloody clothing to obscure involvement in a homicide. Examples of this are socks, belts, hats, glasses, watches, and other jewelry items. Weapons should also be carefully examined for evidence of impact blood spatter.

Blood spatter on clothing should be examined thoroughly, especially with respect to estimation of impact angle and spot size. Fabrics can alter the appearance and size of bloodstains. R.B. White researched bloodstain patterns on fabrics and published an article in 1986 entitled "The Effect of Drop Volume, Dropping Height and Impact Angle." Impact spatter on clothing may be obscured by the presence of other blood patterns. Also, dark clothing, especially blue jeans, may hinder evaluation of small blood spots, as a result of the lack of contrast (Figures 7.15 and 7.16). The use of a stereomicroscope is helpful in many cases to show that impact spatter has penetrated the weave of the material (Figure 7.17).

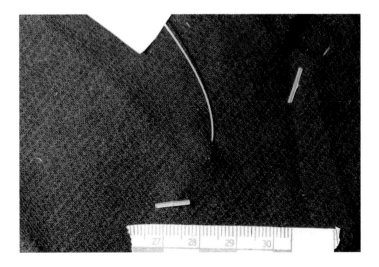

Figure 7.15 Close view of impact spatter on dark, thick, woven shirt.

Figure 7.16 Close view of impact spatter on black denim jeans.

This enables the examiner to differentiate small transfer stains that are on top of the fabric weave (Figures 7.18 and 7.19). Diagrams of clothing and shoes that indicate the location of bloodstains are useful for documentation purposes, especially when the small stains are difficult to visualize without the aid of magnification. This concept can also be applied to the examination of impact spatter associated with gunshot. Because of their relatively small size, care must be taken when observing and measuring impact blood spatters for directionality and impact angles. A pocket microscope or other magnifying device with a scale in tenths of millimeters is very useful.

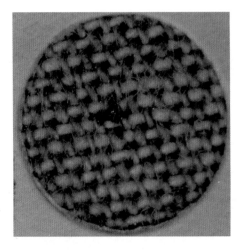

Figure 7.17 Impact spatter on blue denim jeans viewed with a stereomicroscope. (Courtesy of Linda Netzel, Kansas City, Missouri Crime Laboratory.)

Impact Spatter Mechanisms

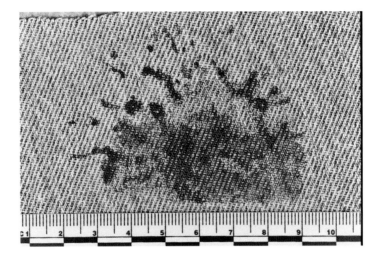

Figure 7.18 Transfer stains appearing to be impact spatter on blue denim that show the stains on top of the weave.

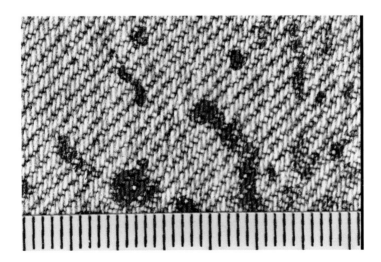

Figure 7.19 Closer view of stains on top of the weave.

Impact Spatter Associated with Gunshot, Explosions, and Power Tools

This type of impact blood spatter is most frequently associated with gunshot injury but also is seen in cases involving explosions and power tools as well as in some automobile accidents. The authors have worked on cases of accident and suicide events involving chain saws, circular and radial saws, as well as vehicular–pedestrian and air bag impacts that have produced spatter. Impact spatter produced as the result of gunshot is associated with the entry wound and/or exit wounds produced by a projectile. The actual mechanism(s) responsible for the production of impact spatter associated with gunshot have been investigated. Studies by Dr. Martin Fackler suggest that the collapse of the temporary wound cavity produced by the projectile is responsible for spatter production (Figures 7.20a and b). In close-range or contact gunshot wounds especially to the head, the discharge of gases that accompany the projectile that often lacerate the entrance wound due to expansion beneath the skin has been studied and proposed as a causative mechanism for back

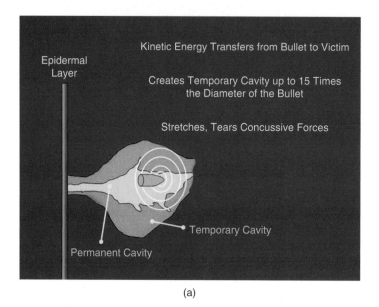

Figure 7.20 (a) Diagrammatic representation of the creation of temporary cavity caused by projectile penetrating into body. (b) Diagrammatic representation of the collapse of the temporary cavity.

Impact Spatter Mechanisms

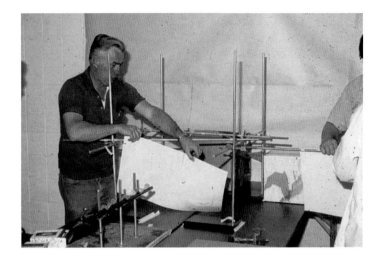

Figure 7.21 Students in basic bloodstain pattern analysis course performing gunshot spatter experiments.

spatter. The concept of conical distribution of gunshot spatter can be demonstrated in laboratory conditions. A basic course in bloodstain analysis includes laboratory exercises to create impact spatter with gunshot (Figure 7.21). These experiments involve shooting a .22 caliber weapon into a blood-soaked sponge with white cardboard or other surfaces such as fabric positioned to receive forward and back spatter. The patterns developed by shooting through blood-soaked sponges usually indicate that the production of forward spatter is generally more symmetric than that produced as back spatter (Figure 7.22). Even with these controlled conditions, the amount of spatter production may vary considerably. In actual casework there are factors that tend to diminish the symmetry and amount of the pattern including the location, size, and shape of the gunshot wound and intermediate

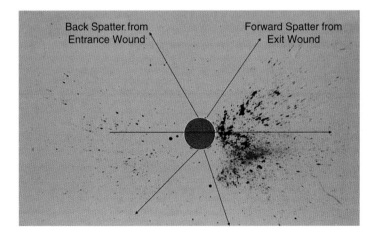

Figure 7.22 Forward and back spatter on horizontal surface produced by shooting through blood-soaked sponge with .22 caliber rifle. The cardboard was positioned below the sponge.

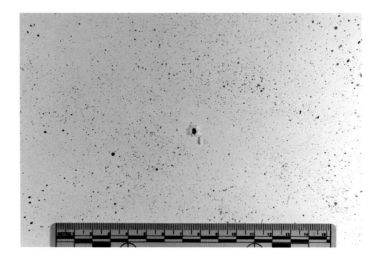

Figure 7.23 Back spatter produced on vertical cardboard placed 6 in. behind blood-soaked sponge.

objects such as hair and clothing. Therefore, the conical distribution is not often seen in casework.

The size of impact spatters resulting from gunshot can be extremely fine in the range of 0.1 mm or smaller. However, the fine spatter when present is usually in association with larger stains that overlap with other spatter mechanisms. Reference is often made to an "aerosol-type pattern" or a "mistlike" dispersion of the stains as a characteristic of spatter associated with gunshot (Figure 7.23 and Figure 7.24). These terms are often overused by some analysts and applied to stain sizes that are clearly larger, and other mechanisms are

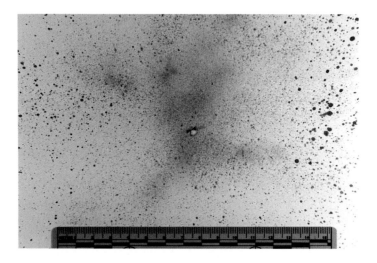

Figure 7.24 Forward spatter produced on vertical cardboard placed 6 in. in front of blood-soaked sponge. Note the true "misting" of the minute stains.

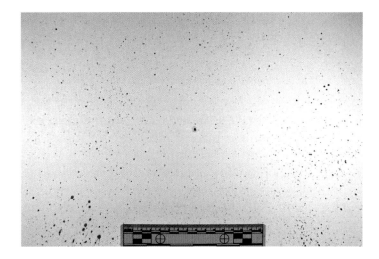

Figure 7.25 Back spatter produced on vertical cardboard placed 12 in. behind blood-soaked sponge.

dismissed. Extremely fine stains can be produced by other mechanisms as well. The small blood droplets produced by gunshot events have low mass and usually travel only a short distance (up to 4 ft) through the air. The extremely fine spatter may often only travel 6 to 12 in. The larger droplets, of course, will travel greater distances owing to their greater mass and may be present without the fine spatter (Figure 7.25 and Figure 7.26). Shooting into exposed blood will often increase the quantity of back spatter, but the distance it will travel will generally be limited to up to 4 ft for the small droplets. When victims have sustained a perforating gunshot injury, there are two types of impact spatter that are possible: back spatter associated with an entrance wound and forward spatter associated

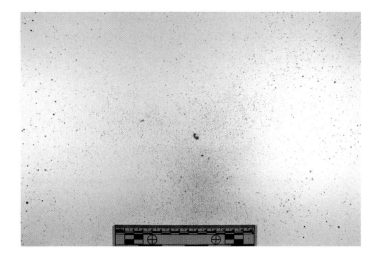

Figure 7.26 Forward spatter produced on vertical cardboard placed 12 in. in front of blood-soaked sponge.

with an exit wound. In the case of a penetrating gunshot wound, the entrance wound would be the only source of spatter and be referred to as back spatter.

Back Spatter

Back spatter results from blood droplets directed back toward the source of energy, which would be the weapon. It may be deposited on any object or surface near the entrance wound relative to the position of the victim at the time of discharge of the firearm (Figure 7.27). Back spatter may also be deposited on the firearm and the exposed hand, shirt cuff, or arm of the shooter holding the weapon if it is discharged at contact or close range (Figure 7.28 and Figure 7.29). In cases of possible self-inflicted gunshot injury, the hands and arms of the victim and the weapon should be carefully examined for evidence of impact blood spatter that may indicate the position of the hands on the weapon at the time of discharge of the firearm.

The amount of back spatter is affected by the type of weapon and ammunition, muzzle to target distance, and anatomic features of the wound site. Stephens and Allen (1983) documented factors affecting amount of back spatter experimentally using .22 caliber weapons. They noted that back spatter may be completely absent with considerable muzzle to target distances. With respect to shotguns and high-powered firearms, a greater quantity of back spatter would be expected, especially when the weapon is discharged at close or contact range. Pex and Vaughan (1987) conducted further experiments with back spatter and applied their findings to actual case investigation where back spatter was identified on the sleeve of suspects. Their work also corroborated the work of Stephens and others. Burnett (1991) conducted research on the detection of bone and bone plus bullet particles in back spatter from close range shots to heads.

Betz et al. (1995) studied a total of 103 suicidal and 29 homicidal gunshot fatalities. They found that in 33 out of 103 gunshot suicides (32%) blood spatters were detectable

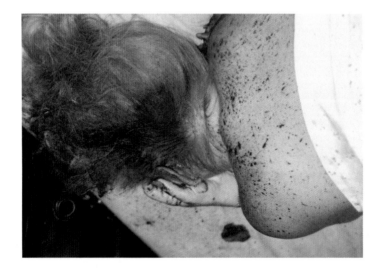

Figure 7.27 Back spatter on exposed back of victim who received contact gunshot wound to back of head.

Impact Spatter Mechanisms

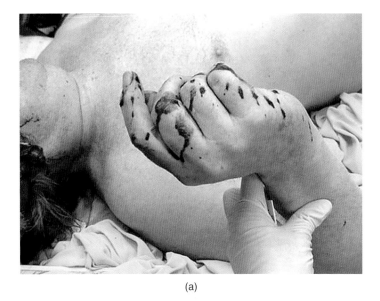

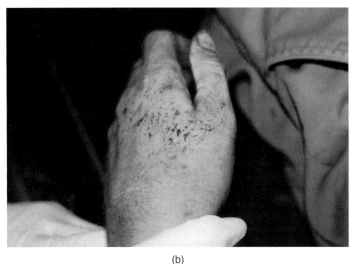

Figure 7.28 (a) Back spatter on right hand of victim who sustained suicidal shotgun wound of the head. (b) Back spatter on left hand of suicidal shooting victim who used a small-caliber weapon. (Courtesy of Todd Thorne, Kenosha, Wisconsin, Police Department.)

on the hands by naked eye inspection. No visible blood spatters were seen on the hands of the 29 homicide victims.

Hueske (1997) studied the frequency of back spatter on .38 and .44 caliber revolvers by shooting into blood-soaked sponges from near contact, 1-in., 3-in., and 6-in. muzzle to target distances. He found that within the limitations of his study, back spatter was present on all weapons at near contact range and on one weapon fired at a distance of 3 inches. Karger et al. (2002) concluded from their research that the deposition of back spatter on the firearm and the hand of the shooter occurred frequently but not regularly as the result of close-range shots. They also noted that back spatter of brain tissue, fat, muscle, bone fragments, skin, and hair has been documented in casework.

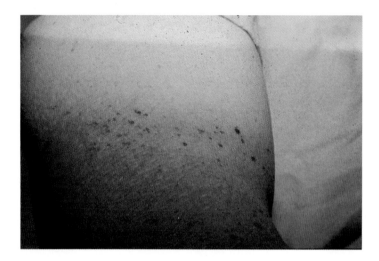

Figure 7.29 Demonstration of back spatter on right arm of shooter. (Courtesy of Todd Thorne, Kenosha, Wisconsin, Police Department.)

Yen et al. (2003) studied five gunshot suicide cases to deduce the position and orientation of the hands and therefore the firearm according to the bloodstain patterns on the hands of the deceased. They stressed the importance of close examination and documentation of the hands of shooting victims for impact spatter prior to the hands being tested for gunshot residue or wrapped for the transport of the body. They noted that wrapping of the hands can smear and destroy the fragile traces of blood. Additionally, the further alteration or washing out of the bloodstains on the hands can occur when the corpse is stored in a cold room by trapping moisture around the wrapped hands.

Forward Spatter

Forward spatter is produced by blood droplets traveling in the same direction as the source of energy and projectile and is associated with the exit wound. An exception to this would be a shoring or furrowing projectile path on a peripheral area of the body where a true exit wound is not apparent. The demonstration of the presence of forward spatter will assist in the location of the victim at the time of discharge of the weapon. When a projectile has reentered a body after passing through a hand or arm, there may be forward spatter around the periphery of the reentry wound, and its presence will help reconstruct the position of the victim at the time of weapon discharge. As previously described, when there is an exit wound the amount of forward spatter will usually exceed that of back spatter.

The quantity and distribution of impact blood spatter, whether it be forward or back spatter, vary considerably depending on many factors. The following is a summary of these factors:

- Caliber of the firearm
- Muzzle to target distance
- Number of shots

Impact Spatter Mechanisms

- Wound characteristics and position of the victim
- Blocking effect of hair, hats, clothing, or other head gear worn by victim

Gunshot injuries of the head may produce significant impact spatter. The study of its location and distribution in conjunction with wound characteristics and path of the projectile(s) can assist with the distribution of this blood spatter, with the reconstruction of the shooting, and with the distinction between homicide, suicide, or accident. A recent case supplied to the authors involved a young man who was determined to have shot himself in the head with a rifle after an argument with his girlfriend, who was at the scene when the shooting occurred. The quantity and distribution of blood spatter, skull fragments, and brain tissue are typical of this type of massive head trauma that essentially evacuates the contents of the skull. The wide distribution of impact spatter and projectile penetration on the ceiling assists with the orientation of the weapon and position of the victim at the time the weapon was discharged (Figure 7.30, Figure 7.31, Figure 7.32, and Figure 7.33). Another case involved projectiles and spattered tissue on a ceiling but no blood spatter. A man shot himself in the face with an upward trajectory while on or near a bed in a motel room with a 12-gauge shotgun that contained a load of #00 buckshot. The pellets formed a somewhat circular pattern on the 14-foot-high ceiling, as shown in Figure 7.34 and Figure 7.35. There were small tissue fragments on the ceiling in the area of the pellet penetrations but no spatters of blood because of the higher mass of the tissue material that was able to travel a further distance (Figure 7.36). The spatters of blood ultimately fell back onto the bed.

Another case where the position of the victim was clear by the location of the impact spatter involved a self-inflicted shotgun wound that occurred in the driver's seat of a van. The shotgun slug entered the right mid chest of the victim and exited the top of the left shoulder. The slug passed through the roof of the van to the rear and left of the driver's seat. The projectile penetration surrounded by impact forward spatter is seen in Figure 7.37.

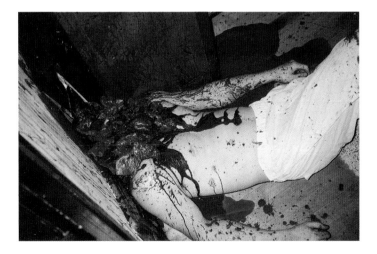

Figure 7.30 View of victim in his bedroom with massive head trauma from self-inflicted gunshot wound to the head. (Courtesy of Detective Jeffrey A. Strobach, Wausau Police Department, Wausau, Wisconsin.)

Figure 7.31 View of wall of bedroom behind and above victim showing impact spatter and projected bloodstains. (Courtesy of Detective Jeffrey A. Strobach, Wausau Police Department, Wausau, Wisconsin.)

Figure 7.32 Distribution of impact spatter on ceiling above victim in bedroom. (Courtesy of Detective Jeffrey A. Strobach, Wausau Police Department, Wausau, Wisconsin.)

Impact Spatter Mechanisms

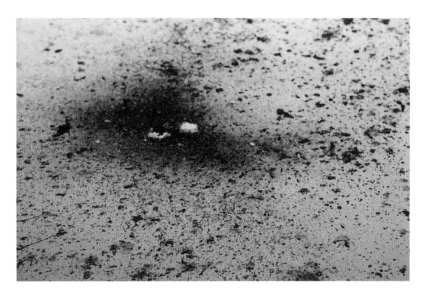

Figure 7.33 Closer view of impact spatter with projectile and tissue (bone) penetration on ceiling. (Courtesy of Detective Jeffrey A. Strobach, Wausau Police Department, Wausau, Wisconsin.)

Occasionally, self-inflicted gunshots involve elaborate apparatus. Figure 7.38, Figure 7.39, Figure 7.40, and Figure 7.41 show an individual who took the time and patience to construct a wooden structure complete with a rope and pulley system in order to fire the rifle by pulling the rope with his hand.

Figure 7.34 View of high-vaulted ceiling with #00 shotgun pellet penetrations.

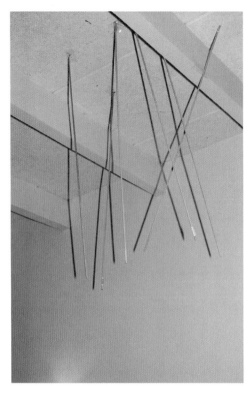

Figure 7.35 View of high-vaulted ceiling with wooden dowels indicating pellet paths and distribution.

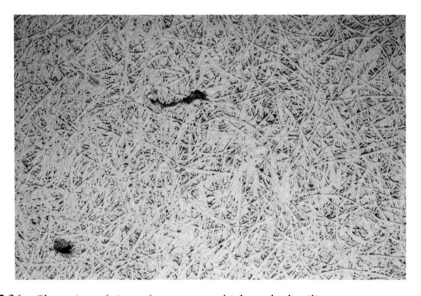

Figure 7.36 Close view of tissue fragments on high-vaulted ceiling.

Impact Spatter Mechanisms

Figure 7.37 View of impact spatter around shotgun slug penetration through roof of van.

Another case involved a man who received two gunshot wounds to the head while driving his pickup truck. The first shot penetrated the windshield as he was approaching the shooter, and the second was fired through the open driver's window as he was passing the shooter. Figure 7.42 shows the projectile penetration through the windshield. There was an

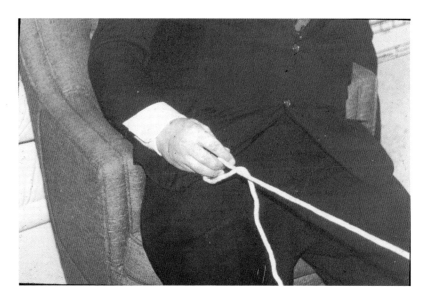

Figure 7.38 View of victim seated in chair with right hand holding rope that discharged rifle. (Courtesy of Martin County Sheriff's Office, Stuart, Florida.)

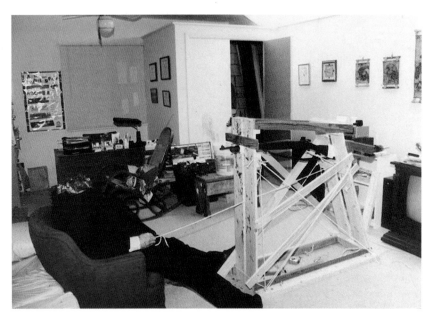

Figure 7.39 View of victim with massive head trauma from gunshot. Rifle is positioned in wooden apparatus. (Courtesy of Martin County Sheriff's Office, Stuart, Florida.)

array of spatters and projected stains on the visor, headliner, and interior roof of the pickup truck, as shown in Figure 7.43 and Figure 7.44. The shot through the driver's window likely produced the spatter seen on the driver's mirror in Figure 7.45.

A man attempting to commit suicide by self-inflicting incised wounds also walked into traffic on a roadway and was struck by a vehicle. The front of the vehicle received impact

Figure 7.40 Closer view of wooden apparatus showing rope attached to trigger that enabled the victim to discharge rifle. (Courtesy of Martin County Sheriff's Office, Stuart, Florida.)

Impact Spatter Mechanisms

Figure 7.41 Impact spatter combined with tissue and hair fragments around projectile penetration on wall behind seated victim. (Courtesy of Martin County Sheriff's Office, Stuart, Florida.)

spatter as the result of impact with a previously bleeding person (Figure 7.46). He survived both attempts at suicide.

Laboratory experiments in basic bloodstain courses have used the principle of shooting through blood-soaked polyurethane sponges to demonstrate forward and back spatter. This creates an amount of impact blood spatter in excess of that produced at many crime

Figure 7.42 Projectile penetration through driver's side windshield of pickup truck. (Courtesy of L. Allyn DeMeo, Arcana Forensics, La Mesa, California.)

Figure 7.43 Impact spatter and projected stains on visor and headliner of pickup truck. (Courtesy of L. Allyn DeMeo, Arcana Forensics, La Mesa, California.)

scenes because the blood source is fully exposed and not covered by head gear, clothing hair, skin, tissue, or bone as previously discussed. James and Eckert (1998) described construction of simulated human heads devised to study experimentally the distribution forward blood spatter and was related effectively to actual casework. Their procedure for construction of the heads was a modification of a technique developed by MacDonell and Brooks for their research of the drawback effect of blood into the muzzle of firearms.

Figure 7.44 Impact spatter and projected stains on interior roof of pickup truck. (Courtesy of L. Allyn DeMeo, Arcana Forensics, La Mesa, California.)

Impact Spatter Mechanisms 147

Figure 7.45 Impact spatter on driver's side mirror of pickup truck. (Courtesy of L. Allyn DiMeo, Arcana Forensics, La Mesa, California.)

Figure 7.46 View of front of vehicle that struck victim who was bleeding from previous self-inflicted incised wounds.

Detection of Blood in the Barrels of Firearms

Another phenomenon associated with contact and close-range gunshot injury is the drawback effect. This is the presence in the barrel of a firearm of blood that has been drawn backward as a result of the effect of the discharged gases accompanying the projectile.

It should be obvious that the examination of firearms for blood both on the outside surface and within the barrel should be conducted prior to handling and test firing of the weapon in question.

Original research in this area was conducted by MacDonell and Brooks in 1977, and it was demonstrated that the depth of penetration of blood into the muzzle of a firearm was a function of the caliber of the weapon and the discharge distance. The research of MacDonell and Brooks determined that a relationship existed between discharge distance and the distance to which blood is drawn back into the barrel of the firearms. Maximum discharge distances at which traces of blood were detected to a depth of 5 mm or greater inside the weapon's muzzle ranged from 1 inch to 1.5 in. for .22 caliber revolvers to 5 in. for 12-, 16-, or 20-gauge standard shotguns. Several general observations were made as a result of their research:

- The larger the caliber or gauge, the greater the depth of penetration into the barrel.
- Recoil-operated auto-loading weapons will produce less depth of blood penetration than a weapon whose barrel does not recoil.
- The use of magnum or similar, higher energy loads will produce more depth of blood penetration than standard ammunition in the barrel of a given firearm.
- When a double-barreled shotgun is discharged at contact, considerable back spatter occurs (up to 12 cm) in the dormant barrel.

Caveats

There are instances where testimony has been given regarding what one would expect to see or what was actually observed in cases involving gunshot spatter. In one case, a pathologist testified that "the gun used to shoot the victim was more that two feet from her head when fired. Both wounds would have produced blood spatter although [I] couldn't say how much." In the same case a forensic scientist testified that the wounds he observed in photographs "would have probably produced enough blood spatter to get on the shooter's clothes." The issue in the case involved the absence of bloodstains (back spatter) on the clothing of the accused. There is no valid scientific basis to relate estimates or establish probabilities for the production of impact spatter resulting from gunshot.

Spatter Associated with a Projection Mechanism

8

Introduction

Projection mechanisms create a variety of bloodstain patterns, as shown in the taxonomy diagram in Figure 8.1. A projected bloodstain pattern is spatter created as the result of a force other than impact. Projected patterns produced with a volume of 1.0 mL or more often exhibit long spines and spatters that radiate from the central bloodstain. The spines indicate the application of force that propelled the blood against the target surface. The edges of the central stain are usually distorted and exhibit spines as compared to the minimal distortion associated with a splashed bloodstain pattern. In actual practice, splashed and projected patterns may exhibit remarkably similar characteristics. More spines will be produced when the target is close to the source of projected blood. At distances in excess of 4 feet, spines are reduced because air resistance causes the volume of projected blood to decelerate (Figures 8.2a and b).

Arterial Mechanisms

Arterial bloodstain patterns are the result of a projection mechanism resulting from arterial pressure that propels blood upward, forward, or downward from a damaged artery, often in a spray-type fashion not unlike a hole in a garden water hose. Distinctive physical characteristics can be observed and documented. The presence of the distinct bright red color of oxygenated arterial blood may also be apparent. A literature search for bloodstain terminology or definitions that addressed arterial patterns indicated that the common phrases for terminology are arterial spurt or spurting, arterial gushing, or both used interchangeably. The common denominator from the sources was the exiting or escape of blood under pressure from a breached artery. Reference was also made in many instances to evidence of spines, pressure fluctuations, or pulsation features of the patterns. A distinction between the terms arterial gushing and arterial spurting was noted in some sources, referring to gushing as large patterns and spurting as a series of individual bloodstains representing less volume and being more distinct in showing the pressure variation. One or more of these series of spurts represented arterial gushing.

The size of individual stains within arterial patterns ranges from less than 1 mm to more than 1 cm and larger in diameter. The stains may exhibit elongated spines and satellite spatters around the central area of the stain. The shape of the stains may vary considerably.

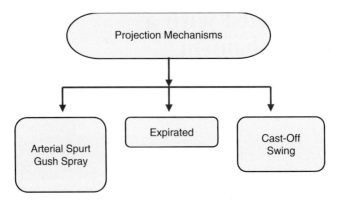

Figure 8.1 Taxonomy of projection mechanisms.

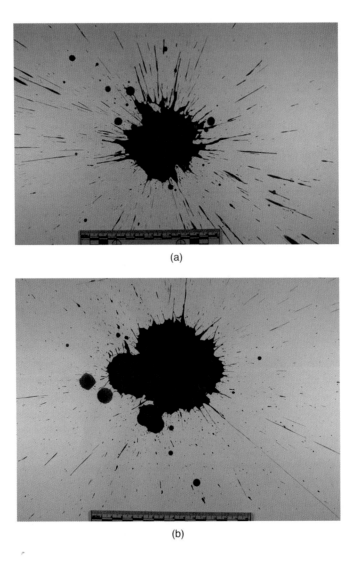

Figure 8.2 (a) Bloodstain pattern produced by 1.0 mL projected downward 40 cm onto cardboard. (b) Bloodstain pattern produced by 1.0 mL of blood projected downward 150 cm onto cardboard.

Spatter Associated with a Projection Mechanism

Figure 8.3 Arterial bloodstain pattern on floor showing numerous spines. Note intense red color consistent with arterial blood.

The central area of the stains may be circular, elongated, or irregular and may be associated with downward flow patterns. Individual stains within the pattern often align with each other in a parallel fashion relative to their direction of travel. These features may be dependent on the nature of the target surface. The alignment of the victim with the surface on which the blood is being projected also can change the shape and directionality of the stains.

The shape of the patterns may vary. The patterns may exhibit fluctuations such as pulsation or undulation and may bear resemblance to an electrocardiogram tracing. Small droplets may be deposited in a spray-type distribution. A variety of common arterial patterns is shown in Figure 8.3, Figure 8.4, Figure 8.5, and Figure 8.6.

Figure 8.4 Overlapping arterial bloodstain pattern showing undulating effect and spray of small spatters around periphery resulting from cut wrist.

Figure 8.5 Arterial pattern overlapping showing numerous stains with downward directionality and flow patterns resulting from cut wrist.

These physical characteristics are influenced by the following:

- Location of artery (deep versus closer to surface)
- Severity of injury (nick or severing)
- Volume of blood dispersed
- Orientation of the surface
- Nature of the surface
- Blocking effect of skin, tissue, or clothing
- Position of victim
- Movement and activities of the victim subsequent to injury including manual compression of the wound site

Figure 8.6 Arterial pattern showing numerous downward flow patterns and small spatters.

Spatter Associated with a Projection Mechanism

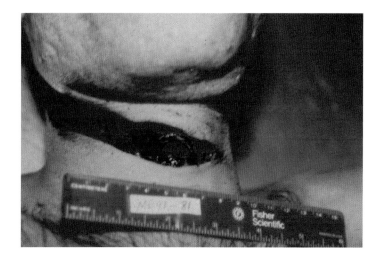

Figure 8.7 Victim who sustained slashed throat with severance of trachea and right carotid artery.

- Sequence of multiple injuries
- Overlapping of patterns created by other mechanisms
- Alteration of stains
- Medical intervention

A victim had sustained a fatal incised wound of the throat that severed his trachea and right carotid artery, as shown in Figure 8.7. The assault occurred on the bed, but the victim was able to leave the bed and ambulate to the area of the bedroom door where he collapsed on the floor by the foot of the bed. A characteristic arterial bloodstain pattern is present on the door and adjacent wall of the bedroom (Figure 8.8).

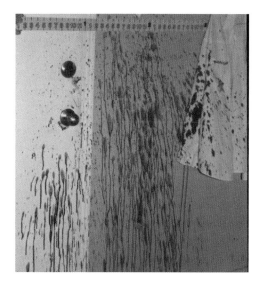

Figure 8.8 Arterial pattern present on door and adjacent wall of the bedroom.

Depending on the relative position of the victim and assailant, the clothing and person of the assailant may receive considerable bloodstaining from arterial sources. In death cases where an accidental fall has occurred, the scene may be extremely bloody as a result of the spurting of blood from a cut temporal artery. The resulting bloodstain patterns may arouse suspicion of foul play unless correctly interpreted. There are cases where the victim has sustained arterial injury inflicted while down on a surface such as a bed, carpet, floor, or ground where the subsequent expelling of blood forms a pool or saturation stain. Typical arterial spurting or gushing patterns are not present. It is important to attend the autopsy or examine the autopsy report to correlate the presence of arterial patterns with evidence of a damaged artery. The presence of arterial bloodstain patterns indicates arterial damage, but their absence does not preclude arterial damage from having occurred.

Medical conditions such as peptic ulcers in the stomach, ulceration in the gastrointestinal tract, or lung cancer can produce massive expulsion of blood through the mouth or rectum. The pathology of the lesion is essentially the same throughout the gastrointestinal tract and is caused by an abnormally large tortuous submucosal artery eruption. An example of this is a man who began hemorrhaging in bed and ultimately collapsed and died in the bathroom (Figure 8.9). A recent case involved the death of an elderly man who was terminally ill with non-small–cell carcinoma of the lungs who had been receiving outpatient hospice care. He was found on the ground outside the rear of his residence. A large projected bloodstain pattern was present on the cement floor of his patio, as shown in Figure 8.10 and Figure 8.11a and b. Prior to exiting his residence, the victim expelled a large amount of blood onto his bathroom floor and into his toilet (Figure 8.12).

Arterial bloodstain patterns have also been the result of therapeutic misadventure such as a malfunctioning arterial/venous shunt, as shown in Figure 8.13, or intentional postoperative disruption of an arterial graft by the patient, as seen in Figure 8.14.

Figure 8.9 Large saturated stain on bed and projection of blood onto bedroom carpet from patient with peptic ulcers.

Spatter Associated with a Projection Mechanism

Figure 8.10 Large projected bloodstain pattern on cement floor of victim's patio.

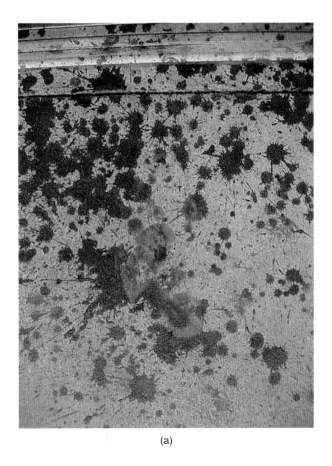

(a)

Figure 8.11a Closer view of area on patio floor near entranceway.

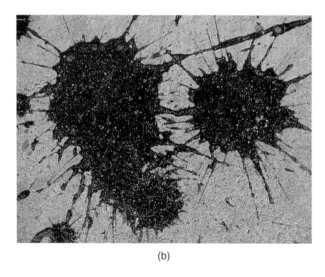

(b)

Figure 8.11b Closer view of large stains with elongated spines; note similarity to Figure 8.2a.

Figure 8.12 View of bathroom where victim expelled a large amount of blood on the floor and into the toilet.

Spatter Associated with a Projection Mechanism

Figure 8.13 Arterial patterns on wall and floor in conjunction with large accumulation of blood from victim with malfunctioning arterial venous shunt.

Figure 8.14 Arterial pattern on bathroom wall resulting from intentional ripping out of arterial graft by patient.

Bloodstain Patterns Produced by Medical Conditions of the Venous System

There are medical conditions that affect the venous system and are capable of causing hemorrhage that may produce bloodstain patterns that appear to be arterial in nature. These conditions may or may not be fatal and require a review of the medical history of the individual in conjunction with the bloodstain pattern interpretation at the scene.

Venous insufficiency syndrome associated with rupture of varicose veins of the lower extremities can produce significant blood loss, and fatalities have been reported in medical literature. Thin-walled varicose veins protrude just under the skin. Bumping or scratching a large varicose vein may cause severe bleeding as a result of the abnormally high pressure within the damaged vein.

Esophageal varices often associated with cirrhosis of the liver can project large volumes of blood from the throat that may be fatal. Projected bloodstain patterns may also result from the rupture of blood-engorged hemorrhoids.

One case involved a 93-year-old woman with extremely poor vision who awakened in the morning and proceeded to the kitchen to make coffee. She noticed a red-colored substance on the kitchen floor and attempted to clean it up. Her husband got up and assisted her; he suspected that the substance was blood and called their son. The son arrived and immediately called the Sheriff's Department. A deputy arrived and observed what appeared to be blood on the kitchen floor and lower walls of the residence. A small amount of apparent blood was observed on the clothing of the couple. They were both taken to a local hospital emergency room as a precaution. Physical examination of the couple indicated no sign of recent bleeding injury and hemoglobin levels were normal. Varicose veins were present on the wife's lower extremities. Samples of the suspected blood were collected and found to be of human origin, and the DNA was a match with the wife.

The bloodstain patterns at the scene consisted of areas of pooled blood and projected stains with some evidence of wiping alterations on the kitchen floor (Figure 8.15). There were

Figure 8.15 Pooled blood and projected stains with evidence of wiping alteration on kitchen floor. (Courtesy of Rex Sparks, Des Moines, Iowa Police Department.)

Spatter Associated with a Projection Mechanism 159

Figure 8.16 Base of doorjamb with laterally oriented projected stains; note downward flow patterns. (Courtesy of Rex Sparks, Des Moines, Iowa Police Department.)

also several areas of small projected bloodstains on the floor. The base of a doorjamb contained a laterally oriented arcing or undulating pattern exhibiting some downward flow patterns (Figure 8.16) that extend onto the adjoining wall approximately 6 in. above the floor (Figure 8.17). This pattern is also located approximately 6 in. above the floor. The lower portion of a doorway exhibited a laterally oriented projected bloodstain pattern (Figure 8.18). These patterns mimic arterial bloodstain patterns resulting from the increased pressure within the engorged vein.

Figure 8.17 Continuation of projected pattern adjoining corner of doorjamb and wall. (Courtesy of Rex Sparks, Des Moines, Iowa Police Department.)

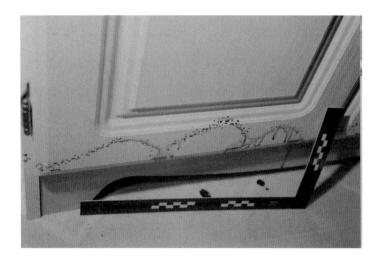

Figure 8.18 Laterally oriented undulating projected bloodstain pattern. (Courtesy of Rex Sparks, Des Moines, Iowa Police Department.)

Expiratory Mechanisms

Expiratory bloodstain patterns are the result of a projection mechanism that propels a volume of blood and air upward, forward, or downward from the nose, mouth, or open wound of the airway or lungs with physical characteristics that can be observed and documented. A literature search for bloodstain terminology or definitions that addressed expiratory patterns indicated that common phrases for terminology are expired, expiratory, or exhaled blood. The common denominator from the sources was the exiting or escape of blood blown out of the nose, mouth, or wound connected to the airway as the result of air pressure. This can occur in a number of different mechanisms:

- Exhalation of blood
- Coughing of blood
- Sneezing of blood
- Wheezing of blood
- Spitting of blood
- Paramedic intervention during resuscitative procedures

The size of individual stains within an expiratory bloodstain pattern ranges from less than a millimeter to several millimeters and in many instances overlaps the size range of impact spatters resulting from beating, stabbing, and gunshot events. Stain size may be dependent on the nature of the target surface. The shape of the stains may be circular or elongated, depending on the angle at which they struck a surface. The shape of the pattern may vary with different mechanisms. Small droplets may be deposited in a spray-type distribution (Figures 8.19a and b), and the pattern may exhibit a conical distribution. Blood expelled from both nares may produce a bilateral pattern, as demonstrated in Figure 8.20, Figure 8.21, and Figure 8.22. The spitting of blood on a computer screen and keyboard is shown in Figure 8.23 and Figure 8.24.

Spatter Associated with a Projection Mechanism

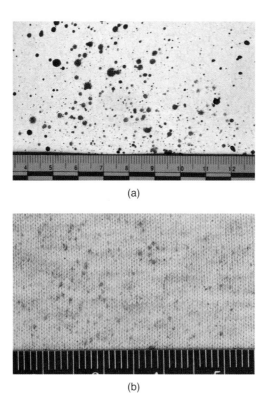

Figure 8.19 (a) Expirated bloodstain pattern produced on vertically oriented white cardboard from approximately 18 in. (b) Expirated bloodstain pattern produced on vertically oriented 100% cotton fabric from approximately 18 in.

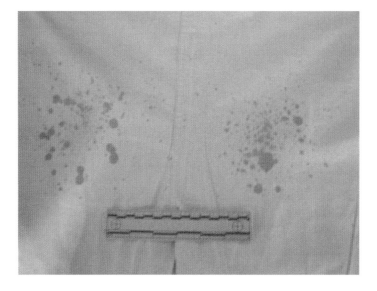

Figure 8.20 Bilateral expiratory pattern of blood on upper thighs of hospital scrubs of nurse, produced by sneeze during bloody nose episode.

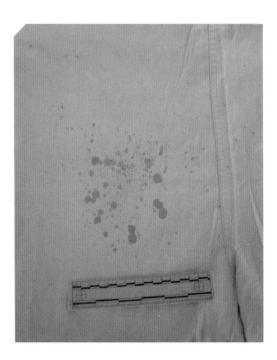

Figure 8.21 Close view of right leg of hospital scrubs.

Expiratory patterns may be intermixed or overlaid with spatter from other mechanisms such as beatings and gunshot.

There are several indicators for the presence of expirated bloodstain patterns. An important consideration is the presence of blood in the nose, mouth, or airway (Figure 8.25 and Figure 8.26). The presence of air bubbles or bubble rings (circular outlines of air bubbles

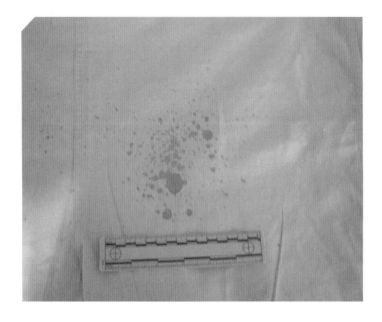

Figure 8.22 Close view of left leg of hospital scrubs.

Figure 8.23 View of computer monitor and keypad showing areas where blood was purposely spit.

in dried stains), as shown in Figure 8.27, Figure 8.28, and Figure 8.29, or the presence of mucous strands within the pattern are also important criteria (Figure 8.30 and Figure 8.31). There may or may not be evidence of dilution of the stains. A high amylase concentration in the bloodstains may be presumptive evidence of the presence of saliva. Amylase is normally present in blood and is elevated in pancreatic disease.

Figure 8.24 Closer view of lower portion of monitor and keypad showing the areas of bloodstains mixed with saliva.

Figure 8.25 Victim of blunt force injury with large amount of blood around nose and mouth with evidence of expirated blood on left arm.

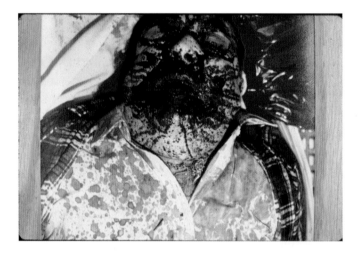

Figure 8.26 Victim of multiple gunshot wounds with a bullet passing through the bridge of nose, producing large amount of blood around nose and mouth and expirated bloodstains.

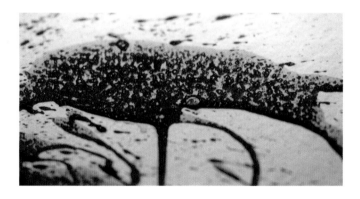

Figure 8.27 Air bubbles and vacuoles in large expirated bloodstain pattern.

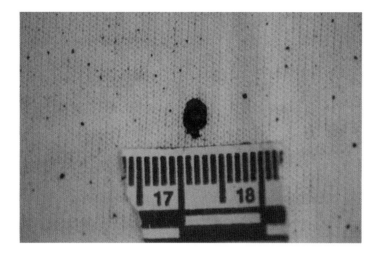

Figure 8.28 Air bubble in recently deposited expirated bloodstain pattern.

If there is no blood in the nose, mouth, airway passages, or lungs, then the bloodstains could not have been produced by expiratory or exhaled blood. It is possible, however, to produce expiratory-like bloodstains from compression of the abdomen when there has been a wound to the intestinal tract that could expel a mixture of intestinal gas and blood. This occasionally occurs during paramedic intervention (Figure 8.32 and Figure 8.33). In medical situations, expiratory bloodstain patterns may be produced by a patient coughing through a tracheostomy tube, as shown in Figure 8.34.

Figure 8.29 Expirated bloodstain experiment performed by student lying down and projecting his own blood from his mouth.

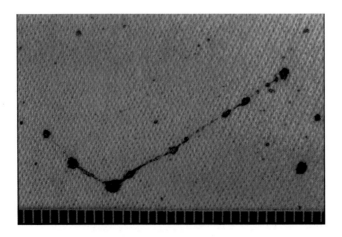

Figure 8.30 Mucous strands within expirated bloodstain pattern.

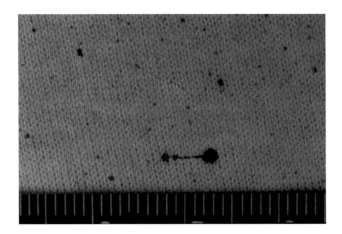

Figure 8.31 Another area of pattern in Figure 8.30 showing additional mucous strands.

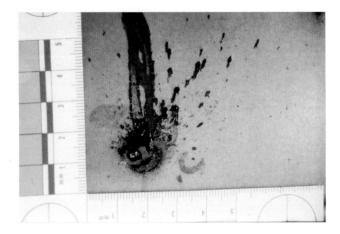

Figure 8.32 Gunshot wound of abdomen producing small spatters of blood due to expulsion of intestinal gas.

Spatter Associated with a Projection Mechanism

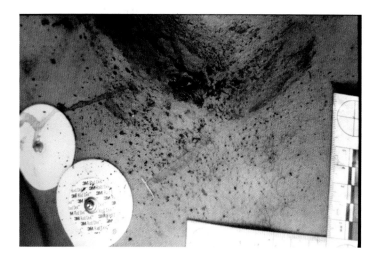

Figure 8.33 Compression of chest by paramedics producing expirated blood on chest and on top of electrocardiogram pads.

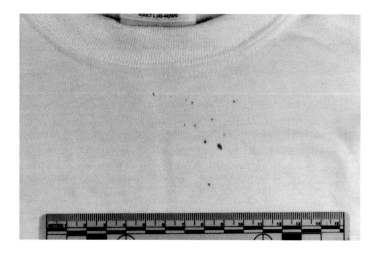

Figure 8.34 Expirated bloodstain pattern produced on T-shirt of attending nurse as a result of patient coughing through his tracheostomy tube in hospital.

Expiratory Bloodstain Experiments

Numerous experiments have been conducted by individuals using their own blood to demonstrate the bloodstain patterns produced by exhalation, coughing, and sneezing blood from the nose and mouth. The results of these experiments clearly showed the similarity of size range of the individual drops within a pattern with impact spatter observed in beating, stabbing, and gunshot events. Extremely fine bloodstains were easily produced. Air bubbles or bubble rings, mucous strands in the stains, or discernable diluted stains were not a frequent occurrence. The results also demonstrated how the quantity of blood expelled could affect the overall concentration and distribution of individual spatters

Figure 8.35 Close view of small spatters on trousers of defendant.

within an expirated bloodstain pattern. In reality, one does not know the amount of blood available to expel during an expiratory event.

One case involved the death of an elderly man who was suffering from terminal throat cancer and spent a considerable amount of time in bed with oxygen supplied from a bedside tank. He was known to have hemorrhagic episodes and in fact had been taken to an emergency room for treatment a few days prior to his death. He was being cared for by a nephew who stated that he heard two shots from his uncle's bedroom and found his uncle slumped over the edge of the bed in somewhat of a sitting posture. The nephew assisted his uncle to the floor and realized that he was dead. The nephew was ultimately arrested for shooting his uncle. One of the major issues in the case was the presence of small stains of the uncle's blood on the sheet where he allegedly was sitting when found by the nephew and small spatters of the uncle's blood on the front of the trousers of the nephew (Figure 8.35). It was interesting that there was no blood on the long-sleeved shirt worn by the nephew. The expert for the state testified that the stains on the sheet and the trousers was "high-velocity mist" that indicated that the uncle was not on the bed but rather sitting or kneeling on the floor when he received the fatal gunshot wounds. Incidentally, there was one projectile in the head, although there was evidence of a second glancing entrance wound. A second projectile was not recovered. The state expert also testified that the small stains on the trousers of the nephew were also high-velocity mist that would place the nephew close to his uncle at the time of the shooting.

To demonstrate an alternative explanation for the production of the small stains in question on the bedsheet and trousers, experiments were conducted by the defense bloodstain analyst using his own blood and coughing and expirating blood onto similar fabrics. Stain patterns were produced that were indistinguishable from those on the trousers and sheet. Figure 8.36 and Figure 8.37 are representative of the experimental results obtained. The nephew was ultimately acquitted by the jury. The case demonstrated the value of performing relevant experiments and the importance of recognizing alternative explanations for the production of small bloodstains.

Spatter Associated with a Projection Mechanism

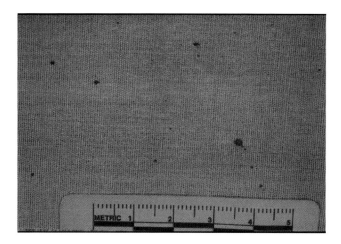

Figure 8.36 Small bloodstains produced as a result of blood expirated onto 100% cotton fabric similar to defendant's trousers.

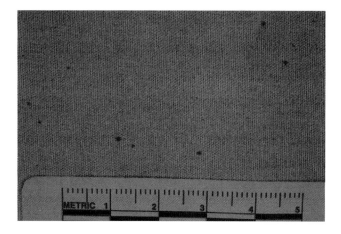

Figure 8.37 Small bloodstains produced as a result of coughing blood onto 100% cotton fabric similar to defendant's trousers.

Cast-Off Mechanisms

Cast-off bloodstain patterns occur when blood is projected or thrown onto a surface from a bloody source or object in motion such as a hand or beating instruments (bat, crowbar, hammer, etc.) owing to a whiplike action and the generation of centrifugal force that overcomes the adhesive force that adheres the blood to the object. Cast-off experiments are routinely performed in courses on basic bloodstain pattern analysis (Figure 8.38). Cast-off patterns are frequently associated with beating events, and cast-off stains may be within an impact spatter pattern and may overlap in size with impact spatter stains (Figure 8.39). The size of cast-off bloodstains is generally in the range of 4 to 8 mm, although they may be larger or smaller depending on the type and size of the bloody object, amount of available blood, and the forcefulness and the length of the arc of the swing. The flicking of bloody fingers can produce

Figure 8.38 Students participating in beating and cast-off experiment in basic bloodstain school.

small stains in the size range of impact spatter (Figure 8.40). Usually the more forceful the swing, the smaller the drops and the resultant stains will be. Cast-off bloodstains are somewhat uniform in size within a given pattern and are often seen as wide or narrow linear or slightly curved trails on the impacting surfaces with the more elongated bloodstains most distant from the source. The term *cast-off bloodstain* should be distinguished from the term *wave cast-off stain*, which is the result of a small blood droplet that has originated from a parent drop resulting from the wavelike action of the blood striking a surface at an acute angle. When a weapon is swung, the overhead or side stroke is subject to a rapid termination as the stroke abruptly stops and the direction is reversed into a forward, injury-producing blow. This abrupt reversal of direction of the swing causes blood to be thrown or projected from the weapon. In many cases of blunt trauma, the weapon is swung at the victim repeatedly. Once blood has been produced or exposed, it will adhere to the weapon in varying quantities, depending on the type of weapon and the quantity of blood at the site of injury.

Figure 8.39 Victim of blunt force injury showing impact spatter and cast-off trails on wall above left shoulder.

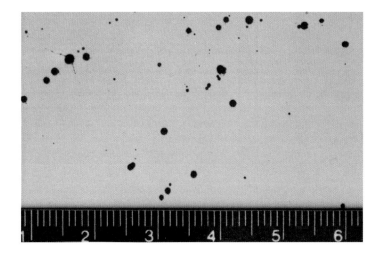

Figure 8.40 Small spatters of blood produced by flicking of bloody fingers.

During the back or side swing away from the victim, the blood adhering to the weapon will be thrown off and will travel tangentially to the arc of the swing; the droplets will impact on nearby surfaces such as walls, ceilings, floors, and other objects in their path (Figure 8.41 and Figure 8.42). Often more blood and smaller drops are cast off during the course of the downward or return swing of the weapon when the back swing arc is complete because more force is directed at the victim. The initial blood drops that are cast off from a weapon during the arc of the back swing may strike a target surface at or near a 90° angle, producing circular to slightly oval-shaped stains on a surface such as a ceiling or nearby wall depending on the plane of the arc (Figure 8.43). Subsequent drops will strike the target at increasingly more acute angles of impact and become progressively more elongated as the drops continue along their path of travel away from the weapon (Figure 8.44).

Another form of cast-off bloodstains are referred to as cessation cast-off, which occurs when a bloody object strikes another object or surface and comes to an immediate stop. The blood droplets continue in the same direction and at the same velocity as did the bloody object

Figure 8.41 Cast-off pattern on ceiling above bed where a victim was beaten.

Figure 8.42 Cast-off pattern on floor of bathroom.

(Figure 8.45, Figure 8.46, and Figure 8.47). The location of cast-off bloodstains provides information relative to the position of the victim and the assailant during the administration of the blows.

When there are multiple cast-off patterns present, a general estimation of the minimum number of swings of the bloody object may be determined. The number of distinct patterns or trails of cast-off bloodstains may equal the minimum number of blows struck plus one

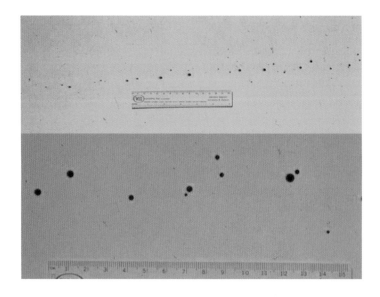

Figure 8.43 Cast-off patterns consisting of circular stains.

Spatter Associated with a Projection Mechanism

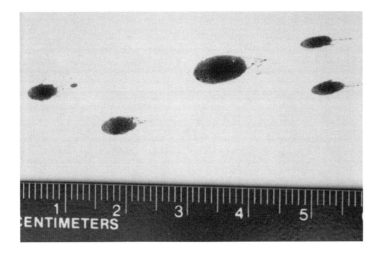

Figure 8.44 Cast-off pattern consisting of elongated stains showing directionality from left to right.

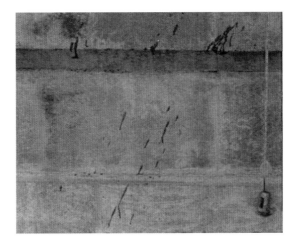

Figure 8.45 Cessation cast-off stains on side of bridge resulting from dismembered thigh and leg of victim striking edge of bridge while being discarded below.

Figure 8.46 Area below bridge where dismembered thigh and leg were found.

Figure 8.47 Cessation cast-off stains on wall resulting from impact with bloody weapon.

because the initial blow generally does not produce sufficient exposed blood on the weapon to produce cast-off bloodstains. Exceptions to this are blows administered with a machete or axe that can become bloody with the first blow. In cases such as that, the number of trails may actually equal the number of blows. If multiple blows were struck on the same plane, the cast-off patterns or trails may overlap, which is another reason why only a minimum number of blows struck may be estimated. However, overlapping cast-off trails may be difficult to interpret as to their number so the total number of blows struck to a victim should be correlated with the autopsy interpretation of the number of injuries. It is also important to remember that cast-off trails may be the result of a swing of a bloody object not related to a blow to the victim.

Occasionally, the location of bloodstain patterns may suggest in which hand a weapon was held while being swung. This interpretation should be made with caution with the possibility of a two-handed or backhanded delivery kept in mind. Cast-off bloodstains on the rear of an individual's shirt or the back of the trouser legs may be produced while swinging the weapon overhead in a leaning or kneeling position. Cast-off bloodstains on the front of an individual's clothing may indicate proximity to a bloodshed event rather than actual participation in the beating or stabbing event (Figure 8.48).

A woman's roommate found her beaten to death on a mattress located on the floor of their living room in front of a couch in the apartment they shared (Figure 8.49). The roommate denied participation in the homicide. The victim had sustained severe blunt force injuries of the head (Figure 8.50). The location of impact spatter on the left edge of the couch and wall behind indicated that blows were struck to the victim while her head was consistent with a sitting position as well as in her final position. Multiple cast-off stains were present on the ceiling over the mattress. A bloodstained aluminum softball bat, which exhibited multiple dents, was located in the bathroom (Figure 8.51). The blood on the bat was determined to be consistent with that of the victim.

Crime scene investigators noticed apparent bloodstains on the clothing and legs of the roommate as she was being questioned. An area on the left sleeve and upper left rear of her T-shirt contained several cast-off bloodstains that were consistent with the blood of

Spatter Associated with a Projection Mechanism

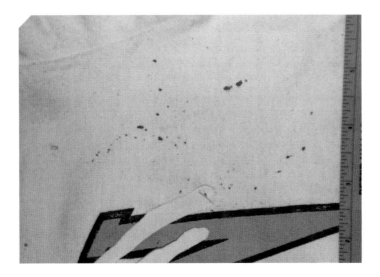

Figure 8.48 Cast-off stains on front of individual's shirt that resulted from him standing near a beating event without participating.

the victim (Figure 8.52). Her shorts contained a single small bloodstain, and a few impact spatters were present on her lower legs. In an attempt to explain these bloodstains she stated that all she remembered was standing over the victim on the mattress with a bat in her hand. She was ultimately convicted at trial.

This case demonstrates that despite significant impact spatter distributed at the scene of beatings, there may be little if any spatter on the assailant. The directionality and force of the blows struck usually dictate the distribution of spatter. In this case the impact spatter was largely directed away from the assailant. However, important bloodstain evidence in

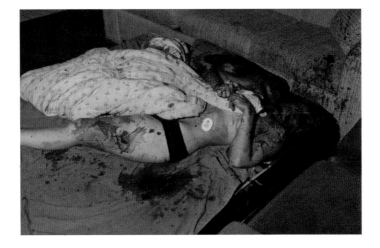

Figure 8.49 Victim of blunt-force injury on mattress on floor of living room showing spattered blood on lower portion of couch.

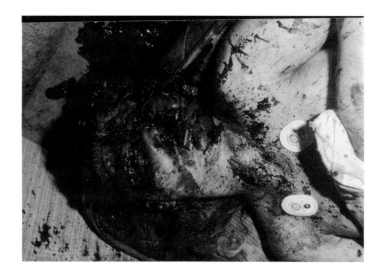

Figure 8.50 Extensive blunt-force injuries to face and head of victim.

this case was the presence of the cast-off bloodstains on the sleeve and rear of the shirt of the roommate that were consistent with her swinging a bloody object overhead.

Overhead cast-off on ceilings is often overlooked by assailants who have attempted to clean and remove bloodstains from a crime scene. Similarly, investigators may fail to "look up" as well. *Look at the ceiling!* A recent case demonstrated the importance of observing the ceiling by confirming that injury to the victim occurred on his bed. A woman called police to report that her husband was missing, and a few days later she claimed that she found him in the closet of their bedroom. He had sustained blunt-force injuries to the head, but she claimed no knowledge of the incident, although she had been sleeping in their bed

Figure 8.51 Bloodstained aluminum baseball bat found in bathroom determined to be weapon used in assault on victim.

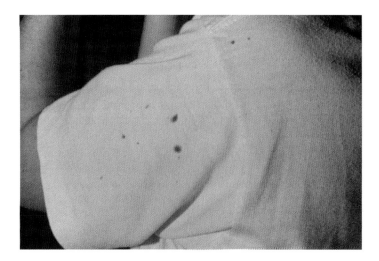

Figure 8.52 Upper-left sleeve and shoulder area of defendant's shirt showing cast-off stains.

since the day he disappeared and the closet was only a few feet away from the bed. Investigators located a large area of blood saturation on the mattress, which was covered by sheets and blankets. The cloth covers on the headboard and footboard had been removed, exposing the underlying material. Some impact spatter was present on the lamp shade on the nightstand and on a picture frame that hung on the wall above the head of the bed. Distinct linear cast-off patterns were observed on the ceiling above the bed. It was clear that the husband was likely beaten to death on his bed and hidden in the closet. The wife eventually pled guilty.

Altered Bloodstain Patterns 9

Introduction

Another segment of our bloodstain taxonomy is the important category of altered bloodstains (Figure 9.1). Actually, any bloodstain or pattern may be subject to an alteration. There are many instances where exposed blood can become altered from its original appearance as a result of physical or physiologic changes that occur. Alterations in the appearance of blood can also occur within the body in the stomach and gastrointestinal tract, and when expelled, exhibit distinct color changes as a result of disease processes. Environmental conditions may subject the blood and bloodstains to a variety of changes. These include diffusion; drying and flaking; dilution resulting from moisture; decomposition; and various types of insect, animal, bacterial, and fungal activities. Often there are cleanup efforts used by the assailant to remove bloodstains, which alter their appearance at the scene.

After bloodshed has commenced, the clotting mechanism is initiated. Bloodstains may become diluted by mixture with other body fluids such as urine and cerebrospinal fluid (CSF). Existing wet bloodstains may be altered and artifactual stains may be produced by activities during and after the assault by the victim or the assailant. Movement of the body by medical and law enforcement personnel can also change the appearance of bloodstains and create additional stains at the scene and on the clothing of the victim. The handling of clothing and other items wet with blood can produce alterations. Packaging of wet bloody items should be avoided.

Diffusion and Capillarity of Bloodstains

Diffusion is defined as the movement of a substance from an area of higher concentration to an area of lower concentration. Capillarity is a function of the ability of a liquid to wet a particular material, such as blood wetting a fabric. Surface tension is responsible for the movement of a liquid through a porous solid.

The dynamics of blood and serum capillarity and diffusion that cause migration on cloth is a chromatographic function of the ability of the liquid components of the blood (mobile phase) to travel by capillary action on a porous surface (stationary phase). Experimentally, it has been demonstrated that the chromatographic separation of blood components on cotton or polyester cloth is not a rapid or efficient chromatographic procedure. The most

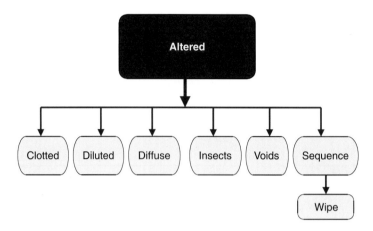

Figure 9.1 Taxonomy of altered bloodstains.

rapid capillary diffusion is seen to occur during the initial 5 min of exposure as a result of the wetting action of the blood with a slow but measurable increase thereafter. The distance traveled by the mobile phase is also a function of the equilibrium in the system. Within a closed chromatographic system, the ultimate height of diffusion is considerably higher. The effects of blood diffusion and capillarity are readily observed when blood saturates bedding and clothing, as demonstrated in Figures 9.2a and b.

Aging of Blood

As bloodstains increase in age, they progress through a series of color changes from red to reddish brown to green and eventually to dark brown and black (Figures 9.3a and b). This change of color is attributable to the drying process and to the loss of oxygen from the oxyhemoglobin in the red cells on exposure to air. Exposure to the sun will hasten the darkening process. A particularly warm and humid environment and the presence of bacteria and other microorganisms during the decomposition process will also affect the sequence and duration of color changes in bloodstains. Postmortem growth of fungus can produce unusual changes to blood as seen in Figures 9.4a and b. This was a scene where the body had begun the decomposition process and the growth of the hairlike fungus was determined to be *Aspergillus* sp. Wet bloodstains that were originally red will usually progress to red brown and then to green within 24 h at a warm temperature as a result of the growth of bacteria and the decomposition process. Medical conditions such as peptic ulcer hemorrhage and vomiting often produce dark "coffee ground emesis" and resulting bloodstains that have the appearance of aged blood. The dark color is the result of the stomach acids denaturing the blood, as described in Chapter 2. Bleeding within the lower gastrointestinal tract can produce loose, black, tarry fecal material that is the result of the breakdown of blood prior to evacuation.

There are instances where the substrate on which the bloodstain is located has an effect on the observed color of the stain. For example, a dark bloodstain on glass where there is transmitted illumination may appear lighter red than on a section of drywall. When

Altered Bloodstain Patterns

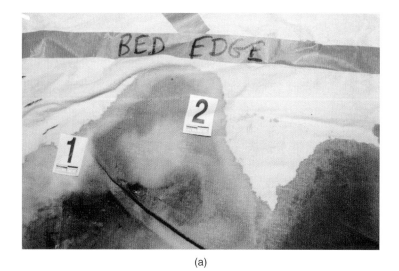
(a)

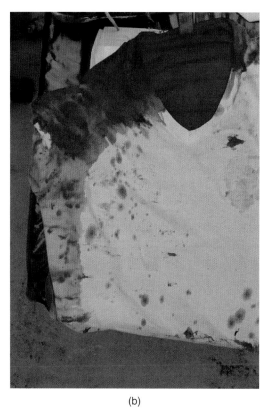
(b)

Figure 9.2 (a) Diffusion of blood on bed sheet. (b) Diffusion of blood on right side of shirt saturated with blood.

examining bloodstains in photographs, the exposure of the photograph must be considered when evaluating slight color differences. It cannot be overemphasized that estimations of the age of bloodstains based on their color and condition at the scene or in scene photographs should be very conservative.

Figure 9.3 (a) Appearance of fresh bloodstain on glass. (b) Appearance of older, darker bloodstain on cardboard.

Drying of Blood

When blood is exposed to an external environment, the drying process on various surfaces is initiated. The drying time of blood is a function of the bloodstain size and volume, the nature of the target surface, and the influences or effects of the external environment. Small stains and spatters, light transfer stains, and thin flow patterns of blood may dry within a few minutes at normal conditions of temperature, humidity, and air flow on nonporous surfaces. Bloodstains of greater size and volume require longer periods to dry under similar conditions (Figures 9.5a and b). The drying time of blood in general is decreased by increased temperature and decreased humidity, the presence of significant air currents such as wind, and the effect of a fan. It follows that humid environments, lower temperatures, and minimal air flow will produce longer drying times. Surfaces that permit the soaking

Altered Bloodstain Patterns

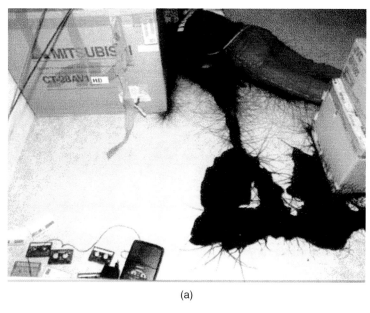

(a)

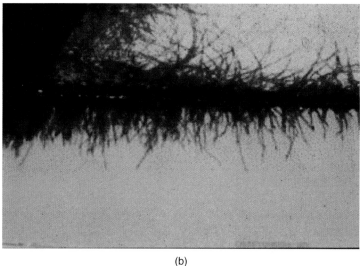

(b)

Figure 9.4 (a) A view of scene where the body has begun the decomposition process; note the hairlike extensions on the pool of blood on the floor. (Courtesy of Martin Eversdijk, The Netherlands.) (b) Closer view of the growth of fungus on the bloodstain. (Courtesy of Martin Eversdijk, The Netherlands.)

of blood into the material such as carpets and bedding usually produce significantly longer drying times (Figure 9.6).

Research by Epstein and Laber published in their manual, *Experiments and Practical Exercises in Bloodstain Pattern Analysis*, published in 1983, contains significant data on the drying time of blood. They studied the effect of temperature, humidity, air flow, and surface texture on the drying time of blood using a single drop, 1-, 5-, and 10-mL volumes of blood to create the bloodstains. Their study can be used as a general guideline. However, when

(a)

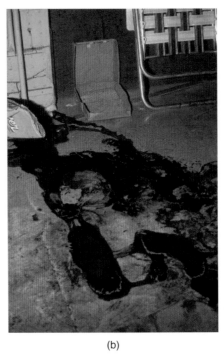

(b)

Figure 9.5 (a) A view of partially dried bloodstain on wood. (b) View of pool of blood at scene 2 days after double homicide occurred.

drying times are to be estimated in casework, they should be very conservative. When comparative experiments are considered, the bloodstain size should be estimated and the surface texture, air flow, and environmental conditions should be duplicated as closely as possible.

The drying of bloodstains is observed initially around the edges or periphery and proceeds inward to the central portion of the stain as the drying process continues. Occasionally the dried central area of a dried bloodstain will flake away, leaving an intact circular rim.

Altered Bloodstain Patterns

Figure 9.6 Blood saturation on bedding and pillow still wet after 3 days.

This is referred to as a skeletonized bloodstain (Figure 9.7). Another type of skeletonized bloodstain occurs when the central area of a partially dried bloodstain is altered by contact or a wiping motion leaving the peripheral rim intact. This can be interpreted as a sequence of activity occurring after the bloodstain began to dry (Figure 9.8). Sequencing will be discussed further later in this chapter.

A useful observation of bloodstains on surfaces other than horizontal is the recognition of the dense zone. When a blood droplet of sufficient volume strikes a nonhorizontal surface, gravitational forces continue to act on the liquid portion of the bloodstain. The lower area or base of the bloodstain will be denser as a result of the continued accumulation of blood. After the bloodstain has dried sufficiently on the surface, this dense zone cannot

Figure 9.7 Skeletonized bloodstain resulting from flaking of dried blood.

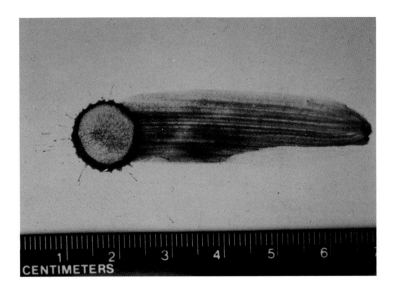

Figure 9.8 Skeletonized bloodstain resulting from wiping through partially dried stain, leaving the peripheral rim intact.

be altered by movement (Figure 9.9). Bloodstains on objects or surfaces that have been moved after the blood has dried may show dense zones inconsistent with their observed positions. This observation would indicate an alteration of the scene.

The significance of the diffusion and drying time of blood was demonstrated in a case that involved the shooting deaths of four members of a family. The father called police at approximately 4:30 AM to report that his oldest son had just shot his wife, infant sister, his brother, and himself. Police arrived within minutes. Each of the victims was shot once

Figure 9.9 Accumulation of blood at lower portion of stain on vertical surface referred to as dense zone.

Altered Bloodstain Patterns

Figure 9.10 View of infant in crib with diffused blood on top of shoulder and area adjacent to upper right arm.

with a .22 caliber handgun. The oldest son was alive but expired later in the day at a local hospital. The three deceased victims at the scene were cold to the touch. An alert detective also noted that the blood on the shoulder of the infant's nightshirt was diffused and dry (Figure 9.10). Experiments were performed with fresh human blood that was allowed to diffuse and ultimately dry on fabric similar to the nightshirt of the infant. The time for this to occur was in excess of 3 h (Figure 9.11). The blood on the nightshirt of the infant likely required a significant amount of time to diffuse and dry, which is not consistent with the version of events given by the father. This fact coupled with additional evidence resulted in the conviction of the father for the four homicides.

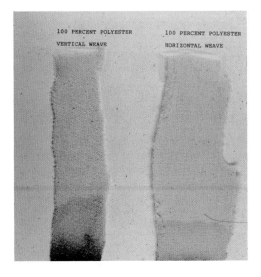

Figure 9.11 Diffusion of blood on 100% polyester cloth with vertical weave strip on left and horizontal weave strip on right.

188 Principles of Bloodstain Pattern Analysis: Theory and Practice

Clotting of Blood

A blood clot is formed by a complex mechanism involving the plasma protein fibrinogen, platelets, and other clotting factors. As discussed in Chapters 2 and 3, fibrin is integral to clotting. For fibrin to be available whenever an injury occurs, it is constantly being transported through the body by the circulating blood. Fibrin must be in a soluble form (called fibrinogen) when it is circulating in the blood, but it must change into an insoluble form whenever clotting becomes necessary. Subsequently, the blood clot begins to retract, causing a separation of the remaining liquid portion, which is referred to as serum (Figure 9.12, Figure 9.13, and Figure 9.14). This serum separation resulting from clot formation and retraction may take

Figure 9.12 Demonstration of formation of clot in fresh blood after 7 min.

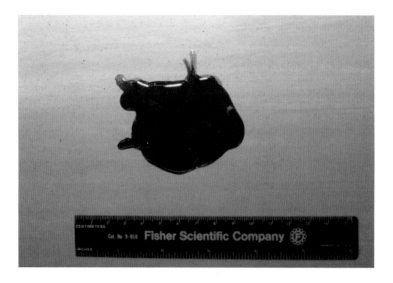

Figure 9.13 Beginning of clot retraction and production of serum around edges.

Altered Bloodstain Patterns

Figure 9.14 Floor of scene showing a dried pool of blood with serum separation; note spatter on floor and lower portion of refrigerator.

an hour or more depending on the original blood volume, the surface or substrate where the blood is present, and environmental factors such as temperature and humidity.

In normal healthy individuals the time between initiation of the bleeding to the appearance of visible clot formation requires approximately 3 to 15 min in a clinical setting, depending on the particular clotting time test used. Normal values vary with the different tests that may be used. There are many clinical factors that will lengthen the clotting time of individuals, including diseases that affect the liver, such as alcoholic cirrhosis of the liver; clotting factor deficiencies, such as occur in hemophilia; and platelet deficiencies that are seen in aplastic anemia and leukemia. Prescribed anticoagulant medications for so-called "blood thinning" that help reduce the risk of recurrent strokes and heart attacks maintain patients at a lengthened clotting time. Aspirin will also increase the clotting time. Many adults are advised by their doctors to ingest a therapeutic dose of 81 mg of aspirin per day to reduce the risk of stroke or heart attack.

In a forensic situation, one must consider external factors that might affect the length of time required to form a clot. Continual disruption, such as a continued beating, would extend the time required. In the author's experience, a victim was beaten about the head and then dunked repeatedly in a toilet in an attempt to elicit a statement. Continually wetting the area of injury would also extend the length of time required for a clot to form.

Cerebrospinal fluid (CSF) is one of the triggers that can accelerate clotting. In years past, CSF was actually used in clinical studies of clotting. Head injuries or injuries to the spinal canal potentially release CSF into the site of injury. As a result, the clotting time in such instances may be shorter.

Blood clots and serum stains surrounding them as well as the degree of observed drying of blood should be recognized as important information at crime scenes. Occasionally, events take place after blood has been shed and has begun the clotting process. Bloodstain patterns produced by partially clotted or clotted blood indicate a time interval between

(a)

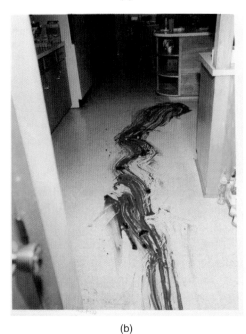

(b)

Figure 9.15 Drag pattern showing clotted blood, indicating movement of victim after significant period of time. (Courtesy of Herbert Leon MacDonell, Corning, New York.)

the bloodshed and the activity producing the pattern (Figures 9.15a and b). This interval may be short or extended, depending on the degree of clotting, source and quantity of blood, and environmental conditions existing at the time. It is also important to obtain a medical history of the victim and a record of medications.

Examples of the significance of partially clotted or clotted bloodstain patterns are as follows:

Altered Bloodstain Patterns

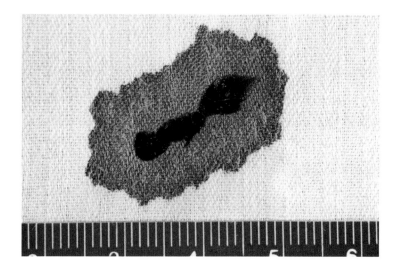

Figure 9.16 Clotted spatter produced in laboratory resulting from impact into partially clotted blood.

- In general, the presence of significant dried and clotted blood on surfaces at a scene indicates a significant time lapse between bloodshed and the observation of the bloodstains.
- Clotted blood spatter on victim's clothing or surrounding surfaces associated with a beating death may indicate a significant interval between blows administered or possibly postmortem infliction of injury (Figure 9.16 and Figure 9.17). Usually small stains of freshly spattered blood will dry before they have the opportunity to exhibit clots.

Figure 9.17 Small clotted spatter on cloth; note dense center and lighter periphery.

- Clotted bloodstain patterns associated with a pedestrian victim on a roadway may indicate an interval between impacts associated with more than a single vehicle.
- Coughing or exhalation of clotted blood by a victim may be associated with postinjury survival time.

An estimation of the degree of blood clotting and drying of pools of blood associated with a victim when used in conjunction with other signs of postmortem change may be helpful in the determination of the postmortem interval or in substantiating postmortem movement of the victim. Estimation of degree of clotting and drying time of blood at crime scenes should be reproduced experimentally using freshly drawn human blood of similar volume placed on an identical surface with similar environmental conditions existing during the experiment as were observed at the scene. Estimates of this interval should be made with extreme caution.

The significance of clotted spatter with the analysis of bloodstains can be demonstrated in the two following cases. The first case involved the beating death of a young woman with a crowbar that occurred on her sofa bed in her apartment. She sustained severe linear lacerations of her head with underlying skull fractures (Figure 9.18). Impact spatters were present on the wall behind the sofa bed with an area of convergence at approximately the level of her head on the bed. Additionally the pillows and a stuffed doll on top of the headrest of the sofa near the pillows received impact spatter. The front of the dress of the doll contained clotted spatter characterized by the dense center and lighter periphery of the stains, referred to as the "fried egg" appearance (Figures 9.19a and b). This finding was consistent with a time interval between blows to the victim later admitted to by the assailant as a means of stopping the sounds made by the victim.

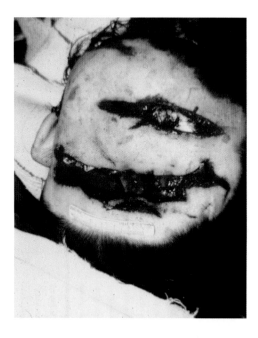

Figure 9.18 View of victim showing severe linear lacerations to the head and depressed skull fractures.

(a)

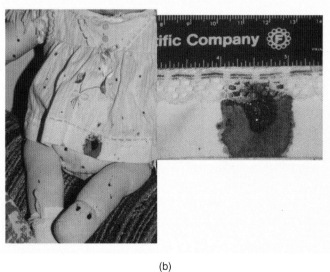

(b)

Figure 9.19 (a) View of doll at head of bed above victim showing clotted spatter. (b) Closer view of clotted spatter on doll.

The second case involved the beating death of a homeless man who lived in a makeshift tent with his friend. The friend called police to the scene claiming that he returned to the campsite and found the victim dead. The victim had sustained massive head trauma likely from a portion of a metal road signpost that was lying across the chest of the victim

Figure 9.20 Homeless man who sustained massive head trauma from impact with portion of metal road sign.

(Figure 9.20). The post contained blood and hair. There was bloodstain evidence that indicated that the attack began inside the tent and proceeded outside into a partially wooded area where there was a large accumulation of blood on the ground. This area was a distance of at least 30 ft from the final position of the victim. The police photographed the friend, whose trousers and shoes were spattered with blood despite his denial of any involvement in the death of his friend. There were numerous clotted spatters within the impact spatter patterns on the trousers and shoes that were consistent with a beating that occurred over a period of time (Figures 9.21a and b).

Diluted Bloodstains

The deposition of bloodstains at crime scenes occurs under many different environmental and physical conditions that may alter their appearance either initially or thereafter. Diluted bloodstains may be present at scenes where excessive moisture is present, such as external environments involving rain or snow, in that the characteristics of the original stains may be altered to the point where analysis is difficult or impossible. Deliberate cleanup attempts often produce evidence of diluted bloodstains. Bloodstained areas often exhibit a lighter, washed-out appearance. A feature of individual diluted bloodstains is a prominent dark outer rim with a lighter center area (Figure 9.22). In extremely cold environments, bloodstain patterns deposited on snow or ice are frequently recognizable and should be preserved photographically as soon as possible (Figure 9.23). The pooling of blood on ice will cause some melting of ice on the surface and create dilution of the original pool. Homicides have occurred where the victims are found in walk-in freezers, and the bloodstains have not presented problems with analysis in the author's experience. In those types of cases (often occurring in fast-food restaurants) the bloodstains may freeze to the surface even before they would have dried, and their characteristics may be less likely to be altered. The maintaining of the cold temperature within these walk-in freezers is critical until the stains

Altered Bloodstain Patterns 195

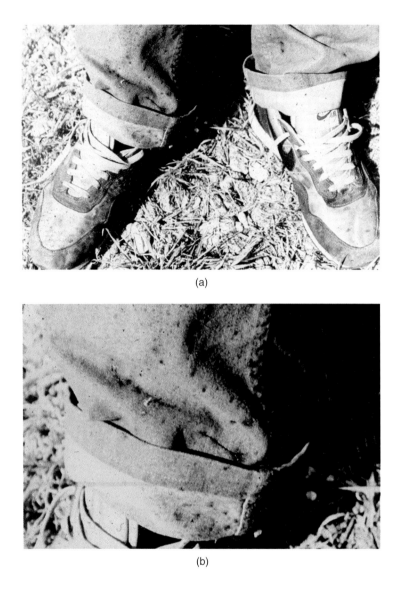

Figure 9.21 (a) View of lower jean legs and shoes of the accused showing numerous clotted spatters of blood. (b) Closer view of clotted spatter on lower legs of jean.

have been thoroughly documented and representative samples have been collected. Allowing the freezer to thaw will cause alteration and possible intermixing of stains as a result of thawing.

Diluted bloodstains may be encountered at crime scenes as a result of mixture with water or other fluids. Damp clothing resulting from excessive perspiration may dilute bloodstains that are present on clothing worn by a victim or assailant (Figures 9.24a and b). In some cases perimortem urination by the victim has caused dilution of blood on their clothing or on a surface such as a couch, bed, or floor. High levels of urea and creatinine in the stained areas can confirm the presence of urine. When bloodstains that appear diluted are suspected to contain saliva, a test for the presence of amylase is recommended, as discussed in Chapter 8.

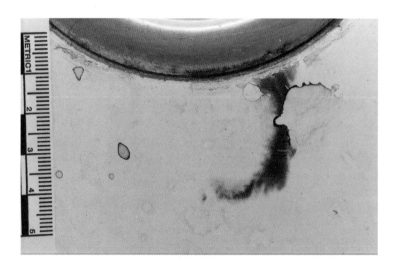

Figure 9.22 View of sink adjacent to drain showing areas of diluted blood with prominent dark outer rim and lighter center area.

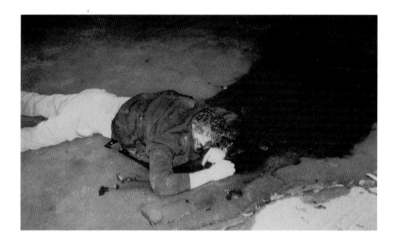

Figure 9.23 View of victim beaten on frozen pond showing extent of blood pooling on ice.

Another form of diluted bloodstains occurs occasionally with massive head trauma involving the combination of blood and spinal fluid. In a recent suicide a young man sat on a bed and discharged a shotgun into his head. Brain tissue and blood were projected and spattered throughout the small room. On the ceiling tiles above and forward of the victim were seen numerous areas of blood tinged and colorless impact spatters of spinal fluid that were easily interpreted (Figure 9.25).

Removal of Bloodstains

Attempts to eliminate bloodstain evidence at a scene or the washing of bloodstained clothing is often encountered during the course of an investigation. The authors have encountered many cases where attempts have been made to clean or obliterate bloodstains from surfaces at the

Altered Bloodstain Patterns

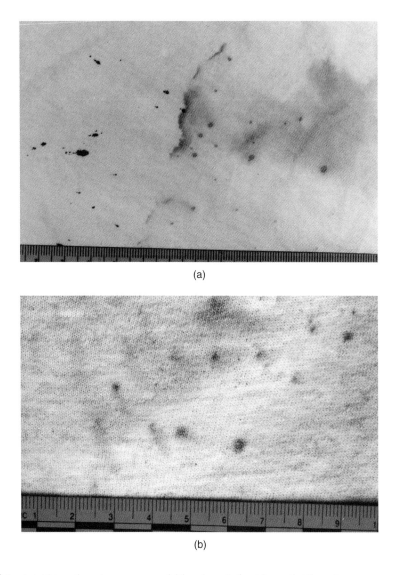

Figure 9.24 (a) Shirt showing spattered blood on left side and diluted spatter on right side. (b) View of shirt showing spatter diluted by perspiration of wearer.

scene and clothing by various methods that result in partial or complete alteration or partial or complete removal of the stained areas. These procedures include the use of detergent and water, bleach, pool acid (HCl), paint, fire, and even gouging. As experience has shown, the cleaning procedures used often do not completely eliminate the presence of bloodstains. Occasionally, recognizable wipe patterns and diluted-appearing projected stains will be visible on nearby surfaces. The use of luminol will detect trace quantities of blood. Cast-off bloodstains and small spatters on furniture, walls, ceilings, and other objects often are not noticed by the person attempting to clean up the scene. There is usually compelling evidence of the use of bleach and caustic agents. In a particular case, an assailant attempted to remove bloodstains from a brown carpet with bleach, which was very obvious (Figure 9.26). The same individual attempted to gouge out small impact blood spatters from a painted sheet rock wall but abandoned this effort because of the large number of blood spatters.

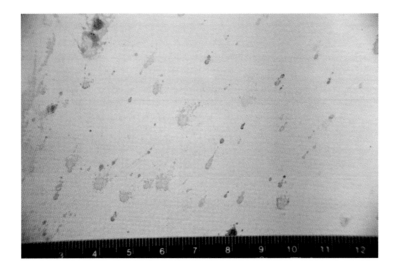

Figure 9.25 View of ceiling tile above and forward of the victim showing areas of impact spatters of blood mixed with spinal fluid.

Blood has been successfully removed from clothing by washing, but this depends on the quantity and age of the bloodstains, the type of fabric involved, and the washing procedure used. Fresh bloodstains are more easily removed by washing in cold water. However, experiments have shown that cold water wash with detergent and rinse do not completely remove drips and transfers of blood on denim material when the bloodstains have been allowed to dry for 72 h. However, small spatters of blood on denim were in fact more effectively removed by this washing procedure after drying for 72 h. Standard dry cleaning procedures that immerse fabric in an organic solvent are not effective for the removal of bloodstains. William Best, a forensic scientist from North Carolina, demonstrated this in a case where a serologist testified that dry cleaning likely removed blood from a jacket that was negative on examination. Best applied his blood to a similar jacket and detected the blood on the jacket with luminol after the dry cleaning was completed.

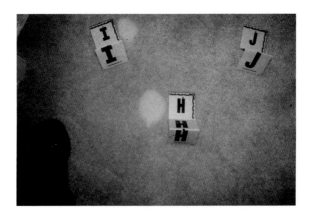

Figure 9.26 View of carpet showing unsuccessful cleanup attempt of bloodstains with bleach.

Altered Bloodstain Patterns

Another interesting case involving the alleged washing of bloody clothing was a triple homicide in which the victims were beaten and shot. A girlfriend of one of the codefendants testified that she observed him washing clothes in the washing machine the day after the homicides. He had apparently used dishwashing detergent, causing soap bubbles to rise in the washer and overflow extensively onto the floor. She maintained that the bubbles were red, indicating that he must have been washing his bloody clothing. At the request of the defense attorney, experiments were designed to determine if an agitated mixture of the same brand dishwashing detergent, water, and blood would produce red suds or bubbles. Initial tests were performed in a beaker on a magnetic stirrer. The results indicated that the blood mixed with detergent and water quickly turned dark brown to black and the foam or suds remained white (Figure 9.27). The experiment was repeated in a washing machine using a heavily bloodstained blanket and the same brand dishwashing detergent (Figure 9.28). There were copious amounts of white foam and bubbles produced that overflowed onto the floor. There was no indication of red bubbles, as maintained by the witness. (Figures 9.29a and b). However, the two defendants in the case were ultimately convicted.

Another case involved the issue of diluted bloodstains in the form of small spatters on the sleeves of a husband's jacket accused of fatally beating and stabbing his three children and stabbing his wife to death. This became important evidence for the defense during his trial. The presence of blood spatters on the sleeves of the jacket was not apparent until a

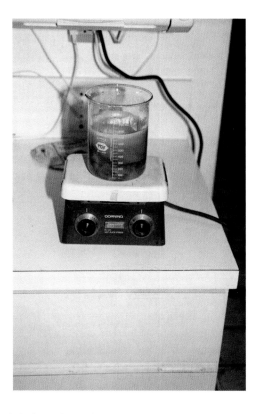

Figure 9.27 Mixture of dishwashing detergent, water, and blood in a beaker on a magnetic stirrer.

Figure 9.28 View of heavily bloodstain blanket used in experiment.

prosecution expert sprayed the jacket with luminol to identify the spatters of blood. The expert testified that the presence of these spatters was consistent with the defendant administering blows or stab wounds to the members of his family. The husband claimed that he had returned to his residence and found the bodies of his wife and children. He admitted carrying one of his children who was barely alive from a bathroom to the living room floor before calling for help. After police and medical personnel arrived, the husband, who had visible blood on his hands, began washing his hands under the faucet at the kitchen sink. A police officer testified that he observed this and directed the husband to stop washing his hands. The scene photographs clearly showed diluted blood in the sink and some blood transfer on a roll of paper towels near the sink. The force of the tap water on the bloodstained hands of the husband certainly could have produced spatters of diluted blood that would deposit on the sleeves of the husband and not be apparent to the naked eye on the jacket. This was presented to the jury in addition to other bloodstain and forensic pathology testimony that resulted in the acquittal of the husband on all charges. The defense presented convincing evidence to show that the wife had beaten and stabbed her children and then stabbed herself in the chest, resulting in a triple homicide and suicide.

Effects of Fire and Soot on Bloodstains

The alteration of bloodstains as the result of the effects of fire and soot is another consideration for analysis. In some cases bloodstains may be covered by soot and be missed entirely at the scene of a homicide that preceded a fire. In many cases where soot is deposited over bloodstains on a surface such as a wall, the bloodstains may appear darker than the surrounding soot-covered surface and lend themselves to analysis. MacDonell, in his 1993 publication entitled *Bloodstain Patterns*, has demonstrated that the outer layer of soot is easily removed with lifting tape. The distinction of soot over bloodstains — or conversely bloodstains over soot — is helpful to determine the sequence of events.

The study of the effects of heat and fire on bloodstains has been studied by David Redsicker and published in *Interpretation of Bloodstains at Crime Scenes* by Eckert and James

Altered Bloodstain Patterns

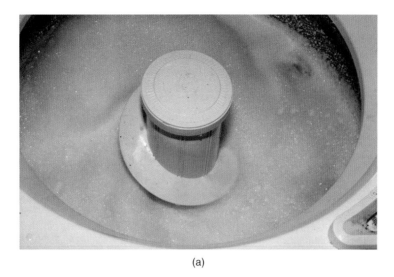

(a)

(b)

Figure 9.29 (a) View of bloodstain blanket agitating in washing machine with dishwashing detergent demonstrating white suds. (b) View of overflow of white suds onto floor from washing machine.

in 1989. Redsicker addressed this subject in conjunction with Sgt. Craig Tomash of the Royal Canadian Mounted Police in Halifax, Nova Scotia in the text *Fire Investigation*, published in 1996. As would be expected, bloodstains undergo a variety of physical changes depending on their proximity to sources of heat, fire, and smoke. Bloodstains may darken, fade (ghosting effect), or be totally destroyed, depending on the intensity and duration of a fire. This was demonstrated by Paulette Sutton in an experiment where blood was distributed on various surfaces in a building that was subsequently set afire for purposes of fire investigator training (Figures 9.30a and b and Figures 9.31a and b). It is also important to recognize alterations to bloodstains at fire scenes that may occur as a result of the diluting effects of water and the activities of fire fighting personnel.

(a)

(b)

Figure 9.30 (a) View of side of stove with bloodstains prior to fire. (b) View of side of stove showing soot-covered area of bloodstains that are darkened but still recognizable.

Altered Bloodstain Patterns

(a)

(b)

Figure 9.31 (a) View of bloodstains placed on wall and couch of room prior to fire. (b) View of room showing obliteration of bloodstains on wall and couch.

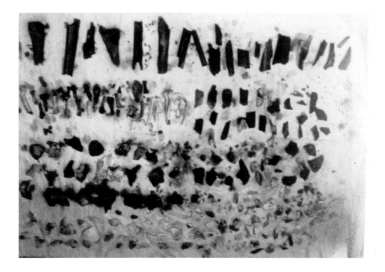

Figure 9.32 View of victim's remains after being burned in crematorium.

Bloodstain patterns may survive the effects of fire, as demonstrated in a case where a funeral home employee was charged with killing his wife and attempting to dispose of her remains in the funeral home crematorium. The charred remains of the victim were discovered in a friend's garage. Despite the suspect maintaining that the remains consisted of animals that he destroyed, charred bones, teeth, dental crowns, and jewelry that were not consumed in the fire were identified as those of his missing wife (Figure 9.32).

The crematorium consisted of a propane-fed furnace with dual jets located on the interior side wall providing outlets for the flames. The temperature within the furnace may reach 1500°F and reduce a body to ashes and desiccated bone fragments (cremains) within a few hours (Figure 9.33). However, in this case, the flame port nearest the furnace door

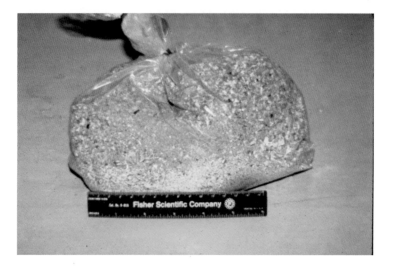

Figure 9.33 View of ashes of normal cremation referred to as cremains.

Altered Bloodstain Patterns

Figure 9.34 View of improperly working flame port in Furnas.

was not working properly and had a delayed ignition of approximately 2 h (Figure 9.34). The pattern of residue on the floor of the furnace indicated that the victim had been placed inside the door of the oven to the right of center and close to the malfunctioning flame port (Figure 9.35). A blackish stained area was present on the wall just below and to the right of this flame port (Figure 9.36). The pattern suggested that a bloody object struck the wall with sufficient volume of blood to create downward flow patterns. Samples taken from this area indicated DNA consistent with that of the victim. Further investigation revealed evidence of attempted cleanup of bloodstains on the floor in the room that housed the crematorium. The husband was convicted at trial.

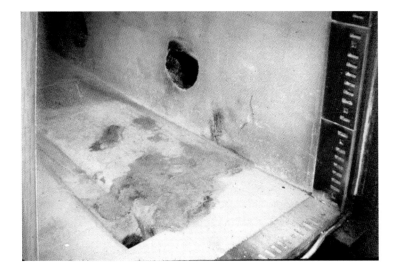

Figure 9.35 Pattern of residue on floor of furnace indicating prior location of victim.

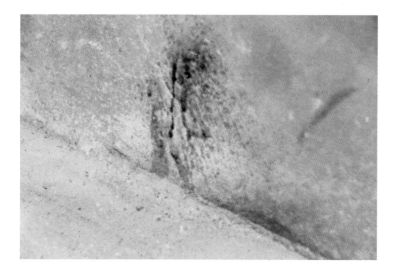

Figure 9.36 View of stained area on wall below and to the right of the flame port.

Fly and Other Insect Activity

The activity of flies at the scene where blood has been shed is another possible source of small stains of blood that may be confused with impact blood spatter (Figure 9.37). Flies will be at a scene if the scene is accessible to them. They will remain at a scene as long as a food source is available or as long as they are trapped. An understanding of the mechanics of flies feeding on blood and decomposing bodies is essential for proper analysis of these bloodstains (Figure 9.38). The horse fly is characterized as a biter, whereas the common house fly is characterized as a lapper and sucker. Flies ingest blood and regurgitate it onto

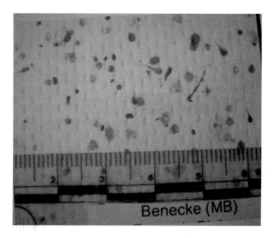

Figure 9.37 View of various types of stains produced by flies. (Courtesy of Larry Barksdale, Lincoln, Nebraska Police Department.)

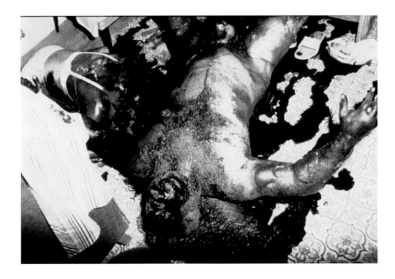

Figure 9.38 View of decomposing bodies with extensive maggot activity.

a surface to allow enzymes to break down the blood. At a later time the flies return to the areas of regurgitated blood and consume a portion of it. The surfaces on which these activities have taken place will contain small spots of blood material, which are often a millimeter or less in diameter with no definite area of convergence or origin. Some of the stains will exhibit dome-shaped craters as a result of the sucking process, and others may show swiping as a result of defecation. These stains may be observed on many surfaces at the scene, especially electric lights, shades, blinds, glass, paper, and ceilings, as well as on the victim and clothing (Figures 9.39a–d). They do not have to be in lighted areas. Their locations are frequently inconsistent with blood spatter associated with injuries sustained by the victim.

In an excellent article published in the journal *Forensic Science International* in 2003 entitled "Distinction of Bloodstain Patterns from Fly Artifacts," Mark Benecke and Larry Barksdale described their research and established criteria for distinguishing fly artifacts from human bloodstains (Figure 9.40). The following is a summary of their criteria:

- Are any of the geometric patterns similar to the "snake, tadpole," "double wave," or "sperm cell" standards?
- Is there a lack of directionality that would indicate a logical convergence?
- In any teardrop pattern, is the ratio of the length of the tail divided by the length of the body greater than one?
- Is there a lack of spiny edges on circular stains on a rough surface texture?
- Is there a logical relationship of stains to the scene (i.e., evidence of decomposition, flies present, maggots present)?
- Is there a logical relationship of stains to events (i.e., nature of wounds to the victim, stains in areas not related to investigative events)?

Other insects may also create small artifactual bloodstains. Haglund and Sorg described what they termed "pseudo blood spatter" in their text, *Forensic Taphonomy: The Postmortem*

208   Principles of Bloodstain Pattern Analysis: Theory and Practice

Fate of Human Remains. They referred to a case where blood spatters were trailing in the wrong direction, which had created some confusion with the bloodstain pattern interpretation of the scene. The issue was resolved when a large cockroach was observed crawling through a pool of blood and then depositing stains on a nearby surface identical to those originally observed.

The impact of insects on vehicles can also produce geometric patterns similar to human bloodstain patterns. As a result of the morphology of an insect spatter, human blood may flow around an insect spatter. Insect spatters may become relevant during motor vehicle accident investigations as well as other crimes.

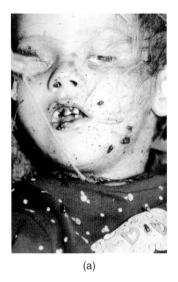

(a)

(b)

Figure 9.39 (a) View of small spatter-like stains of blood on face of victim as a result of obvious fly activity. (b) View of toilet with numerous small spatter-like stains from fly activity (c) Small spatter-like stains on mug as a result of fly activity. (d) Small spatter-like stains on denim cloth as a result of fly activity.

Altered Bloodstain Patterns

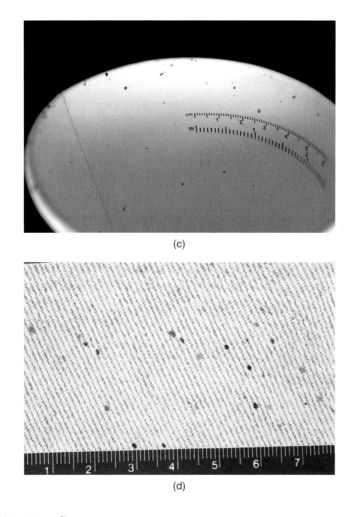

(c)

(d)

Figure 9.39 (Continued).

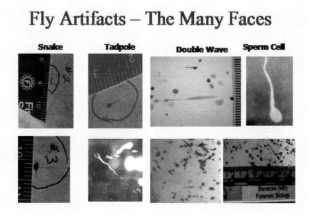

Figure 9.40 Classification of fly artifacts established by Benecke and Barksdale. (Courtesy of Larry Barksdale, Lincoln, Nebraska Police Department.)

Void Patterns

A void pattern, a form of alteration, is the absence of bloodstaining in an otherwise continuous bloodstained area. The void is a result of the presence of intervening objects or people. If the object or person is shifted in position or completely removed, the resultant unstained void pattern remains (Figures 9.41a and b). The void may be the recognizable outline of an object or may just show that someone or something interrupted the bloodstain pattern. Frequently, the position of a victim can be reconstructed by observation of a complete or partial outline of the victim or within a bloodstained area (Figures 9.42a and b). Void areas within spatter patterns may assist with the location of the assailant. Void areas may be present on the clothing of victims that may indicate folding of the material or

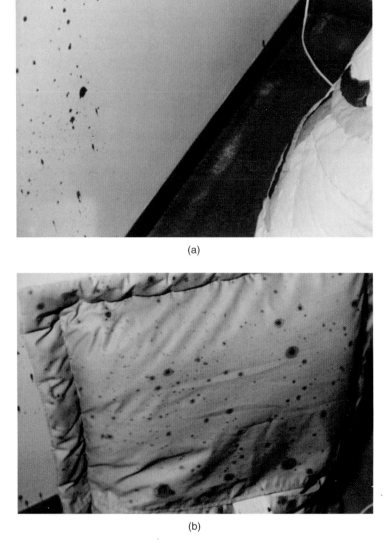

(a)

(b)

Figure 9.41 (a) Void area within pattern of spattered blood on wall. (b) View of pillow with intercepted spatters of blood that was against wall.

Altered Bloodstain Patterns 211

(a)

(b)

Figure 9.42 (a) View of victim on floor prior to removal. (b) View of void area on floor after removal of victim. (Courtesy of Martin Eversdijk, The Netherlands.)

position of arms or hands or legs at the time of bloodshed (Figures 9.43a and b). Folding of clothing and other fabrics while wet with blood often creates additional voids and additional stained areas, as shown in Figures 9.44a and b.

The repositioning of bloodstained objects to their original location that have been moved after bloodshed in a cleanup attempt by the assailant based on the presence of voids can be demonstrated by a case involving the gunshot death of a young woman. It was determined that the woman was not shot in the location where she was found. The woman's boyfriend denied knowledge of the shooting and permitted police to search the house he shared with the victim. Initial examination of the living room of the residence did not reveal apparent bloodstains until the throw rugs near the coffee table were removed (Figures 9.45a and b). A bloodstained hassock was found hidden in the garage. Based on the location of dried pools of blood on the living room carpet and flow patterns on the sides of the hassock, the hassock was able to be positioned in its original position. This area was originally partially covered by the coffee table (Figures 9.45a and b). The victim originally was lying across the hassock while bleeding. The boyfriend finally admitted to an argument with his girlfriend that resulted in the shooting as they struggled over a shotgun. He eventually pled guilty to manslaughter.

Another example of a void assisting with the positioning of a shooting victim is shown in Figure 9.46. The victim discharged a shotgun close to his chin with the wound track furrowed through his left cheek and forehead. Birdshot pellets were embedded in the ceiling near the edge of the bed. The edge of the bed showed a void pattern that indicated that

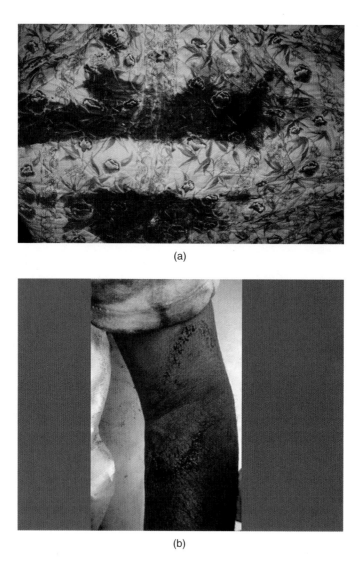

Figure 9.43 (a) Void created by folding material during the accumulation of blood on cloth. (b) Void on arm indicating that arm was folded upward at prior time.

the victim was sitting on the edge of the bed when the weapon discharged. He survived his injury after remaining in his bedroom for 2 days until found by a friend.

Sequencing of Bloodstains

Sequencing — the placing of an order of events producing alterations of bloodstains — is possible in some cases. This could involve a significant amount of time with importance to the overall bloodstain reconstruction of the scene. With greater periods of elapsed time, a wipe through a partially dried stain will originate more from the center of the stain than from the edges. If skeletonized stains exhibit these types of differences, it may be assumed that more than one event took place and that these events were separated by a significant lapse of time (Figure 9.47).

Altered Bloodstain Patterns 213

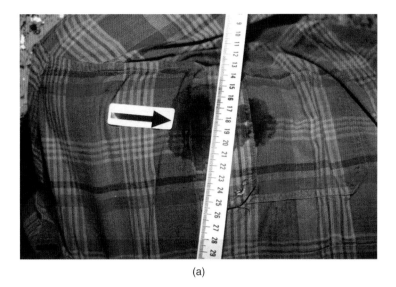

(a)

(b)

Figure 9.44 (a) Bloodstained area shirt prior to removal of victim. (b) Appearance of shirt after removal of victim. (Courtesy of Martin Eversdijk, The Netherlands.)

Sequencing may also be demonstrated by the observation of wipes through previously existing transfer stains (Figure 9.48) or overlapping stain patterns, including spatters on top of transfers and spatters of blood on top of a palm, shoe, or footprint. (Figure 9.49 and Figure 9.50). In a recent beating case a partial bloody palm print consistent with the

Figure 9.45 (a) View of living room showing bloodstains with void area near coffee table. (b) Repositioning of hassock to corresponding void area in living room.

Figure 9.46 Edge of bed showing void pattern where victim was sitting when weapon discharged.

Altered Bloodstain Patterns 215

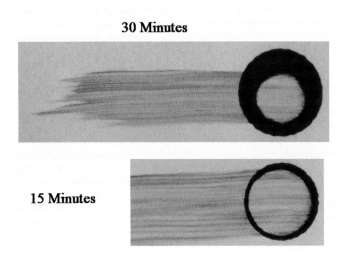

Figure 9.47 Wipes through partially dried bloodstains showing increasing peripheral rim relative to time interval.

victim's blood and matching a suspect was found on a wall above the head of the victim. The suspect stated that he was indeed at the scene but after the victim was dead. He must have touched the victim and transferred the print in that manner. However, impact blood spatters appeared on top of the ridge detail of the print, suggesting that the suspect deposited the bloody print at an earlier time than stated. Physical activity that produced the blood spatters occurred afterward, consistent with the suspect's presence at the scene during the beating of the victim. The accused later pled guilty.

Experimentation by Hurley and Pex in 1990 has shown that the differentiation of blood spatter that has dried prior to being stepped on from a dried blood print with blood spatter on the top is not always possible. Experiments conducted by the authors and students in

Figure 9.48 A wipe through a preexisting transfer stain showing an example of sequencing.

Figure 9.49 Wall at scene showing projected and spattered blood on top of a previously deposited transfer stain.

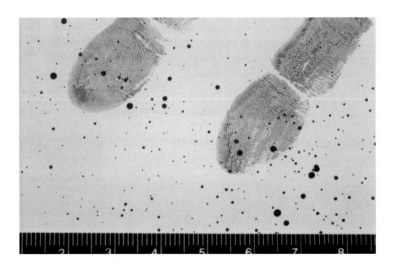

Figure 9.50 View of fingerprints in blood with spatter on top of ridge detail.

their advanced bloodstain pattern analysis classes have confirmed the study of Hurley and Pex. Caution should be exercised, especially if this is attempted from photographs.

Conclusion

In summary, the alterations of bloodstains by physical and environmental effects should always be considered when interpreting bloodstain patterns because they often affect the overall reconstruction of a case. If the appearance of the stains is out of the ordinary, then physical or physiologic alteration should be considered.

Determination of the Area of Convergence and Area of Origin of Bloodstain Patterns

10

Introduction

The goal of bloodstain pattern analysis is to determine the what, when, and where of bloodshed events. The "what" is answered by examining the size of the preponderance of the stains, the overall appearance of the stains, and the pattern distribution combined with the case particulars. The "when" may be answered by alterations such as clotting and sequencing. The "where" requires the analyst to determine the area of convergence and area of origin that corresponds to the location of the source of the blood. These determinations are often referred to as being the location of the victim, but more precisely they refer to the location of the blood produced by the victim's injury at the time a pattern was created.

A Pattern

The first step in determining the victim's location is recognition of a pattern. A pattern is a group of individual spatters generated by the same impact or applied force. The majority of the spatters in a pattern will fit within the acceptable size range for a given impact category. In its simplest form, a pattern would be created by a single impact of uniform velocity. In that instance, the individual spatters will radiate away from the site of impact in a fan-shaped or radial distribution. The overall pattern is influenced by a number of factors including, but not limited to, the velocity of the impacting force, the directionality of the applied force, the physical characteristics of the wound providing a blood source, surface textures, and obstructions presented by the external environment.

Area of Convergence

Once a pattern has been identified, determining the area of convergence is a relatively simple process. Several individual, well-defined stains are selected from all "sides" of a pattern. A string, or a line drawn using a straight edge, is aligned with the long axis of the stain and then extended backward, or 180° opposite of their individual directions of travel. Each of these lines approximate the path along which each of the drops traveled prior to impact — obviously ignoring the perpendicular axis. As the process continues, the line will cross in a generalized area equating to the location of the blood source within the scene — the point or area of convergence. This procedure will demonstrate a geographic location. In other words, it is a two-dimensional solution.

The convergence demonstrated by the crossing strings or lines in Figure 10.1a and b describe an *area* as opposed to a more precise location implied by the term "point of

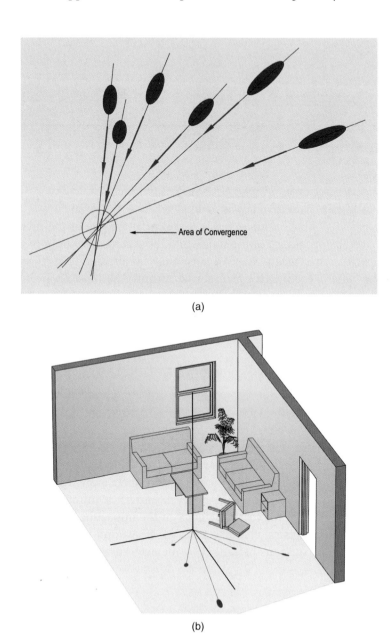

Figure 10.1 (a) Two-dimensional area of convergence of individual stains within a pattern. (b) Two-dimensional area of convergence of a stain pattern inside a typical room. The lines shown on the floor describe the area of convergence. (Courtesy of Alexei Pace, Malta.)

convergence." Very few wounds are pinpoints, and impact spatter should be anticipated from the entire space encompassed by the wound *plus* from any areas on which blood has subsequently been deposited. For this reason, many analysts use the term "area of convergence" instead of "point of convergence." Although "area of convergence" is considered more accurate, in reality most analysts use the two terms interchangeably.

Area of Origin

Determining the area of origin combines the two-dimensional area of convergence plus the angle of impact determination for each of the selected stains. The angle of impact adds the third dimension to the point of convergence determination, creating a spatial representation of the location of the blood source. The area of convergence shows a victim location in a given room or area, and the area of origin provides information about their relative posture: standing, kneeling, sitting, or lying down (Figure 10.2).

There are several different methods available to determine the area (or point) of origin. The method chosen is often a matter of personal preference or it may be dictated by limitations imposed by the individual scene. For any method used to determine the victim's location coupled with posture, the angle of impact for the individual stains is a necessary piece of data. To determine angle of impact, an understanding of basic trigonometry is needed. This will be followed by some principles regarding stain selection, followed by the step-by-step process used by the various methodologies.

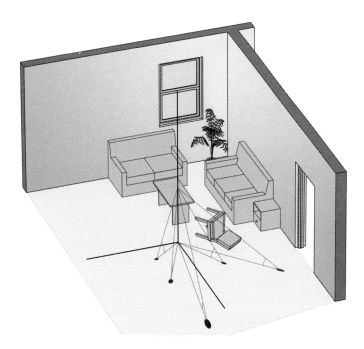

Figure 10.2 Three-dimensional area of origin inside a typical room. The lines shown on the floor describe the area of convergence. By adding the angle of impact for each of the stains, the area of origin is shown by the elevation of the original lines of convergence. (Courtesy of Alexei Pace, Malta.)

Basic Trigonometry of a Right Triangle

A right triangle is any triangle containing a 90° angle, which is also called a right angle. By definition, a triangle has three sides and three angles. The beauty of right triangles is that if we know the value for two parts of the triangle, and at least one of those known values represents the length of one of its sides, it is possible to determine the value for the remaining pieces of the right triangle. Because of the presence of the 90° angle, there are certain proportional relationships that exist between the angles and the sides of a right triangle. These relationships are represented by trigonometric functions.

Before discussing trigonometric functions, we must first define the parts of a right triangle and how the parts are ascribed a designation relating to the trigonometric relationships. Referring to Figure 10.3, the location of the 90° angle is indicated by the perpendicular lines. Obviously there are two other angles, and it must first be determined which of these two angles is to be studied. The terms that will be designated for the sides (or legs) of the triangle are determined by their location in relationship to the angle under scrutiny.

Again referring to Figure 10.3, angle A will be used as the point of reference. The leg of the triangle directly opposite angle A is designated as the leg "opposite." The leg closest to angle A is designated as the leg "adjacent." The remaining leg is designated as the "hypotenuse." The hypotenuse is always the leg opposite the right angle.

The sine of an angle = length of the leg opposite that angle divided by the length of the hypotenuse. This is usually abbreviated as sin = opp/hyp. Trigonometry uses six different functions, each defined by differing ratios between two sides of the triangle. In addition to the sine function, there is tangent (tan), secant (sec), cosine (cos), cotangent (cot), and cosecant (csc). The definition of each of these functions is shown in Figure 10.4.

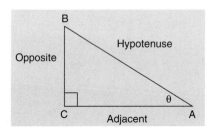

Figure 10.3 A basic right triangle.

$$\sin 2 = \frac{\text{opposite}}{\text{hypotenuse}} \qquad \cos 2 = \frac{\text{adjacent}}{\text{hypotenuse}}$$

$$\tan 2 = \frac{\text{opposite}}{\text{adjacent}} \qquad \cot 2 = \frac{\text{adjacent}}{\text{opposite}}$$

$$\sec 2 = \frac{\text{hypotenuse}}{\text{adjacent}} \qquad \csc 2 = \frac{\text{hypotenuse}}{\text{opposite}}$$

Figure 10.4 Trigonometric functions.

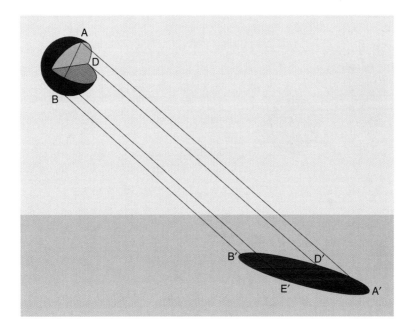

Figure 10.5 Sphere projected to an elliptical stain. (Courtesy of Alexei Pace, Malta.)

Angle of Impact

The impact angle for an individual stain is the internal angle formed between the flight path of that drop and the target surface the drop strikes, as shown in Figure 10.5. As such, 90° is the largest possible angle of impact for bloodstain pattern analysis and describes a drop falling directly downward onto a target from a perpendicular source. A 90° angle of impact produces a stain that is equal in width and length.

The geometric shape of a drop of blood is spherical for the overwhelming majority of its time in flight, and hence is considered to be spherical at the time of impact. Once the drop impacts a target surface at an angle other than 90°, it becomes an elongated stain. As Pizzola states in *Blood Drop Dynamics — I*, the distortion is limited to the lower hemisphere of the drop. "As the drop continues its travel, it gradually collapses downward with respect to the target surface, accompanied with little change in the shape of the upper hemisphere."

In Figure 10.5, the line D'E' corresponds to the width of the sphere of blood prior to its impact with the surface. As Pizzola noted, there is some radial displacement of the blood but the cohesive forces of surface tension counter the lateral spreading. For this reason, the width of the resultant stain at a crime scene is equated to the width of the original spherical ball of blood when it is in flight.

The length of the ensuing stain changes proportionally with the change in angle of impact and is represented as the line A'B' in Figure 10.5. Another way to think of the long axis of a stain is as the product of the drop skidding along the target surface (assuming a smooth, hard surface). Stains are less elongated when the angle of impact is closer to perpendicular because there is less skidding along the target surface. When the angle of impact is more acute (smaller), the stain will be more elongated. A drop striking at a 10° angle of impact

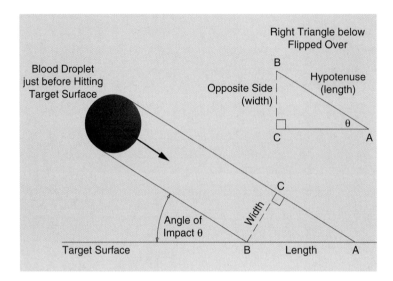

Figure 10.6 Width to length of bloodstain equated to a right triangle. (Courtesy of Alexei Pace, Malta.)

will skid a great distance. Stains produced on target surfaces of differing characteristics — for example, glass as opposed to concrete — can visually appear very dissimilar. This variation in their visual appearance is the result of the effects of the target surfaces and does not affect the angle of impact calculation. Certain target surfaces, such as absorbent cloth, carpet, etc., can make the determination of the angle of impact for the individual stains difficult, if not impossible, but theoretically at least the ratio of the width to the length remains constant. It is this relationship between the width and the length of stains that provides the basis needed to determine the angle of impact.

If one were to view Figure 10.5 from the side, it would become two-dimensional, as shown in Figure 10.6. Because a spherical drop of blood is equal in width and length prior to impacting the target surface (i.e., AB = ED in Figure 10.5), the line projected from the bottom of the drop to point B and the line extending from the top of the drop of blood to point A are parallel lines. The angle of impact is the angle between the line extending from the bottom of the ball of blood to point B. Because the two lines are parallel, the angle of impact θ as indicated in Figure 10.6 is equal to the angle A. Again, because the spherical shape of a drop of blood in flight has identical width and length, one can drop a perpendicular line from B to C, as shown in Figure 10.6. A right triangle ABC is thus created, with the 90° angle at C.

As discussed earlier, we want to determine the angle of impact by determining angle A. The opposite side of the right triangle represents the width of the bloodstain, and the length is represented by the hypotenuse. If the width is divided by the length, in essence the leg opposite of angle θ has been divided by the hypotenuse. This defines the sine of angle θ. Because our desire is to know the angle, the arc sin of angle θ is determined.

$$\frac{\text{Width}}{\text{Length}} = \frac{\text{Opposite}}{\text{Hypotenuse}} = \text{sine of angle } \theta$$

$$\text{Angle of Impact} = \text{arc sin } W/L$$

The historic method of determining the angle of impact from stain measurements was to calculate the ratio of the width divided by the length. That numeric ratio was then searched in a trig table for sine value and matched to the appropriate angle for that ratio. Handheld calculators have made this task much easier and much quicker. Once the ratio between the width and length has been determined—a simple division process—the arc sine or inverse sine of that ratio equals the angle of impact. Examples of this calculation are as follows:

$$\frac{\text{Width}}{\text{Length}} = \text{sine of angle}$$

For a circular bloodstain:

$$\text{Width (W)} = 1.0 \text{ unit}$$
$$\text{Length (L)} = 1.0 \text{ unit}$$
$$\text{W/L ratio} = 1.000$$
$$\text{Arc sin of } 1.0000 = 90°$$

For an elliptical bloodstain:

$$\text{Width (W)} = 0.5 \text{ unit}$$
$$\text{Length (L)} = 1.0 \text{ unit}$$
$$\text{W/L ratio} = 0.500$$
$$\text{Arc sin of } 0.500 = 30°$$

Measuring Bloodstains

Under ideal circumstances of velocity, impact angle, and target surface, a drop of blood would create a perfectly uniform, oval stain upon impact. Ideal circumstances rarely, if ever, exist and occur with even less frequency in actual casework. The distortion created in response to the landscape of the target surface will prevent the stain from being a perfect oval. If the volume of the drop is great enough, the blood that has not been stabilized by deposition on the surface will continue along the path of travel, creating even more deformation from the ideal elliptical stain. All of these causes of distortion are the result of conditions *other* than the angle of impact. For that reason, any deformation at the terminal edge of a stain must be negated when the length of the stain is measured for impact angle determination.

Stains are "rounded" to reapproximate this ideal oval or ellipse as closely as possible before measurement. This may be accomplished by attempting to visually match the terminal edge of the stain to the leading edge, as shown in Figures 10.7a–d.

Some analysts prefer to eliminate the distortion from the terminal edge by measuring the distance from the center of the widest point to the leading edge of the stain. This value is then doubled to arrive at the overall length of the undistorted stain.

The device used to measure stains also varies among analysts. Some prefer a magnifier loupe with a built-in scale, and others prefer to use calipers to measure stains. As with any other similar situation, one's measurements can only be as accurate as the measuring device used. Although the use of a 6-inch ruler and a keen eyeball seems popular, it is not an accurate method.

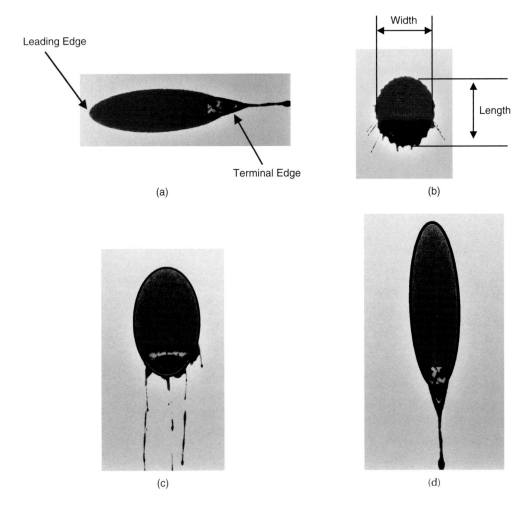

Figure 10.7 (a) Location of the leading and terminal edges of a bloodstain. (b) Measurement of the width and length of a near-circular bloodstain. (c) Measurement of the width and length of an elongated bloodstain by fitting the best ellipse. (d) Measurement of the width and length of a more elongated bloodstain by fitting the best ellipse.

Selection of Stains

There is no "magic number" of stains to be selected for the determination of area of convergence or area of origin. One should certainly use a sufficient number of stains to have a representative sampling and to comprise an accurate area of convergence. This process becomes very complex when you have a significant amount of spatter on a wall that was created by multiple impacts in multiple locations. The stains selected should be well defined in shape and should consist of stains from all primary zones surrounding the generalized point of convergence (i.e., from the area in front of the point of convergence and from all surrounding planes).

Stains lying closest to the area of convergence will give a more accurate solution than stains lying a greater distance away. For the same reason, smaller stains are better choices

Determination of the Area of Convergence and Area of Origin

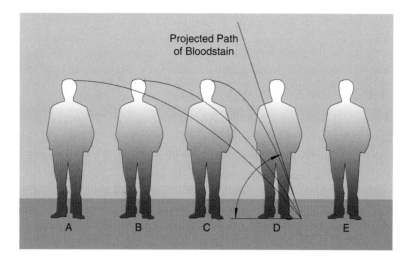

Figure 10.8 Possible versus impossible victim position. (Courtesy of Alexei Pace, Malta.)

than larger stains. Drops of blood follow a parabolic flight path from the source of creation until they are deposited onto a target surface. The use of lines or strings to depict what we know to be a parabolic path has an inherent margin of error. The further a drop has traveled away from the source, the more its path has assumed the parabolic arc. If a line is placed across a parabolic curve, the line will necessarily be forced high.

The area of origin as shown by any method using a straight line to depict the parabolic flight path of a drop has limitations that must be recognized in the final interpretation. Because we know that the true path taken by the drops was parabolic, the use of straight lines to recreate the area of origin will result in a finding that is artificially high. The area of origin represents an approximation of the victim's location at the time of impact, but the actual location of origin is *at or below* the area demonstrated by the straight-line method (Figure 10.8). In other words, if the area of convergence were a location on the floor of a room, then the area of origin shown by the strings would be the maximum height above the floor or the upper limit. This error is inherent in the process. The magnitude of this error depends greatly on the care and attention given during the selection of stains, the initial stain measurements, and recreating the pattern if physical strings are used. In most instances, pinpoint accuracy is not necessary. To be able to place an upper limit on the vertical distance is sufficient in most cases — for example, was the victim lying on the floor, sitting in a chair, or standing upright when struck in the head?

Methods of Area of Origin Determination

There are four basic methods in use to determine the area of origin:

1. Stringing method
2. Trigonometric method
3. Graphic method
4. Computer programs

The use of computer programs to determine the area of origin will be discussed in Chapter 11. This chapter will include only the remaining three methods listed. Before one can proceed with the actual determination of the area of convergence or the area of origin, documentation must be completed. For this reason, the documentation of the bloodstains and the stringing method will be intermixed.

Documentation

Equally important to establishing "where" the blood source was located when the spatter pattern was created is the documentation of the data upon which this determination is made. The documentation of the area of convergence and origin of a spatter pattern occurs simultaneously with the process itself, with the goal being to create a clear, concise, and accurate record of the size, shape, distribution, and location of the pattern and the spatters that comprise the pattern. The best method of documenting an area of convergence and origin determination is through notes and photographs. By using a data worksheet created specifically for recording the measurements taken during this procedure, a clear record of the measurements is created. The information on the worksheet along with photographs may be peer reviewed at a later date. An example of this worksheet is shown in Figure 10.9.

The size, shape, distribution and location of the spatters that compose the pattern should be discussed in your notes prior to any formal area of convergence or origin determination. The location of the spatter pattern within the scene as well as on a specific surface should be carefully noted. Photographs should be taken of the overall pattern prior to any markings being made on the surface containing the spatters. If a large piece of furniture (sofa, bed, etc.) is located in front of a spatter pattern on a wall, document the overall location of the furniture with notes and photographs and move the furniture away from the wall prior to conducting the origin determination. This will facilitate the documentation of the pattern, and the origin determination will become more manageable. By creating an outline of the furniture on the wall with a permanent marker, prior to moving it, the marker outline will appear in the photographs of the spatter pattern. This will assist the jury in understanding how the origin determination interrelates with the original furniture location. In addition, collect one of the other nearby spatters that exhibit similar characteristics as those used in the origin determination for a serologic sample. If sufficient spatters exist, it is also practical to perform a presumptive blood test on one of the pattern's spatters prior to initiating a formal origin determination. Record the location of the samples in your notes as well as through photography. If a presumptive test for the presence of blood is conducted, the results need to appear in the analyst's notes.

The following steps should be taken when documenting the area of convergence and origin of a spatter pattern.

- The individual spatters to be used should be labeled (A, B, C, etc.), and a scale should be placed beneath each spatter. Take an overall photograph of the pattern with these markings.
- An individual macrophotograph of each of the labeled spatter should be taken. It is necessary that you use a method to depict the spatters' orientation of the surface (i.e., floor/ceiling or N, S, E, W coordinates on a horizontal surface). On a vertical

Determination of the Area of Convergence and Area of Origin

```
                          ORIGIN DATA WORKSHEET
Scene Location: _____  Date: _____

Pattern Location: _____

Analyst _____  Case Number _____
```

Stain	Width (mm)	Length (mm)	Angle of Impact	Distance to Area of Convergence	Tangent Distance	Direction of Flight Angle	Coordinates (N, S, E, W) ( )	( )

Location of Origin: _____

Comments: _____
_____

Figure 10.9 Origin data worksheet used to collect numeric data associated with origin determination.

surface, this may be accomplished by allowing the line of a plumb bob to extend beside the stain while being photographed or by placing a "level-line" directly on the surface under the spatter with a level, as shown in Figure 10.10.
- The "level-line" method works best when a colored marker, such as blue, is used consistently for all the "level-lines." The "level-line" should always be underneath the spatter. When using the plumb bob method, you should place arrows on the surface or some other notation to depict the location of the floor. From these

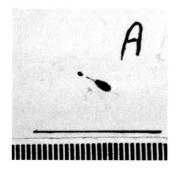

Figure 10.10 Macrophotograph documenting an individual spatter used in a three-dimensional origin determination. The blue line under the spatter is the "level-line."

photographs, another bloodstain analyst may verify the angle of impact and the direction of flight angle.
- Fill in the origin data worksheet with the letters signifying the spatters you have selected along with the spatters' width, length, and angle of impact.
- Record the location of each spatter on the wall or surface. If the surface is a wall, select two corners and use the coordinate method for documenting the location of each of the selected spatters. Measure from the leading edge of each spatter, using the same two measuring termination points (wall corners) for all coordinate measurements including the area of convergence location. These data can be recorded on the origin data worksheet. Specify on the worksheet the location of the two measuring termination points (wall corners).
- A photograph should be taken after the area of convergence is established either by drawing on the surface or by using elastic cord (Figure 10.11 and Figure 10.12). For

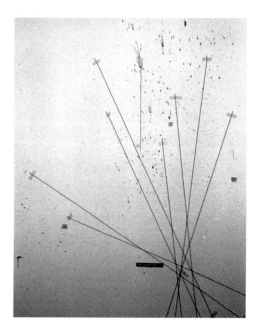

Figure 10.11 Area of convergence determination using black elastic cord.

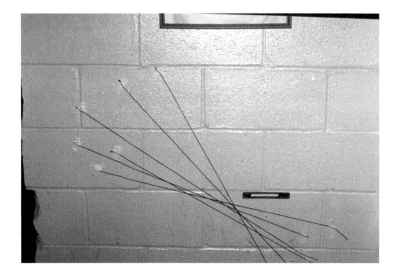

Figure 10.12 Area of convergence determination using black elastic cord.

photographic purposes, the color of the elastic cord selected should provide a contrast between the cord and the background. Nonelastic string or cord is more difficult to keep taunt. Elastic cord can be purchased in small packages from fabric retail stores in either white or black (Figure 10.13). The small individual packages prevent waste and contamination that could occur with larger spools that would be used at more than one scene.

- Measure the distance from the leading edge of each spatter back to the area of convergence previously established by drawing on the surface or by strings, and record this measurement on the origin worksheet under the "Distance to Area of Convergence" column.
- Calculate the "Direction of Flight Angle" of each selected spatter by extending the level line until it intersects the area of convergence cord. Some analysts choose to draw a line through the long axis of each spatter, skipping over the top of the actual spatters (Figure 10.14). The "Direction of Flight Angle" is established by placing the zero point of the protractor at the intersection of the level line and the area of convergence path. This angle is then recorded on the origin worksheet under the "Direction of Flight Angle" column. A note should also be made indicating the

Figure 10.13 Elastic cord for area of convergence and origin determinations.

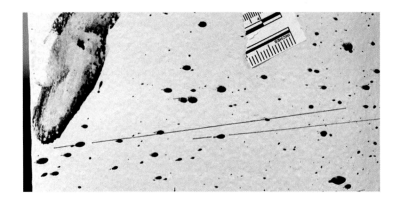

Figure 10.14 Illustration of how drawing through the long axis of the stain right on the wall can assist with area of convergence determination.

compass direction designated as 0°. For example, when the stains are on a wall, the ceiling would be designated as 0°; if the stains are on a floor, 0° might be the north wall or the south wall. This measurement can also be verified using the individual spatter photographs previously discussed.
- Record the coordinates of the actual area of convergence on the origin data worksheet.

Depending on the method of origin determination being used, the documentation may be complete at this time, with the exception of collecting one of the spatters actually used in the origin determination for DNA purposes. Signify in your notes which spatter was collected for this purpose. If the string method is being used to establish the origin, then you will need to document the location of the origin away from the wall or surface. When pulling the elastic cords away from the surface to represent the angle of impact, it is imperative that a protractor with a "zero" baseline is used. It can be difficult to document photographically (Figure 10.15, Figure 10.16, and Figure 10.17). Once the cords are pulled away from the surfaces to depict the angle of impact, you are now dealing with a three-dimensional

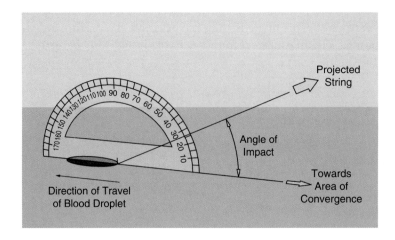

Figure 10.15 Positioning of a protractor at stain. (Courtesy of Alexei Pace, Malta.)

Determination of the Area of Convergence and Area of Origin

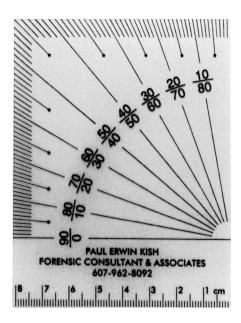

Figure 10.16 Example of custom-made protractor for the purpose of bloodstain pattern analysis.

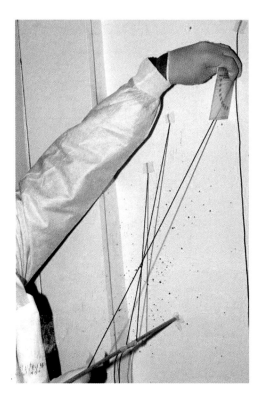

Figure 10.17 Analyst using protractor with "zero baseline."

Figure 10.18 Overall and origin determination using string method.

model versus a two-dimensional diagram. Trying to document any three-dimensional model on a two-dimensional plane is not practical. At a minimum, take an overall photograph of the three-dimensional strings depicting the distance between the bloodstained surface and the origin (Figure 10.18 and Figure 10.19).

The documentation of the origin determination needs to be completed in a manner that will allow another analyst to use your documentation materials to arrive at the same opinion as to the origin of the blood spatter pattern. If documented in a sufficient manner, you should be able to reasonably reconstruct the spatters within the pattern that you used for your origin determination.

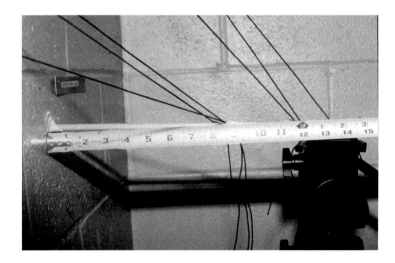

Figure 10.19 Determination of the distance the origin is located away from the wall.

Alternative Methods

The data collected using the procedure outlined earlier for proper documentation of area of convergence or origin will also allow the analyst to use alternative methods to determine the area of origin. Using the alternative methods substitutes *only* for the actual stringing.

Trigonometric Method

If the area of origin for a pattern were viewed from the side, it would very much resemble Figure 10.20. Examining each stain individually, one can visualize a right triangle. The string is the hypotenuse of the right triangle, and the angle of impact is the acute angle θ. The length of the string/line from the base of the stain to the area of convergence is the leg adjacent to the acute angle. This conceptualization facilitates the use of trigonometry to determine the length of the leg opposite the angle of impact. That leg represents the distance above the plane of the X and Y axes.

Referring to the trigonometric functions in Figure 10.4, the tangent of θ was defined as the length of the opposite leg divided by the length of the adjacent leg. The point described by the two-dimensional point of convergence has defined the location for the intersection of the vertical and the horizontal axis. The third dimension or the Z axis is added by determining the distance above horizontal where the string is attached (i.e., the leg opposite). The leg opposite is determined by multiplying the tangent of the angle of impact by the distance (d) from the base of the stain to the two-dimensional area of convergence. The resultant value (Z) is the location of the area of origin in space, or the three-dimensional location. An example of this calculation is as follows:

$$\text{Angle of impact} = 30°$$

$$\text{Tangent of } 30° = 0.5773$$

Distance of stain from area of convergence (d) = 15 units (length of string from each point)

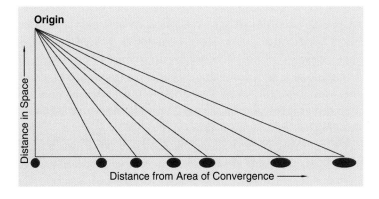

Figure 10.20 Area of origin with X and Y axes. (Courtesy of Alexei Pace, Malta.)

Applying the formula:

$$Z = \tan \text{ of the angle of impact} \times d$$

$$Z = 0.5773 \times 15$$

$$Z = 8.66 \text{ units} = \text{Area of origin from surface}$$

Graphic Method

The graphic method also makes use of the concept that each string is making a right triangle and the distance above the plane of X and Y equates to the leg opposite the angle of impact. Because the geographic location of the area of convergence has determined the location for the intersection of the horizontal and vertical axes, determining the area of origin equates to locating the point above horizontal.

Necessary data:

1. Determine angle of impact for selected stains.
2. Determine distance from the base of each individual stain to the area of convergence.
3. Using a piece of graph or grid paper, scale the horizontal and vertical axes. The horizontal axis should correlate with the measurements taken when measuring the length of the string from the base of the stain to the area of convergence. The vertical axis will be used to determine the values for the distance perpendicular to the X and Y axes.
4. For each stain, mark the horizontal scale at the distance that corresponds to the length of that string from the base of the stain to the area of convergence.
5. Place the zero of a protractor at the mark on the horizontal axis and make a second mark corresponding to the value for that particular angle of impact.
6. Extend a straight line from the value marked on the horizontal axis through the value marked for the angle of impact, and continue the line until it intersects the vertical axis.
7. The value indicated by the point where the line intersects the vertical axis is the distance above the plane of X and Y (Figure 10.20).

"Using" Surfaces

As anyone who has ever used the stringing method to determine the area of origin will readily attest, this is a long, arduous process. In many instances, careful observation can provide an adequate answer without the need to run strings. For example, if impact spatter is observed on the underside of a table and none is on the tabletop, it should be readily apparent that the source of the spatter was below the tabletop. This solution has not required any measurement, calculation, strings, etc. The only requirement is observation and analytical thinking.

There are many instances where it is not necessary to pin down a precise location for the area of origin. Often, it is only necessary to have a general idea. In Figure 10.21 the spatters shown on the wall are traveling in a generally downward direction. Lines were drawn to demonstrate the area of convergence. Purely by the location of the area of convergence, it is obvious the victim was "upright," not lying down on the floor or kneeling. The additional information to be added by calculating the area of origin would only bring the place where these lines intersect outward into the room. The first important consideration would be the

Determination of the Area of Convergence and Area of Origin 235

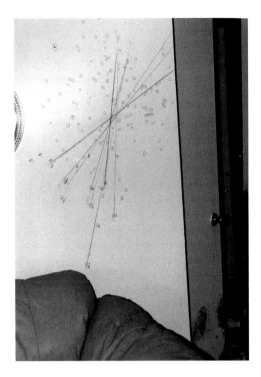

Figure 10.21 Area of convergence of stains on wall with downward directionality.

size of the spatters. In this instance, they were very small and therefore would not physically be able to travel great distances to reach the wall. This knowledge alone will provide valuable information as to the distance perpendicular to the wall. The second factor to be considered is that only a short distance from the wall, 90° stains were observed on the floor. All of this must then be combined with the transfer stains on the arm of the sofa when the gunshot victim fell. These simple observations have basically provided an area of origin without any further data being necessary.

Voids can also be particularly telling in determining the victim's location. The void to the left of the door facing in Figure 10.22 can only be created if the victim is to the right of the door facing when the impact occurs. The area of convergence is demonstrated by the proximity of the void, and again no calculations are needed. Many common surfaces within the home have contours that can be exploited to provide valuable information about the victim's location or direction of movement. In Figure 10.23 the finger transfers can be said to be traveling from left to right as a result of the two narrow yellow strips void of staining. Any observations that can help determine a victim's location, either two dimensionally or three dimensionally, should be considered for inclusion in the overall bloodstain pattern analysis.

Three-dimensional objects in close proximity to a spatter-producing event often assist the analyst in establishing the general location of the spatter event by simply evaluating distribution and location of the spatter on these objects. In cases where the room is covered with spatter as a result of a devastating gunshot wound to the head, the areas void of staining can be very useful in establishing the general location of the blood source.

Figure 10.22 Void to left of door facing.

Figure 10.23 Bloody finger transfers traveling from left to right showing movement from left to right.

Determination of the Area of Convergence and Area of Origin 237

Figure 10.24 Overall view of the basement laundry area where the assault occurred. (Courtesy of Detective Anthony Dinardo and Detective Paul Mergenthaler, New Castle County Police, Delaware.)

The significance of using the location and distribution of impact spatter on three-dimensional objects can be illustrated in a case where the accused beat his mother to death with a "dead blow" hammer (hammer reportedly used for auto-body work). The beating occurred in the basement of the dwelling in the vicinity of a laundry area (Figure 10.24). The location and distribution of impact spatters on the dryer were of particular interest. On examining the dryer the exterior side of the dryer door was void of impact spatter (Figure 10.25), yet impact spatter was identified on the interior side of the dryer door as well as within the dryer drum (Figure 10.26 and Figure 10.27). The overall location and distribution of the impact spatter on and within the dryer was consistent with the victim being at or below the height of the dryer opening, with the door partially open, when she was beaten.

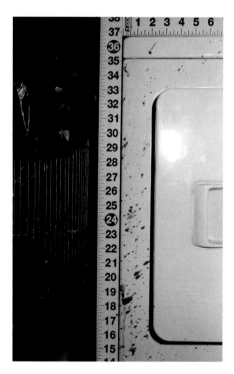

Figure 10.25 Closeup of spatter distribution on corner of the dryer. Note the absence of spatter on the exterior of the dryer door. (Courtesy of Detective Anthony Dinardo and Detective Paul Mergenthaler, New Castle County Police, Delaware.)

Figure 10.26 Impact spatter located on the interior dryer door phalange. (Courtesy of Detective Anthony Dinardo and Detective Paul Mergenthaler, New Castle County Police, Delaware.)

Determination of the Area of Convergence and Area of Origin 239

Figure 10.27 Impact spatter located within the drum of the dryer. (Courtesy of Detective Anthony Dinardo and Detective Paul Mergenthaler, New Castle County Police, Delaware.)

Directional Analysis of Bloodstain Patterns with a Computer

11

ALFRED L. CARTER

Introduction

In this chapter we will describe a procedure called the directional analysis of bloodstain patterns, whereby photographs of individual bloodstains on vertical surfaces are analyzed with the help of a computer and the BackTrack Suite[1] of programs. The object of the analysis is to discover how many blows or bloodletting events took place and where these events occurred (e.g., was the victim standing or lying on the floor?).

The procedure is solidly based on the scientific principles of mathematics and physics. Assumptions regarding unknown quantities, such as the sizes and speeds of the blood droplets and the parabolic arcing of their flight paths, are not necessary. The only data required are the measured positions of the individual stains, their directions, and the shapes of the stains as recorded by the photographic images.

The analysis yields numeric results for the location of each blow and in some cases the standard deviation errors. This result is scientifically valid and based on measured quantities only and therefore can be defended with confidence in any court of law.

Details of directional analysis have been published[2] in a refereed journal. The method will be described here only to the extent that is appropriate for this book.[3]

Some Basic Theory

The laws of physics are clear and simple when applied to freely moving blood droplets. Only two forces can influence the movement of the center of mass points of the blood droplets.

---

[1] Distributed by Forensic Computing of Ottawa (see www.bloodspattersoftware.com).
[2] "The Directional Analysis of Bloodstain Patterns, Theory and Experimental Validation," *Journal of the Canadian Society of Forensic Science*, Vol. 34, Dec. 2001.
[3] See also James, S. H. (1998). *Scientific and Legal Applications of Bloodstain Pattern Interpretation*, CRC Press, Boca Raton, FL, Chapter 3.

1. The force of gravity, which always points downward and remains constant in magnitude. It is the dominant force.
2. The force of air resistance, which has a direction that is exactly opposite to direction of motion. This force is generally much less than the gravity force and depends on the size and speed of the droplet. The air resistance force increases or decreases in a manner that is proportional to the cross-sectional area of the droplet; at the same time it increases or decreases in a manner that is proportional to the square of the speed of the droplet.

Two other forces common to fluids are surface tension and viscosity. Surface tension forces govern the droplet formation. They also give rise to the spherical shape of the blood droplets at low speeds and an oblate shape at high speeds. Blood droplets created by an impact to a pool of blood can carry away internal energy that produces oscillating changes in the diameters of these droplets. However, the oscillations quickly dissipate as a result of the action of the internal viscous forces. It is important to note that these internal forces always add up to a net force equal to zero and therefore have no direct effect on the flight path.

The following example describes the behavior of a blood droplet that is created by an impact causing it to fly upward with an approximate direction of 45° to the horizontal. Let's suppose that it happens to avoid impact and continues to move freely in air. The solution to the equation of motion of such a blood droplet, with only the gravity force present, is a parabolic path concave downward. Adding the air resistance force to the equation of motion causes changes to the flight path. These changes are best described from the point of view of the kinetic energy of the droplet, $1/2\ mV^2$, where m and V are the droplet's mass and speed.

The air resistance, being a dissipative-type force, does work at the expense of the kinetic energy of the droplet. The decrease in kinetic energy causes the speed to decrease, resulting in an increase in the downward curvature of the path until the horizontal component of the motion is zero and the droplet falls straight downward. The velocity of the droplet, now directed downward, continues to increase until the upward air resistance force cancels the downward gravity force. The downward speed is now constant and is known as the terminal velocity of the droplet.

This describes the possible flight path of a blood droplet free of impact with another object. In reality it could impact with a surface at any point along its path. We describe here a procedure for backtracking the flight paths to find the location of the blood source. We limit our discussion to impacts with vertical surfaces. These impacts are the most problematic.

Locating the Blood Sources

The well-known laws of motion predict that the flight paths of each blood droplet, assuming negligible air currents, define unique vertical planes. The blood droplets move up and down, not left or right. They also predict that both the source of the blood droplets and the resulting bloodstains are located in the same vertical plane. Viewed from above, the bird's eye view, these planes appear as straight lines joining each bloodstain position with its source position.

The directions of the blood droplets at the instant of impact with the vertical surface can be computed from the impact angle and the directionality angle of the bloodstain. These directions when viewed from above are identical to the flight paths and are called, in this work, *virtual flight paths*. The experienced reader will recognize these virtual flight paths as the traditional string directions. The laws of physics predict that they always pass directly over the source locations.

This means that top view of the virtual flight paths for a number of bloodstains that have a common blood source will show a convergent pattern that converges on the horizontal position of the source. Each intersection of two virtual flight paths, viewed from above, is a possible source location (triangulation). By averaging all the X and Y coordinates of probable intersections, one can compute an average value for CPx and CPy, where CP stands for the convergent point.

The laws of motion also help us to estimate an *upper limit* for the height of the blood source CPz. The virtual flight paths or strings are straight lines, whereas the real flight paths are curved parabolic arcs of unknown curvature. However, we know that strings must pass directly over the flight paths and the source. Therefore, by averaging the heights of the intersections of the strings with CPx or CPy values we can compute an upper limit for the value of CPz.

An Example of Virtual Paths Compared to Actual Flight Paths

The string direction or virtual flight path for a bloodstain is defined as the direction of the blood droplet at the instant of impact. These string directions are computed from the shapes and directions of the bloodstains. They are used to locate the blood sources by the directional analysis procedure, as explained earlier (top-view analysis).

Figure 11.1 is a screen capture from the program Tracks.[4] It shows the flight paths of three droplets launched from a point source toward a vertical surface. The vertical surface is marked off with a 50-cm grid. The strings or virtual flight paths are red straight lines that run from each bloodstain to a point above the point source.

The laws of physics dictate that the virtual flight path and the corresponding true flight path both lay in the same vertical plane. This means that the top view of the virtual flight paths shows a convergent pattern that is identical with that for the actual or true flight paths. The virtual flight paths or strings when viewed from above converge on the location of blood sources. This is depicted by the bird's eye view in Figure 11.1, where the three red lines have one intersection located exactly at the source position, X = 50 cm, Y = 250 cm. In a real crime scene there are experimental errors of measurements. Therefore, the top view would show a convergent pattern with $3 \times 3 = 9$ possible intersections. The average of these intersections would then be the best estimate of the horizontal source position.

The side view shows the true flight paths connected to the source position and the three red string lines passing directly over the source position. The investigators at the crime scene, of course, will not be able to reconstruct the true flight paths. However, they will be able to construct the red lines showing the virtual flight paths. Also they will be able to estimate the value of CPx, the distance from the wall, by means of a top-view analysis. An upper limit for the value of the height of the blood source above the floor, called CPz, can now be obtained by averaging the intersections of the red lines with a vertical line positioned at the value of CPx.

---

[4] Tracks is a program that creates bloodstain patterns by launching droplets of varying sizes, speeds, and directions toward a vertical surface. It computes the flight paths with air resistance taken into account. It also computes the strings or virtual flight paths. The flight paths are shown in black, and the strings are shown in red.

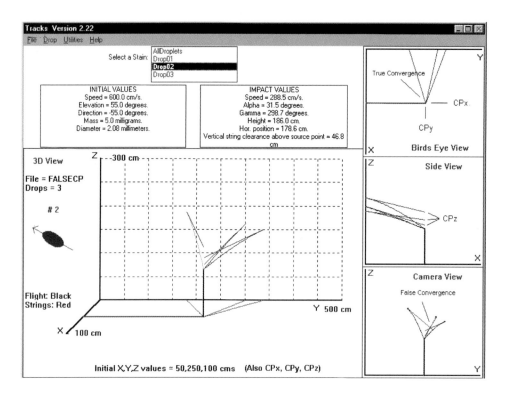

Figure 11.1 Three blood droplets have been launched toward the target wall from the initial point X = 50 cm, Y = 250 cm, and Z = 100 cm. Flight paths are normally black, except for the selected droplet (#2). The virtual flight paths or strings are in red. The strings, in the top view upper right, converge onto the blood source.

The end view or camera view shows that a convergence of the red string lines in this projection can be wrongly interpreted as a source position (e.g., we see here a definite convergence of the red lines occurring above and to the right of the correct position, which is located at the top of the blue line). This convergence of the red lines is labeled as a false convergence. This misleading property of the strings occurs because the flight paths do not project onto the wall as straight lines and therefore, in general, cannot be represented as strings. The *directions* of the strings and the directions of the flight paths coincide only at the end point of the flight paths — that is, the instant of impact. At this point the string or virtual flight path is tangent to the true flight path. This problem does not exist for the top view of the strings.

The Two Impact Angles (α and γ) Depend on the Shape of the Bloodstain

Simple experiments with free-falling blood droplets onto inclined planes show that the angle of the inclined plane alpha (α) and the width (W) and the length (L) of the bloodstain satisfy the empiric formula $\sin(\alpha) = W/L$ (Figure 11.2).

The impact angle (α) of the blood droplet is just one of the two angles that is necessary to determine the string direction. The second angle is the directionality angle gamma (γ)

Directional Analysis of Bloodstain Patterns with a Computer

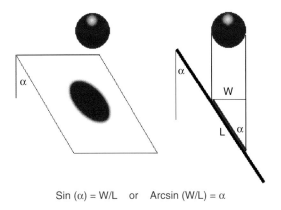

Sin (α) = W/L   or   Arcsin (W/L) = α

Figure 11.2 Using an appropriate value for L, the length of the bloodstain, this formula is reasonably accurate over α values ranging from 10 to 60°.

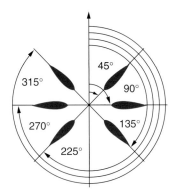

Figure 11.3 The directionality angle gamma is measured clockwise from the upright direction.

for the bloodstain. Figure 11.3 shows how this angle is defined for bloodstains found on a vertical surface.

Figure 11.4 shows a bloodstain on a wall with a gamma (γ) angle equal to 45°. The blue arrow pointing at the stain is the vector V representing the direction and magnitude of the impact velocity. This direction is also the string direction or virtual flight path direction. The three components — V_x, V_y, and V_z — form three interlocking right-angle triangles containing the impacting angles α, β, and γ. A new angle, beta (β), is introduced by means of this diagram. It is the angle between the Y-axis and the projection of the flight path onto the floor, shown above as a dotted line in the X-Y plane. These projections represent the top view of the flight paths.

It can be shown[5] that the three angles — alpha (α), beta (β), and gamma (γ) — are related to one another through the following equation:

$$\mathrm{Tan}(\beta) = \mathrm{Tan}(\alpha)/\mathrm{Sin}(\gamma) \tag{11.1}$$

---

[5] See footnote 2.

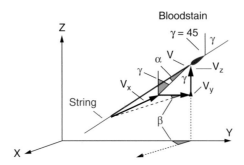

Figure 11.4 The relationships between the angles α, β, and γ and V_x, V, and V_z which are the three components of the impact velocity V.

Solving for the impact angle alpha:

$$\alpha = \text{Arc tan } (\text{Tan}(\beta) \text{ Sin}(\gamma))^{[6]} \qquad (11.2)$$

In the laboratory, bloodstain patterns are produced and studied as useful training exercises. From the known position of the blood source, Tan (β) can be computed for each bloodstain that is selected for study. Sin (γ) for each of these stains can be found by direct measurements of the gamma values. Therefore, we can predict or compute the values for the impact angles of selected stains.

We will use alpha values predicted by Equation (11.2), to demonstrate that ellipses, with width over length ratios equal to Sin (α), can be fitted to bloodstain images with good results.

Using Ellipses to Analyze Bloodstains

A computer program was created that accepts images of bloodstains taken by a digital camera. It is called BackTrack/Images.[7] Figure 11.5 is a sample screen from this program showing the image of a bloodstain that is being fitted with an ellipse.

This particular stain is part of a set of six stains that is supplied as a tutorial for BackTrack/Images. Each stain of the set shows a yellow ellipse that is prefitted with the impact angle that is predicted by formula (2). This ellipse serves as a template while the new user selects appropriate values of the impact angle α and the directionality angle γ. The final results of this fitting procedure are the α and γ values of each stain. Once these values are known, the virtual flight paths or strings are displayed as attached to each stain in the pattern.

Figure 11.6 summarizes the result of the analysis of the six stains. The first column is the string number, which is assigned by the program. The second column lists the stain ID number, which is assigned by the user. The third column is the status number, which

---

[6] Because Tan (β) approaches infinity when β approaches 90°, the formula is unreliable for stains directed vertically upward or vertically downward within ±15° of the vertical direction.
[7] Distributed by Forensic Computing of Ottawa. See www.bloodspattersoftware.com.

Directional Analysis of Bloodstain Patterns with a Computer

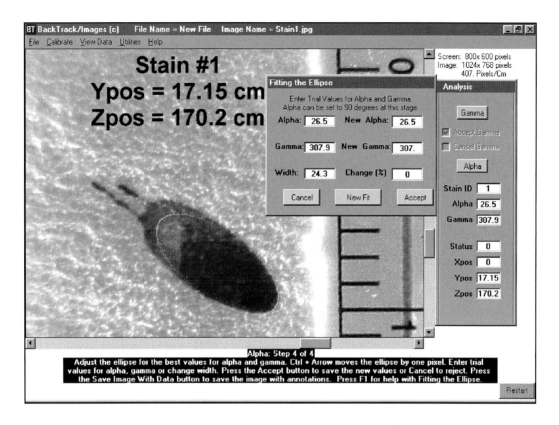

Figure 11.5 Stain #1 is part of a tutorial included with the BackTrack Suite. The width/length ratio for the yellow ellipse agrees with $\alpha = 26.5°$, the value that is predicted by formula (2).

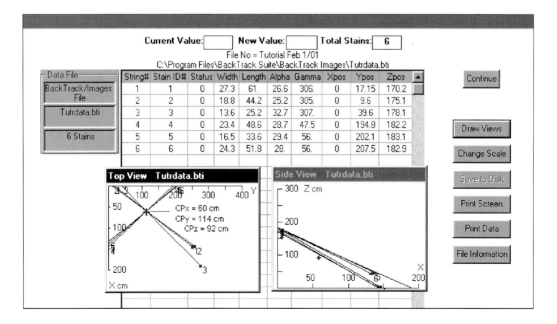

Figure 11.6 A screen grab from BackTrack/Images summarizing the analysis of the six stains from the tutorial.

identifies the surface bearing the stain. The top view and side views of the strings are provided at this stage for guidance. A data file called, in this instance, "Tutrdata.bti," is also created for further analysis with the program BackTrack/Win. The latter program is a sophisticated string analysis program that is capable of analyzing up to 100 stains located on 12 different vertical and horizontal surfaces. An example of the use of BackTrack/Win is given later in this chapter.

More About Ellipses

The gamma angle (Figure 11.7) shows an image of a bloodstain obtained as part of a laboratory setup. The width of the stain is 1.6 mm, and the length as defined by the ellipse is 3.6 mm. The impact angle (α) was computed by Equation (11.2) to be 29°, and the gamma angle (γ) was found by best fit to be 54°. Two more ellipses are shown with gamma values 53 and 55° (i.e., 1° greater and 1° less than the best fit). One can see that a sensitivity of ±0.2° or better is readily achieved by the fitting procedure.

The impact angle alpha (α) (Figure 11.8) shows the same stain with three ellipses in place. The three ellipses have the same value of 54 degrees for the gamma value but show three different values for the impact angle alpha (α). The correct computed impact angle is 29 degrees. Ellipses suitable for 27 and 31° are also shown for comparison.

It turns out that there is no infallible rule for choosing an appropriate length for the ellipse. One is advised to carry out laboratory trials to gain experience. By computing and fitting ellipses using Equation (11.2), one can learn to better estimate the length of the ellipse that is suitable for each stain.

Overestimating the length of the ellipses does result in underestimating the average value of CPx. This has often been observed to occur in laboratory measurements. It should also be noted that the value of CPy is not as sensitive as the value of CPx to the choice of the length of the ellipses. The measured CPy values, in our experience, have consistently been measured with better accuracy than the CPx values.

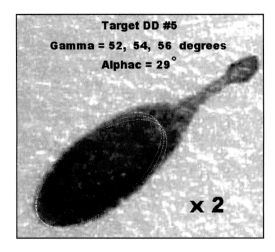

Figure 11.7 The fitting procedure for γ is seen to have a sensitivity of better than ±0.2°.

Directional Analysis of Bloodstain Patterns with a Computer

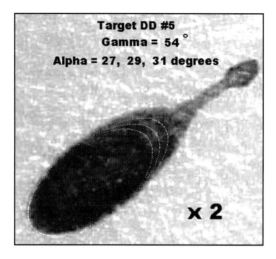

Figure 11.8 Here the gamma value is fixed at 54° and the α values varied.

Laboratory Study of a Double Blow with the BackTrack Software

The directional analysis is now described of 32 selected bloodstains produced by two blood spatter events performed in the laboratory. This work was carried out as part of an exercise during the Advanced Math/Physics Course for Bloodstain Pattern Analysis that was held in Edmonton, Alberta in 2003.[8] The object of the exercise was to locate the positions of the two events using the computer programs BackTrack/Images and BackTrack/Win.

The experimental arrangement consisted of a box sitting on the floor in the corner of a room. A photo of the arrangement can be seen in Figure 11.10. The top, side, and end surfaces of the box are labeled with the letters T, S, and E. A total of 32 stains were select for the analysis. There are 16 stains residing on the main wall, 6 stains on the end surface of the box, and 10 stains on the right wall.

Each of the 32 stains were photographed by the author, with a Nikon Coolpix 4500 digital camera that was equipped with a macro cool-light SL-1. Typical resolutions of about 800 pixels/cm were obtained with this arrangement. These images were analyzed, also by the author, by fitting ellipses with the program BackTrack/Images to produce the data file depicted in Figure 11.9. The final analysis of the virtual flight paths or strings was done with BackTrack/Win. Figure 11.10, Figure 11.11, and Figure 11.12 are composite images of the computer screens depicting the top, side, and end views of the string patterns.

The first column is the String number, which is assigned by the program. The second column is the Stain ID number, which is a unique number that is assigned by the user. The Status is a number from 0 to 11 that identifies which of the 12 possible surfaces contains the bloodstain. The width and length refer to the ellipse that was fitted to the bloodstain during the analysis by BackTrack/Images. These numbers are given in units of 0.1 mm (e.g., stain 32, which lists a width of 33.4, has an actual width of 3.34 mm).

Column 11, labeled Visible, which has a yellow background color, shows the letter "Y" for each stain. This entry can be changed to the letter "N," which renders that string invisible

---

[8] Royal Canadian Mounted Police, K Div, Edmonton. See Credits at the end of this chapter for more information.

String#	Stain ID#	Status	Width	Length	Alpha	Gamma	Xpos	Ypos	Zpos	Visible	String Place
1	1	0	18.4	30.5	37.	313.	0	89.	172.8	Y	Main Wall
2	2	0	30.5	43.1	45.	313.	0	99.5	159.6	Y	Main Wall
3	3	0	17.1	23.3	47.	319.	0	106.4	158.8	Y	Main Wall
4	4	0	20.7	28.3	47.	330.	0	114.4	163.3	Y	Main Wall
5	5	0	18.6	28.3	41.	339.	0	114.4	186.8	Y	Main Wall
6	6	0	18.5	22.6	55.	339.1	0	117.2	168.4	Y	Main Wall
7	7	0	27.5	39.5	44.	347.	0	122.2	178.3	Y	Main Wall
8	8	0	40.	46.2	60.	350.	0	130.9	154.8	Y	Main Wall
9	9	0	37.4	62.1	37.	358.	0	132.2	191.1	Y	Main Wall
10	10	0	27.2	45.2	37.	4.5	0	139.2	197.5	Y	Main Wall
11	11	0	26.2	40.8	40.	14.	0	145.4	192.1	Y	Main Wall
12	12	0	55.1	75.4	47.	15.	0	146.2	169.	Y	Main Wall
13	13	0	58.5	85.8	43.	24.5	0	155.7	174.	Y	Main Wall
14	14	0	32.	57.2	34.	18.	0	156.5	197.1	Y	Main Wall
15	15	0	26.1	37.6	44.	51.	0	169.5	162.4	Y	Main Wall
16	16	0	43.	71.5	37.	39.5	0	182.7	189.3	Y	Main Wall
17	17	1	20.8	26.7	51.	305.	42.	190.9	40.5	Y	MW Offset
18	18	1	18.8	28.1	42.	332.	42.	196.5	52.4	Y	MW Offset
19	119	1	43.6	74.1	36.	11.3	42.	217.5	53.7	Y	MW Offset
20	20	1	27.9	47.4	36.	22.	42.	223.8	56.4	Y	MW Offset
21	21	1	15.7	34.6	27.	26.	42.	230.2	66.3	Y	MW Offset
22	22	1	19.	35.8	32.	33.	42.	233.	59.1	Y	MW Offset
23	23	2	22.3	49.	27.	330.	39.8	244.	81.4	Y	Right Wall
24	24	2	19.1	36.1	32.	326.	45.5	244.	69.2	Y	Right Wall
25	25	2	28.5	46.2	38.	335.	54.3	244.	64.2	Y	Right Wall
26	26	2	28.6	45.4	39.	335.	55.8	244.	76.8	Y	Right Wall
27	27	2	0	0	36.	347.	64.	244.	59.7	Y	Right Wall
28	28	2	0	0	41.	6.	75.7	244.	59.7	Y	Right Wall
29	29	2	0	0	32.	34.3	95.1	244.	99.3	Y	Right Wall
30	30	2	45.4	90.7	30.	28.	99.8	244.	85.9	Y	Right Wall
31	31	2	37.4	72.6	31.	40.5	107.5	244.	78.3	Y	Right Wall
32	32	2	33.4	68.9	29.	37.	110.3	244.	90.2	Y	Right Wall

Figure 11.9 The data screen from BackTrack/Win. All the pertinent information concerning the bloodstains is summarized here in a spreadsheet format.

to the user and the program and also excludes that string from any of the computational routines. This is a useful operation.

When calculating the values of CPx and CPy, using the automatic routines, it was found necessary to treat the stains found on vertical surfaces separately from those found on horizontal surfaces.

One can have up to 100 individual bloodstains residing on 12 separate surfaces. These surfaces include the four walls, the floor, the ceiling, and six more parallel surfaces inside the room, which are called offset surfaces. Each stain has a status value that identifies the surface containing the stain and a unique ID number that is assigned by the user.

Six colors are used to identify the strings from a given surface or offset. Strings from offsets are drawn as dotted lines.

The top views of the strings are, in principle, identical to the top views of the actual flight paths. This means that strings attached to stains from a common source will show a convergent pattern. In Figure 11.10 we can see two convergent patterns. One is produced

Directional Analysis of Bloodstain Patterns with a Computer

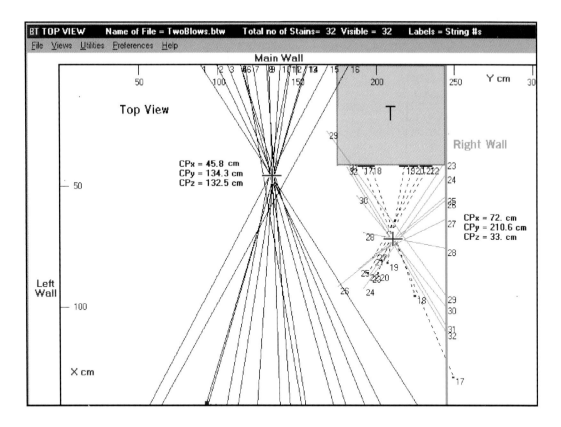

Figure 11.10 Sixteen stains reside on the main wall, ten stains on the right wall, and six stains on the end surface of the box. This is a composite, showing the top view of the two source positions. The letter T identifies the top surface of the box.

by the 16 strings, black in color, emanating from the main wall, and the second is produced by the green strings coming from the remaining 16 stains residing on the right wall and the end surface of the box.

BackTrack/Win has special automatic routines to identify the string intersections that are possible source positions. These routines will find all the possible intersections and compute average values for the X and Y coordinates of these intersections. These average values are then equated to CPx and CPy.

In addition to these automatic routines, which are useful and even necessary, when the number of strings becomes very large, there are manual routines whereby the investigator has full control of which intersections are to be included in the analysis. These manual routines are invaluable for checking the results.

An upper limit, CPz, for the height of the blood source above the floor is computed from the side or end views of the strings.

This view of the 16 strings from the main wall was used to estimate an upper limit for the height of the first blow. The procedure uses a routine called "Manual CPz." This routine draws a vertical red line at the X-position given by CPx = 45.8 cm. The investigator then expands the display and using the mouse selects the z-coordinate where each string intersects the CPx value. These 16 Z-values are averaged to yield a best estimate of CPz = 132.5 cm.

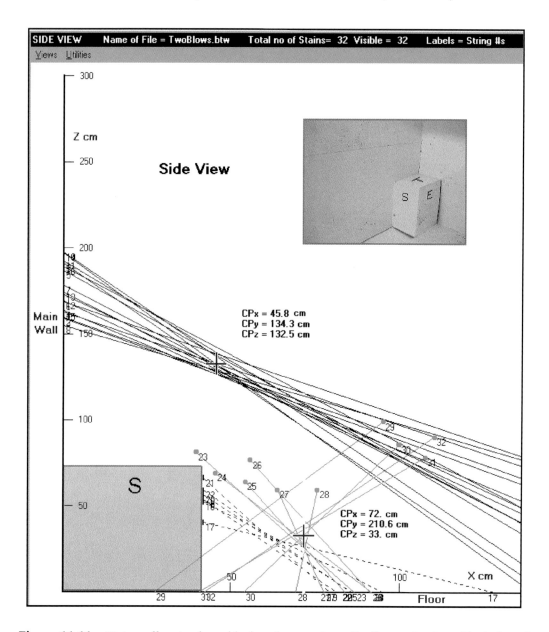

Figure 11.11 Main wall stains have black strings, main wall offset stains have black dotted strings, and the right wall stains have green strings.

It should be emphasized that the information necessary to compute the vertical details of the flight path are unknown quantities (droplet speed and size). However, the laws of physics dictate that, normally, the strings must pass directly over the source positions. This means that the value of CPz found by the previously explained procedure will be an *upper limit* for the true value of the height.

This view was used to estimate the upper limit for the CPz value of the second blow. For strings 17 to 32, the Z-coordinates of their intersection points with the CPx value of 72.0 cm was averaged and set equal to CPz. The result is shown as 33.0 cm.

Directional Analysis of Bloodstain Patterns with a Computer

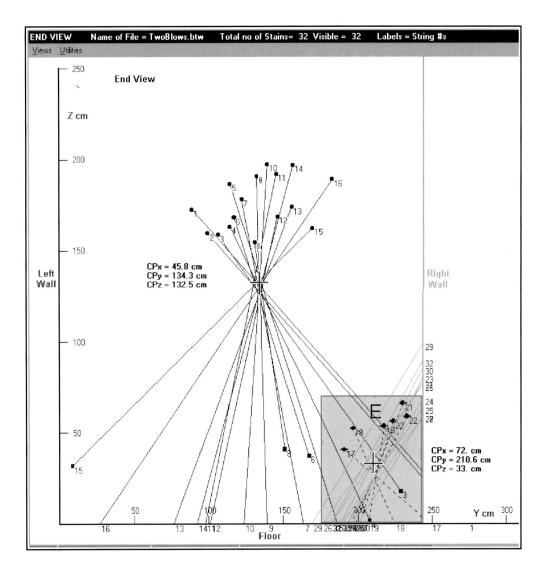

Figure 11.12 The bloodstains are colored circles, black for the main wall and its offset surface. The strings are continued until they exit the crime scene, where they are marked with blue squares.

Summary

A procedure known as the directional analysis of bloodstain patterns has been described for a laboratory exercise involving two blood spatter events called the first blow and the second blow. This procedure uses a standard PC computer running the blood spatter analysis software known as the BackTrack Suite.

The three coordinates of the estimated location of a blow are designated as CPx, CPy, and CPz. Table 11.1 compares the results of the analysis with the true values. The horizontal coordinates, CPx and CPy, of the second blow have been measured to within ±2 cm — a remarkably accurate result that is not uncommon with this procedure. However, the horizontal coordinates, CPx and CPy, of the first blow do not show the same degree of

Table 11.1 CPx, CPy, and CPz are the X, Y and Z Coordinates of the Positions of the Blood Sources or the Locations of the Blows (the results of the analysis and the true values are tabulated for comparison)

	First Blow Analyzed	True Value	Second Blow Analyzed	True Value
CP_x	45.8 cm	51 cm	72 cm	73 cm
CP_y	134.3 cm	139 cm	210.6 cm	212 cm
CP_z	132.5 cm	124 cm	33 cm	27 cm

agreement with the true values. It is tempting to speculate about possible systematic errors. The bloodstain pattern in this instance displays a left–right symmetry when viewed from above. Therefore the 10% difference between the CPx value and the true value can be explained by a systematic underestimating of the proper lengths of the ellipses. However, the 3.6% difference between the CPy value and true value cannot, in the author's opinion, be blamed on any obvious systematic errors and therefore may be interpreted as a random error. The two estimated values for CPz are predicted by the theory to be upper limits for the true values. Comparing the estimated results for CPz with the true values, we see good agreement with that prediction.

A Case Study by Detective Craig C. Moore of the Niagara Regional Police Service, Ontario, Canada

In 1999 a homicide beating occurred in a small two-bedroom bungalow. When the victim's brother contacted police, he led the investigators to believe he had found the decedent in the main floor bathroom seated in a tub full of water. Police noticed that the decedent had suffered blunt trauma blows to the head and upper chest area. The distribution of bloodstain patterns throughout the house and the degree to which the dining room furniture had been overturned and broken indicated the victim met a violent death.

The analysis of the bloodstain patterns found at the scene indicated that the victim suffered numerous blows to the body while lying on the floor against the north wall of the dining room. Figure 11.13 shows the projected spatter bloodstains on the lower portion of the north wall. These stains revealed directionality that indicated the blood droplets were ascending when they impacted with the wall.

Well-defined blood smears to the dining room carpet indicated that the victim was dragged from this north wall area across the floor, into the hall, and into the nearby bathroom. Blood was smeared across the ceramic-tiled bathroom floor, coming to an end at the bathtub where the decedent was found. One uninterrupted individual drag mark in blood measured 3 meters.

Figure 11.14 shows the dining room table returned to its normal position. Projected spatter patterns with good directionality could be seen across the tabletop and on the edge of the table facing north toward the large bloodstained wall/floor area.

The two patterns on the table and the pattern on the north wall/floor area were analyzed using the BackTrack computer program (Figure 11.15 and Figure 11.16).

The analysis of the blood spatter on the north wall indicates that the victim was close to the floor and to the wall. The pattern is consistent with multiple blows. The pattern spattered across the tabletop is from a blood source above the table at a maximum possible

Directional Analysis of Bloodstain Patterns with a Computer 255

Figure 11.13 The north wall bloodstain pattern is located just below the window curtain.

height of 1.3 m from the floor. The blood source for the spatter pattern located on the edge of the table had originated from below the height of the table.

From the various bloodstain patterns found throughout the home and the computer analysis of the projected stains found in the dining room, it was possible to carry out a reconstruction of the probable events that occurred at the crime scene.

Figure 11.14 The table has been restored to its normal position, and the selected bloodstains are located with position markers.

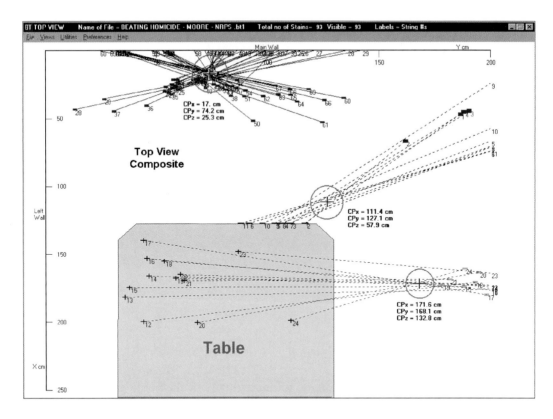

Figure 11.15 The main wall is the north wall of the crime scene. Three bloodstain patterns have been analyzed with BackTrack/Win. The top views of the string patterns are shown here.

- A physical altercation between the two men started in the bedroom. The two combatants made their way into the dining room next to the bedroom.
- The victim received a blow to his head. This blow projected blood onto the tabletop. BackTrack analysis determined that the blood source was at a maximum vertical height of 1.3 m.
- The victim moved a few feet northward into the dining room. Here, on his hands and knees, he takes another blow to his head, which projects blood onto the edge of the table.
- The assailant lifts the end of the table up off the floor and the column leg is broken off, where the table rests standing in an upright position. The victim is now on the floor near the north wall, where he is struck repeatedly, causing blood to be spattered onto the north wall.
- The victim was also struck with the vacuum cleaner. Cast-off bloodstains on the east wall indicate it was probably swung in a golf clublike manner, lower left to upper right fashion, facing north toward the body.
- The unconscious, possibly lifeless body was then dragged to the bathroom and put into the bathtub, where it was found by the police.

Directional Analysis of Bloodstain Patterns with a Computer

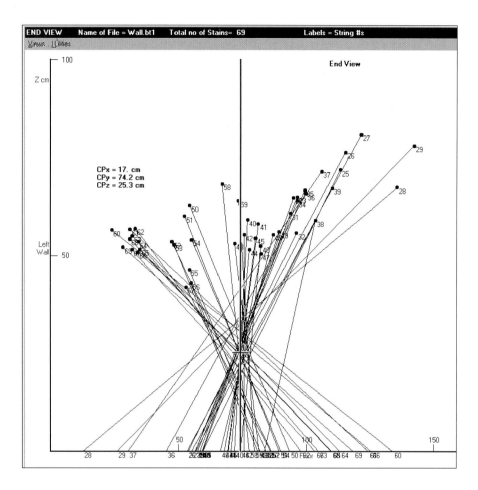

Figure 11.16 This is a screen grab of the end view of the pattern found on the north wall near the floor. The convergences suggest that multiple blows took place here.

The suspect was charged with 2nd degree murder, and after the preliminary inquiry, he pled guilty to manslaughter. He was sentenced to time served plus 40 months.

Two Case Studies by Sgt. Mike Illes of the Ontario Provincial Police, Peterborough Detachment

Case #1 Multiple impacts in a limited space

A convenience store clerk was robbed and confined within the store in the early morning hours. The victim was held hostage within the store for several hours until a Special Weapons and Tactics Unit entered the store, shooting the suspect. The victim had been badly beaten in a narrow stairwell, which was located between the main floor and the basement of the store, as illustrated in Figure 11.17.

258 Principles of Bloodstain Pattern Analysis: Theory and Practice

Figure 11.17 The narrow stairwell where the victim was badly beaten.

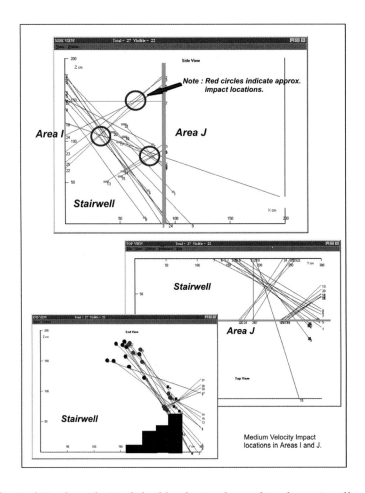

Figure 11.18 BackTrack analysis of the bloodstains located in the stairwell.

Directional Analysis of Bloodstain Patterns with a Computer

There was a minimum of three impact patterns within the stairwell. The stairwell was long and narrow, which created difficulties in processing the scene. The stairwell width was only 90 cm and therefore could not be easily photographed. A photographic mosaic could have been used; however, this would have taken hours of setup for each picture on uneven stairs. The use of the string method to calculate the bloodstain origins was impossible because of the close confines of the area. The three impact locations were easily located using BackTrack/Win, and the results illustrated in Figure 11.18 were accomplished in approximately 2 h.

Case #2 Expirated bloodstaining

An inmate who was wielding a shiv attacked another inmate within a federal institution in Canada. The homemade knife was shoved into the victim's chest and throat areas, causing immediate life-threatening bleeding. The victim moved from his bunk area within his cell toward the range hallway, apparently looking for help.

On examination of the scene, the bloodstain evidence confirmed this scenario. Several small mistlike stains were observed on the front of a drawer within the victim's cell (Figure 11.19). The pattern had the appearance of a very small impact pattern; however, it became apparent that these stains where created by expirated blood. The victim coughed or sneezed while moving past this area in his attempt to get help.

Several of these stains were measured, and BackTrack/Win was used to see if they came from a singular event and to determine the blood source location. A single blood source was located directly in front of the middle drawer; this indicated a continuity of the events throughout the alleged scenario. Figure 11.20 and Figure 11.21 depict the scene and BackTrack/Win results.

Figure 11.19 Showing the bloodstain pattern that was found on the desk drawer.

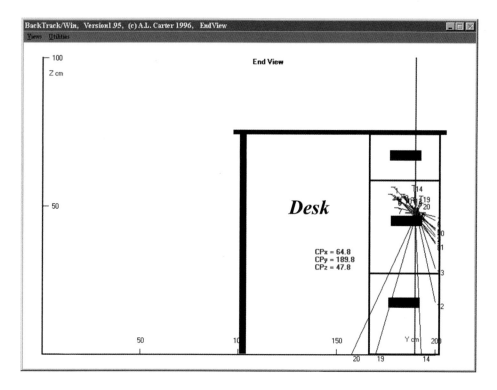

Figure 11.20 The directional analysis by BackTrack/Win of the pattern on the middle drawer.

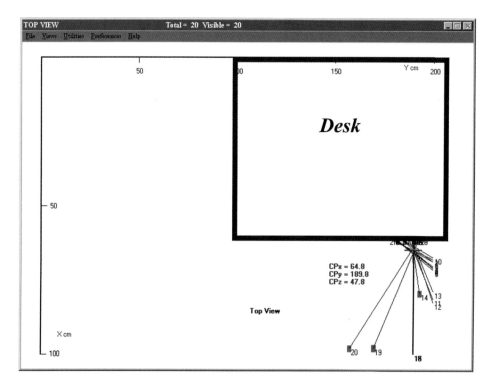

Figure 11.21 Top view of the directional analysis of the pattern on the middle drawer.

Credits

The author is grateful to NCO I/C S/Sgt Bruce MacLean of the RCMP in Edmonton for his continued support of the Math/Physics Advanced Course in Blood Spatter Analysis. Also, for their assistance, Sgt. Jon Forsyth-Erman, Sgt. Serge Larocque of the RCMP, and Cst. William Basso of the Lethbridge Police Service.

Dr. Brian Yamashita of the Research and Review Section, Royal Canadian Mounted Police, Ottawa, taught the mathematics of Blood Spatter Analysis with his usual skill and good humor.

Special acknowledgments go to the three members of the 2003 class in Edmonton who participated in the exercise and actually selected the bloodstains that were used in this analysis. They are Johanna van der Meij of the Netherlands Forensic Institute, Netherlands; Sgt Mark Reynolds, Western Australia Police; and Sgt. Jim Hignell, RCMP, Vancouver BC.

I am also grateful to Sgt Mike Illes of the Ontario Provincial Police and Det. Craig Moore of the Niagara Regional Police for contributing to the case studies.

Documentation and Examination of Bloodstain Evidence

12

A photograph is worth a 100 years.

P. Kish

Introduction

Physical evidence is defined as any and all materials or items that may be identified as being associated with a crime scene, which, by scientific evaluation, ultimately establishes the elements of the crime and provides a link among the crime scene, the victim, and the assailant. The proper recognition, documentation, collection, preservation, and examination of physical evidence are crucial to the successful reconstruction of a crime scene and litigation in a criminal proceeding. Physical evidence may be present at the crime scene whether it be indoors or outside, on a vehicle, on the victim, on the assailant, and/or his environment. The Locard Exchange Principle is foundational for the explanation of the transfer of evidence—that is, it is not possible to enter a location without changing it in some manner, by either leaving something behind or taking something away. Frequently, blood and trace evidence are transferred between the victim and the accused, victim and scene, and the accused and the scene.

Bloodstain pattern analysis is completed at the crime scene as well as the laboratory. Whether the analysis is conducted at the crime scene or at the laboratory, the purpose behind the documentation of bloodstain patterns remains the same: to create a clear, concise, and accurate record of the bloodstain patterns' size, shape, distribution, location, and overall appearance as well as all case facts that may influence the appearance and/or formation of the bloodstain patterns.

The accurate documentation of bloodstains identified within a crime scene or on an article(s) of physical evidence will assist in the following:

- The reconstruction of the bloodshed events
- Maintaining the "chain of custody" of items of evidence
- A peer review of the bloodstain pattern analyst's opinions

Often the extent to which a bloodstain pattern analysis can be completed is dependent on how well the scene and evidence was recognized, documented, and preserved at the onset of the case. Prior to discussing the methods for documenting bloodstains, we need to first evaluate the types of information the bloodstain pattern analyst needs to have regarding the bloodstains. At a minimum, the bloodstain pattern analyst needs to document the following data regarding the appearance of bloodstains:

1. Size of overall stain patterns and individual stains
2. Overall shape of the bloodstains
3. Location of bloodstains
4. Overall distribution of the bloodstains
5. Physical appearance of bloodstains (dry, wet, etc.)

In addition, any case-related factors that may have had an effect on the formation of the bloodstain need to be documented. These factors would include but are not limited to actions taken by the first-responding officer, actions taken by emergency medical personnel, temperature, air flow, poor handling of wet clothing, etc.

Methods of Documenting Bloodstain Patterns

Bloodstain patterns are documented with notes, sketches, diagrams, photography, and video. Each of these documentation methods has its own advantages and disadvantages, which make it essential to use more than one method when documenting bloodstain patterns. Only one opportunity exists to create an accurate record of the stain patterns. Once the crime scene has been released or the stain has been removed for DNA testing, your opportunity to view this evidence in an unaltered condition is gone. At a minimum, notes and photographs must be taken of the bloodstain patterns. The overlapping documentation methods should limit the loss of any potentially valuable information in regard to the bloodstain patterns. *Too much information is better than too little.*

Notes

Documentation by note form is initiated at the onset of the case and will continue through the completion of the case. Anything written or typed regarding any case should be retained within the case file as case notes. Notes taken during the examination of the bloodstain patterns should be a detailed and accurate narrative record of the physical characteristics of the stain patterns (size, shape, distribution, location, and overall physical features). The notes taken at the crime scene should have at a minimum the following general information:

- Case identifier
- Date and time of arrival and departure
- Who was present on arrival
- Geographic location of scene
- Weather conditions and environmental factors

The times when photographs and observations are noted may prove to be beneficial if a drying time issue should arise later in the case. Recording accurate environmental conditions within a scene is often complicated when you are not one of the first people to respond to a scene. Factors such as open windows or doors, running ceiling or floor fans, wet sinks, running appliances (especially those with timers), and lighting should all be recorded in your notes. Case factors that may have influenced their formation should also be noted. Because of the diversity in the materials examined, loose-leaf paper works well for bloodstain pattern analysis notes. It allows the examiner the choice between lined or unlined paper, graph paper, and generic diagrams of clothing. By using a light-colored paper for our original handwritten notes, it makes file organization easier, and it photocopies more clearly for later discovery requests. Notes should be maintained in chronologic order. Each page of notes should have the date, case identifier, and page number and should be maintained in chronologic order.

All potential bloodstains observed at the scene or on items of evidence should be noted as "bloodlike," "visually consistent with blood," "red-brown stains," or "stains" until serologic testing is completed. Some analysts write a caveat in the beginning of their notes indicating each time the word blood or bloodstain is used they are referring to stains that are "visually consistent with blood."

Sketches and Diagrams

Good photography is essential for the proper documentation of bloodstain evidence at crime scenes. However, photographs may not always depict relative distances between objects and other details that are best demonstrated with sketches. A finished sketch is the graphic illustration of the crime scene and items within the scene that are pertinent to the investigation. The primary purpose is to provide a good orientation of items of evidence within the scene. A sketch will contain only essential items, whereas photographs may be overcrowded with detail. It should show the overall view, geographic location, size, and position of objects at the crime scene. A good sketch of a crime scene should complement the photographs and together provide the investigator, and eventually a jury, a clear picture of the scene. Some good general rules for crime scene sketches are as follows:

1. Get a general, visual idea of the layout before collecting data and measurements. Use photographs for reference if available.
2. Sketch an outline of the area first and then record objects.
3. Note position of magnetic north.
4. Include only essential items of furniture, fixtures, and so forth to avoid clutter in the sketch.
5. The person reading the tape measure should record ALL the measurements.
6. Measure from finished wall to finished wall and record dimensions of the room.
7. Note position of doors, windows, and light fixtures.
8. Note direction in which doors swing, windows open, etc.
9. Measurements of the location of movable objects should be noted in reference to at least two fixed objects so they may be relocated to their original position at a later time.
10. If there is more than one body at the scene, number each body.

There are two types of drawings: rough sketches and finished diagrams. The rough sketch is usually completed at the scene, whereas the finished or final diagram is prepared to be used in court. The rough sketch or sketches should be retained within the case file along with your notes. Finished sketches are often completed with computer-aided design (CAD programs). Figures 12.1a, b, and c show representative examples of CAD-generated diagrams.

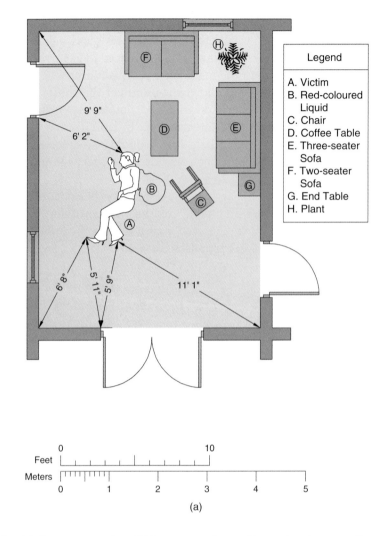

(a)

Figure 12.1 (a) A representative CAD-generated two-dimensional scene diagram using the triangulation method. (b) A representative CAD-generated two-dimensional scene diagram using the transecting baseline method. (c) A representative CAD-generated three-dimensional scene diagram. (Courtesy of Alexei Pace, Malta.)

With complex cases, multiple sketches of specific areas or walls may be necessary to adequately document the stain patterns and their locations. When multiple overlapping sketches are created, specify in detail how they relate to each other and the scene. Sketching outdoor scenes where there is significant movement of the blood source, producing blood trails, can be very difficult and may require the use of surveying equipment. Global positioning system (GPS) is useful for extensive outdoor scenes.

The basic materials needed for sketching at crime scenes are as follows:

1. Tape measures (100 ft, 25 ft)
2. Assorted rulers and triangles
3. Graph paper and sketching paper
4. Appropriate pencils, pens, and eraser

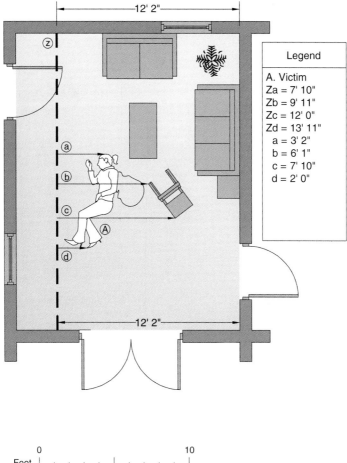

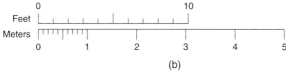

(b)

Figure 12.1 (Continued).

5. Templates (building interiors, body, etc.)
6. Compass

Sketches are also routinely used during the laboratory examination of items of physical evidence. During the examination of items of evidence, sketches are created simultaneously with the analyst's notes to record stain and sample locations.

Photographs

Photographs are the best way to document bloodstain patterns. Bloodstain pattern analysis is a visually intensive forensic discipline. Hence, the need for high-quality color photographs including scales cannot be overemphasized. When a crime scene or item of physical evidence is correctly photographed, each photograph should overlap with another, as shown in Figure 12.2, Figure 12.3, and Figure 12.4. Overall, mid-range as well as closeup photographs should be taken

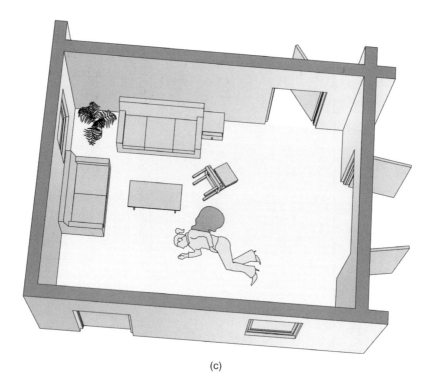

(c)

Figure 12.1 (Continued).

of bloodstain patterns. Photographs of the evidence should be taken prior to any alteration or collection of the bloodstain evidence. Figure 12.5, Figure 12.6, and Figure 12.7 depict a laboratory perspective of documenting with overall, mid-range, and closeup photographs.

It is highly recommended that the bloodstain pattern analyst have a background in photography attained either through formal educational programs or on-the-job training.

Figure 12.2 Overall crime scene photograph.

Documentation and Examination of Bloodstain Evidence 269

Figure 12.3 Mid-range scene photograph of spatter pattern on wall and heat register.

The quality of photographs depends on the quality of the photographic equipment, but, more importantly, it is dependent on the knowledge and skill of the photographer.

The 35 mm, single-lens reflex camera is a good choice for documenting bloodstain patterns when comparing image quality to its relative size, cost, and versatility with interchangeable lens, flash units, and film types. Digital photography has also become a commonly used image-capturing medium. The advancement of the digital technology, along with the increasing quality of the digital images, has resulted in many agencies converting completely to digital photography. A major advantage to digital photography is being able to confirm that an acceptable image was captured prior to altering the scene or the removal of samples. A digital photograph needs to be a fairly high resolution with a camera capable of at least 3 megapixels or greater.

Figure 12.4 Closeup scene photograph with a scale of the spatter pattern on the wall and heat register.

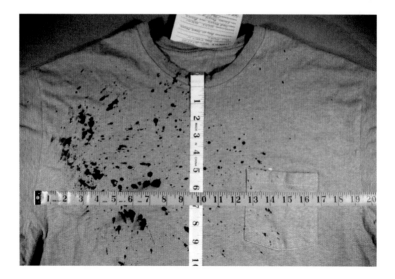

Figure 12.5 Overall laboratory photograph of spatter pattern on the chest region of a T-shirt.

Photographic images are dependent on the use of light and time. Sufficient lighting at crimes scenes and during evidence examinations is not only valuable for photographic purposes but also for bloodstain recognition purposes. Flash photography is probably the most common method for documenting scenes. An inherent problem with flash photography is the location of the flash unit on top of the camera body. This often results in the flash "washing out" the image in mid-range to closeup photographs, as shown in Figure 12.8 and Figure 12.9. Using a remote flash attachment and removing the flash unit from the top of the camera or using photoflood lights can solve this problem. If external lights are used, the color temperature of the film and the temperature of the lights need to be matched in order for proper color. A ring-flash (Figure 12.10) is highly recommended for taking photo macrographs and is ideal for documenting the individual spatters used for origin determinations.

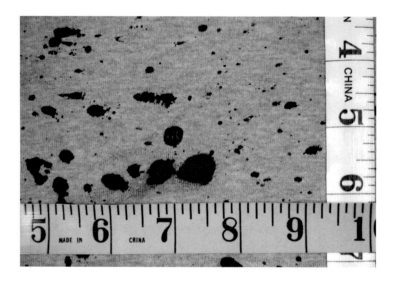

Figure 12.6 Mid-range of T-shirt depicted on Figure 12.5 narrowing in on an area of interest.

Documentation and Examination of Bloodstain Evidence

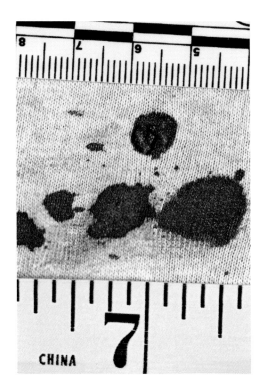

Figure 12.7 Closeup photograph of a spatter with tissue/bonelike material adhering to the spatter.

Figure 12.8 An arterial bloodstain pattern on the side of a vehicle in which a portion of the pattern was "washed out" by the camera flash. (Courtesy of Joseph Cocco, Pennsylvania State Police.)

Figure 12.9 The portion of the pattern in Figure 12.8 that was "washed out" by the camera flash. (Courtesy of Joseph Cocco, Pennsylvania State Police.)

General points to remember when photographing bloodstain patterns:

1. Overall photographs of bloodstain evidence should occur prior to any type of alteration or collection of bloodstained items.
2. Photograph stain patterns prior to and after using *any* blood enhancement techniques.
3. The camera lens must be perpendicular to the plane of the bloodstain for mid-range and macrophotographs. With overall photographs, it will not be possible to have every item within the frame perpendicular to the camera lens.
4. Use external light sources to properly illuminate the area or items being photographed.
5. Use appropriate lighting and/or lens filtration to match the color temperature of your film.
6. Use scales as well as reference points within the photographs (Figure 12.11).

Figure 12.10 Canon™ ring flash used for macrophotography of small bloodstains.

Figure 12.11 Overall scene photographs where areas of interest have been marked out with numbered cards. Record the location of each letter region in your notes. Refer to the region letter "Region A," "Region B," etc. as you describe the bloodstains within those regions.

7. When collecting bloodstained evidence at crime scenes, take photographs of the surface underneath the items being collected.
8. Use a photograph log. This is especially critical with photomacrographs as well as photomicrographs.

Photomicrographs are simply photographs taken through a microscope. They are used to document what was observed during microscopic examinations of bloodstains on items of evidence. Photomicrographs can be very useful when describing the morphology of a bloodstain on an item of clothing (Figure 12.12). The images viewed through a microscope may also be captured via a video attachment.

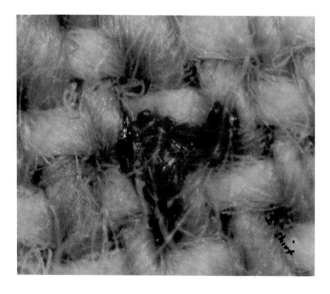

Figure 12.12 Photomicrograph of impact spatter on blue jeans. (Courtesy of Linda Netzel, Kansas City, Missouri Crime Laboratory.)

Basic Photographic Equipment List

Basic photographic equipment includes the following:

1. High-quality 35 mm SLR camera body
   a. Lens: normal, wide angle, macro
   b. 35 to 105 mm or 28 to 80 mm with an f/3.6 to f/16 or f/22
   c. 28 mm wide-angle lens
   d. 50 mm macro lens
   e. 90 to 100 mm macro lens (preferably one with 1:1 capability)
2. Filters
3. Remote shutter release
4. Flash unit
5. Ring flash unit
6. Film (assorted speeds)
7. Camera tripod
8. External lighting/light stands
9. Extension cords
10. Assorted scales, markers, labels, evidence placards
11. Equipment case, batteries, battery chargers

Photographs should be treated as items of evidence no matter the format. It is recommended that the negatives from wet film processing and/or archived versions of the digital images be secured in your evidence control area.

Video

Video is normally taken at the onset of processing a crime scene no matter what type of evidence is being documented. Video is a good overlap for scene photographs and may occasionally catch information that was inadvertently not photographed. They also assist in documenting scene layouts and the spatial relationships between stain patterns and items of evidence. Video cameras are useful in capturing lighting conditions, door and window positions, and outdoor crime scenes where weather and lighting conditions may change rapidly, as well as irregularly shaped items that may be difficult to document photographically.

Following is a list of issues to consider before and while using video to document crime scenes, evidence, or experiments:

- Become aware of how to operate the camera, especially focusing.
- Use adequate lighting. The use of external lighting is often necessary.
- Leave the camera on (recording) throughout the entire scene video.
- Use overall and zoom functions when documenting bloodstains.
- Alter your position to assure your camera is 90° to the bloodstain patterns.
- Clear the scene of all personnel prior to videotaping, and plug the microphone to prevent recording any extraneous noise.
- DO NOT edit the original videotape.
- Log in the videotape as an item of evidence.
- Indicate in your notes when (date and time) and who took the scene video.

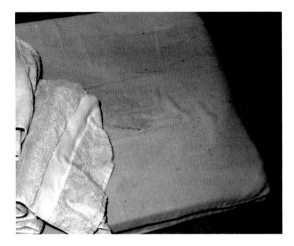

Figure 12.13 Scene photograph of a couch cushion that was located next to the decedent who received a single gunshot wound to the head. (Courtesy of Detective Gregory Coughlin, New Castle County Police, Delaware.)

Documentation of the Crime Scene

It is important to recognize that the study and analysis of bloodstains at a crime scene should be integrated within the systematic approach to the examination of all types of physical evidence and crime scene reconstruction. Bloodstain evaluations should be coordinated with the overall documentation, collection, preservation, and examination of the other types of physical evidence that may be present at the crime scene. The major time-consuming details of bloodstain analysis can usually be accomplished after the crime scene has been processed for other types of physical evidence. In most cases, the search and collection of trace evidence, such as hairs, fibers, and fingerprints, can be given priority over evaluation of the bloodstain patterns. In those instances, where trace evidence coexists with bloodstain evidence, consultation and communication between investigators and laboratory personnel at the scene will minimize unnecessary contamination, alteration, or destruction of physical evidence.

In actual practice, the crime scene environment can be indoors within a room or rooms of a house or other structure, it can be outdoors, or it can involve an automobile or other type of transportation vehicle. Frequently, all these environments may be involved in the commission of the same crime and all may be sources of physical evidence, especially in cases where a victim has been killed in one location and transported to another. The subsequent actions of the assailant may produce additional sources of physical evidence involving disposition of a weapon or other incriminating materials (Figure 12.13, Figure 12.14, and Figure 12.15). The accused's clothing is a valuable source of bloodstain evidence and trace evidence, especially when it is acquired within a reasonable time and has not been subjected to extensive washing.

Indoor Crime Scenes

The indoor crime scene is for the most part protected from the elements and easily preserved for extended periods. Usually there should be no need to hasten the processing of the scene. Occasionally, complications arise with the examination of a victim and crime scene in public

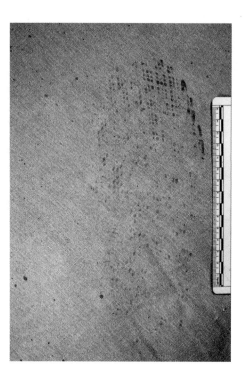

Figure 12.14 Closeup of the sole pattern made on the couch cushion depicted in Figure 12.13. This footwear pattern was made with the deceased's blood.

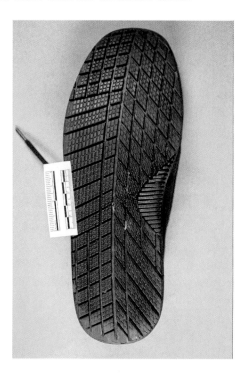

Figure 12.15 Depicts the shoe sole design of the accused's shoe. The transfer pattern on the couch cushion in Figure 12.14 exhibited similar characteristics as this sole, and the deceased's blood was identified on this shoe sole.

places such as restaurants, stores, and airports because of the necessity to clean up the scene as quickly as possible. In all situations, unauthorized persons should be denied access to the scene. Those entering the scene should be cautious so as to minimize the unnecessary tracking of wet blood and thereby not altering significant bloodstain patterns and producing artifacts. Documentation must occur prior to any appreciable alteration of the scene. When the blood is still wet on processing, great care should be used not to create artifacts.

Video usually is used at the beginning of the scene processing to document the overall appearance of the scene on discovery. Notes, sketches, and photographs are taken simultaneously while the crime scene is being processed.

Small bloodstains, spatters, and thin smears of blood will dry rapidly at the scene and will usually remain intact on a surface. There are, however, instances where flaking and ultimate alteration of bloodstains will occur after relatively short periods of time. This phenomenon may occur when there is excessive air movement as created by a fan and when the target surface is wet as a result of various heating devices including heated waterbeds (Figure 12.16). Certain surfaces such as waxed floors, greasy surfaces, or plastic bags (Figure 12.17) may not hold bloodstains very well. In these circumstances, slight disturbances can cause bloodstains to become dislodged. Special attention should be paid to these kinds of surfaces at the crime scene. Merely collecting them with the idea of examining them in greater detail at a later time may, in fact, be a poor decision because moving and handling the items will often cause the stains to flake off completely. If you

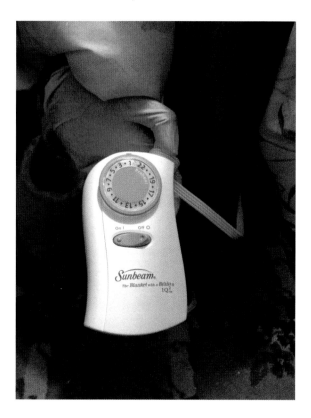

Figure 12.16 Documentation of a bed warmer control. The photograph depicts the exact location of the heat setting and the illuminated indicator light.

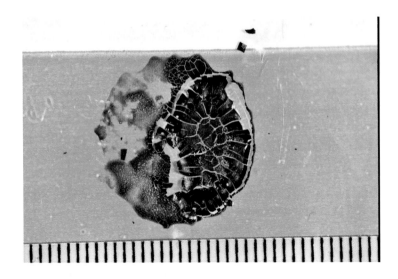

Figure 12.17 Bloodstain flaking from the surface of a plastic bag.

should encounter bloodstains on these types of surfaces, it is highly recommended that you immediately examine and photographically document the actual stain patterns.

People rendering assistance to a bleeding victim can create a lot of bloodstain patterns. The investigator should document activities at the scene during medical intervention. In cases where medical attention is given to a surviving victim, the scene may be subject to unavoidable alteration. Often it is necessary for a victim to be moved to a more suitable position or location for the administration of medical treatment and emergency procedures. Alteration of bloodstains and creation of new bloodstains and other artifacts may be produced as a result. It is important to recognize these alterations at the crime scene. When a victim has been transported to a hospital, it is important to retrieve physical evidence, including victim's clothing, projectiles, and so forth, from the emergency room before they are grossly altered or lost. The sheets or blankets used on the victim during transport in an ambulance should be collected because these items may contain valuable trace evidence that became dislodged from the victim.

Presumptive testing of blood may be made on dried stains after they have been properly documented and photographed. This may be accomplished with the use of leucomalachite green, phenolphthalein, or an equivalent reagent. This technique is especially helpful when the stains are suspect in nature. In many household crime scenes, reddish brown stains may be easily mistaken for blood, and elimination of these stains permits concentration on the actual bloodstains present. The results of all presumptive tests for blood must be recorded in your notes along with a description of where the stain (sample) originated within the scene. Consult with laboratory personnel prior to conducting field tests for the presence of blood in situations where only a limited amount of staining is available.

A diligent search for inconspicuous bloodstains at a crime scene should be made using a good light source such as photo floodlights. Many surfaces may be dirty and discolored or the patterns or texture of the surface may blend in with the bloodstains that are present. There also may be areas of the crime scene where there has been an attempt to wipe up the blood or otherwise clean the surface. Blood may still be present, especially in cracks in floors and walls.

It is important to keep in mind the possible activities of an assailant at the scene not only during the assault but also after the incident has taken place. Assailants may produce bloodstain patterns with the victim's blood or in some instances with their own blood if they have sustained an injury during the altercation with the victim. Foot and footwear patterns made in blood are commonly encountered at crime scenes. These transfer patterns need to be documented with overall photographs that illustrate their general location within the scene, and then they need to be individually photographed (Figure 12.18, Figure 12.19, and Figure 12.20). A bloody knife, for example, may be wiped on a surface and leave a visible pattern (Figure 12.21). Various items may be used to wipe blood from the assailant's hands including towels, tissues, furniture, rugs, and drapes. Partial blood-transfer handprints or fingerprints may be deposited on door edges, knobs, handles, and light switches. Bathrooms should be thoroughly examined for bloodstain evidence. Assailants may deposit or transfer blood on sinks, toilets, shower stalls, or tubs during their attempt to clean up prior to leaving the scene. Sink traps and tub traps should be examined and water samples should be taken to be tested for the presence of blood. A thorough search should be conducted for discarded weapons and bloodstained clothing. An assailant may leave bloodstained cigarette butts or other personal articles such as glasses at the scene.

Figure 12.18 The number placards depict the overall location of a trail of bloody foot patterns created by the accused. (Courtesy of Detective Anthony Dinardo and Detective Paul Mergenthaler, New Castle Police, Delaware.)

Figure 12.19 Number placards depict the location of bloody foot transfer patterns on the stairway depicted in the background of Figure 12.18. (Courtesy of Detective Anthony Dinardo and Detective Paul Mergenthaler, New Castle Police, Delaware.)

Figure 12.20 Closeup photograph of one of the bloody foot patterns identified on the stair treads in Figure 12.19. Note that the evidence placard remained in the closeup image for identification purposes. (Courtesy of Detective Anthony Dinardo and Detective Paul Mergenthaler, New Castle Police, Delaware.)

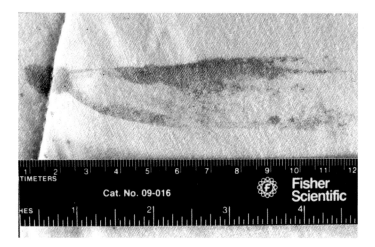

Figure 12.21 A closeup view of a transfer pattern identified on a bed comforter in a case where the victim died as a result of multiple stab wounds and the knife was not recovered at the crime scene.

Outdoor Crime Scenes

The same principles of recognition, documentation, collection, preservation, and examination of physical evidence that are used for crime scenes indoors apply to crime scenes outdoors. Exterior scenes are not as likely to be altered or cleaned by the assailant, but the nature of the terrain and environment including weather conditions may significantly alter the appearance of such physical evidence as bloodstains and cause difficulty in their recognition. Bloodstains and patterns may be absorbed into soil or otherwise altered by wind, rain, snow, or ice. It is important to consider the existing weather conditions and their effect on existing bloodstains and their analysis. Historical weather data will be needed in some cases. Outside target surfaces are more likely to consist of rougher textures than those inside. The analysis of bloodstains on rock, concrete, and grassy areas may be more difficult with respect to directionality and angle of impact. Bloodstains that have been subjected to moisture may undergo alterations in appearance and appear diffuse and diluted. Blood-flow patterns will spread on wet surfaces such as ice and appear to be of greater volume as the blood is diluted. In freezing temperatures, bloodstains and spatters have been observed to retain their characteristics fairly well and remain quite suitable for interpretation.

Photography and documentation of outdoor crime scenes should be accomplished as soon as possible for the obvious reason that unexpected weather changes may later alter the scene considerably and obliterate physical evidence. Bloodstained items should be processed and removed before these changes occur. It may be necessary to videotape or photograph the scene at night with the use of strong light sources. The use of a ladder or truck with a boom is useful for overall photographs of the outdoor scene. Local fire departments can be of assistance when additional exterior lighting is needed. The weather may not hinder additional photography the following day, but if weather change becomes a detrimental factor the night photography is crucial.

It may be necessary to collect soil samples for the demonstration of bloodstain locations. Bushes, leaves, and various other items of outdoor debris should be examined closely for evidence of bloodstains. In certain cases, the absence of significant bloodstain evidence may be associated with the dumping of a victim subsequent to the fatal injuries having occurred elsewhere. If sufficient time has elapsed for blood on the body to dry, the existing blood-flow patterns and lividity patterns may not be consistent with the present position of the victim.

Prompt examination of road surfaces for evidence of bloodstain patterns during the investigation of pedestrian–vehicular accidents may reveal impact sites and directionality of blood spatter that can assist in the repositioning of the victim and reconstruction of the accident. In addition to examination of road surfaces for bloodstain patterns in pedestrian–vehicular accidents, the vehicle's exterior often provides substantial bloodstain evidence in the area of impacting surfaces and undercarriage of the vehicle. The undercarriage of the vehicle should be examined while the vehicle is on a hydraulic lift in a garage.

Documentation and Examination of Motor Vehicles

The examination of vehicles for bloodstain evidence may provide valuable physical evidence in various types of investigations. The interior of vehicles should be examined in a systematic manner and should be divided into sections including the trunk. When photographing bloodstains on windows, mask off the opposing side with paper to decrease background images and reflection. Each area should be photographed and examined separately. The trunk or interior of a vehicle may reveal bloodstain evidence relating to the transportation of a victim, bloody clothing, or objects. The assailant may have transferred their own or the victim's blood onto door and window handles, steering column, gear shift or dashboard controls, seats, floor pedals, and the like.

In cases of motor vehicle accidents involving injury to driver and/or occupants, the evaluation of bloodstain evidence may help to resolve the issue as to the actual driver of the vehicle and the seat location of passengers. This may be accomplished by the characterization of bloodstain patterns and impact sites relative to specific types of injuries in addition to blood grouping and testing of genetic markers.

Documentation and Examination of Bloodstains in the Laboratory

Laboratory standard operating procedures set forth within the laboratory's accreditation standards often dictate the manner in which the analyst records notes in the laboratory setting. The laboratory analyst in many situations is performing more than a bloodstain pattern analysis function; he or she often is either conducting a trace analysis or a serologic analysis along with bloodstain pattern analysis.

When examining articles of evidence for the presence of bloodstain patterns, the time, date, case identifier, item number, and a general description of the item should be included in your notes along with your bloodstain pattern observations. Any anomalies with the appearance of the evidence packaging or inconsistencies with labeling as compared to the contents of the packaging should be recorded in your notes and addressed with the submitting agency. Examine the inside of the packaging on removal of the items. Occasionally, you will find transfer bloodstains on the inside of the packaging, which will give you a valuable insight into the condition of the item when packaged.

Notes and sketches are normally created simultaneously while examining physical evidence within a laboratory setting. Rather than sketching items of evidence, some analysts are now taking digital photographs, printing them in color on plain paper, and writing and drawing directly onto the image. Essentially, any item that can be removed from a scene can be submitted for laboratory analysis. The most common items examined for the presence of bloodstains are articles of clothing and weapons.

List all items contained within the evidence packaging including items found within items, such as contents of pockets. The size, shape, distribution, location, and overall appearance of all bloodstain patterns identified on the items being examined should be included in your notes. The overall appearance should include characteristics such as wet, dry, worn, cut/torn, dirty, etc. The location and appearance of the individual spatters used for serologic testing or for an origin determination should be discussed in detail within your notes. Include the locations of samples that were removed as well as the results of any presumptive test conducted on suspected stains.

The methods used to examine the items of evidence for the presence of blood should also be noted: visual examination, microscopic examination, chemical searching, blind swabbing of the item, etc.

Examination of Bloodstained Clothing

A large portion of our time is spent examining articles of clothing for trace evidence, including hairs, fibers, dirt, debris, residues, and biological materials to include blood. Usually the clothing items submitted will fall into one of the following categories:

1. Victim(s)
2. Accused(s)
3. Witness(es)
4. Miscellaneous garments from scene

As previously discussed, the clothing of the accused may provide valuable bloodstain evidence. Evaluation of the bloodstain patterns associated with the victim's blood on the accused's clothing will in many cases indicate the type of physical activity that is necessary to produce them. Correct analysis of bloodstains on the accused's clothing can help to establish the probability of his or her presence at the scene and involvement with the assault on the victim. Often the reason offered by the accused for his or her bloodstained clothing can be refuted.

The careful examination of bloodstained clothing and footwear often provides valuable information for accurate reconstruction of a violent crime. Bloodstain patterns on the clothing of a victim and accused may represent the activity, position, and movement of each during and subsequent to an attack or struggle after blood has been shed. Examination of bloodstained garments may be difficult and stain patterns may be complex and sometimes partially obscured on blood-soaked material, but a diligent examination is often quite rewarding. Blood found on the clothing of a victim is usually that of the victim, but do not assume this always to be so. Difficult-to-explain features of bloodstain patterns and inconsistent directionalities of individual bloodstains on the victim's clothing may represent blood of the assailant rather than that of the victim. Bloodstain analysis of patterns on an accused's clothing will help confirm or refute explanations offered by the accused concerning the reason for his or her bloodstained clothing. For optimal results certain procedures and precautions should be observed:

- The bloodstained clothing of a victim should be carefully removed after initial photography, and it should be preserved to avoid contamination and the production of

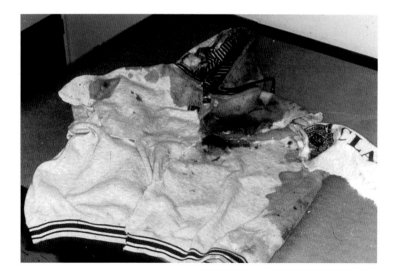

Figure 12.22 Overall photograph of a sweatshirt that was cut off of a victim at the crime scene during resuscitation efforts.

artifacts. Consideration should be given to whether the victim should be disrobed at the crime scene. You will need to establish whether the stain patterns were the result of the violent event or artifactually created during resuscitation or disrobing efforts.
- Clothing that is haphazardly cut from a victim causes many problems for the bloodstain reconstruction. If clothing must be cut from the victim, it should be done in an orderly fashion, avoiding bloodstain patterns, if possible. Perforations, tears, and other defects in the fabric should not be altered during the removal process (Figures 12.22 and 12.23).
- Wet garments should not be folded together or packaged in a damp condition. Blood transfers may occur as well as possible bacterial contamination of the bloodstains,

Figure 12.23 Closeup view of the front of the sweatshirt depicted in Figure 12.22 after reapproximating. Note that the saturation pattern on the right side does not extend across where the garment was cut, indicating the stains were created after the garment was removed from the victim.

which may hinder blood grouping procedures. It is best to hang and air-dry bloodstained clothing over clean paper in a secure location prior to packaging in paper bags. This procedure should also be followed with all wet bed linens and fabrics.
- Trace evidence collection and other special tests and procedures such as examinations for gunshot powder patterns and residue testing should be conducted before extensive bloodstain pattern examination is undertaken.
- Bloodstain analysis and serologic studies on clothing should be coordinated and appropriate sites should be chosen for serologic and DNA testing. If certain bloodstains must be removed, they should be carefully photographed and their exact location should be documented. One experienced in bloodstain analysis can suggest grouping and individualization of bloodstains in specific areas that will assist the reconstruction. Remember that the source of all blood on victim's clothing may not be that of the victim.
- If the clothing of the victim was cut prior to removal at autopsy, it can be extremely helpful to restore it by sewing or taping to conform to its original configuration before bloodstain pattern study is begun.
- The use of a mannequin or model is helpful in orienting the location of bloodstains as they were while the victim or accused wore the garments (Figure 12.24).
- The investigator should study the original scene photographs of the victim at the time the clothing examination is being conducted. This will alert him or her to the possibility of artifacts being produced during removal and subsequent handling of the garments prior to examination.

Figure 12.24 Placement of clothing on a mannequin assists with the orientation and evaluation of the bloodstains.

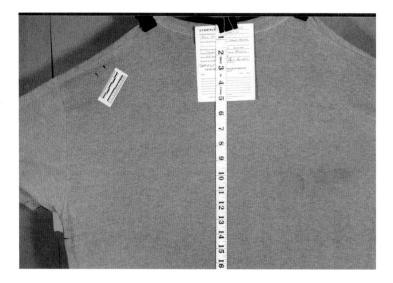

Figure 12.25 Overall photograph of the upper back of a gray T-shirt previously worn by the accused in a beating case. The scale on the upper left shoulder/back region depicts the location of two spatters, which are consistent with cast-off stains.

- Clothing should be adequately described in conjunction with the observed bloodstain patterns. Significant bloodstains should be measured and sketched showing their directionalities, relative sizes, shapes, and appearances. Be cautious with the analysis of physical appearance and directionality of bloodstains on fabrics. Overall and closeup photographs should be taken with a measuring device in place (Figure 12.25 and Figure 12.26).
- A good light source is essential, especially when small spatters are being examined on dark clothing and denim materials.

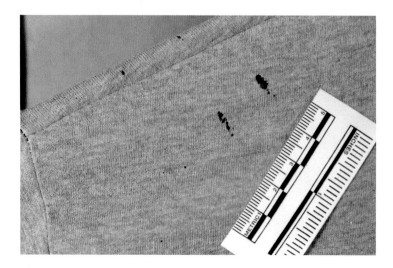

Figure 12.26 Close-up of the spatter identified on the back of the T-shirt depicted in Figure 12.25.

Documentation and Examination of Bloodstain Evidence

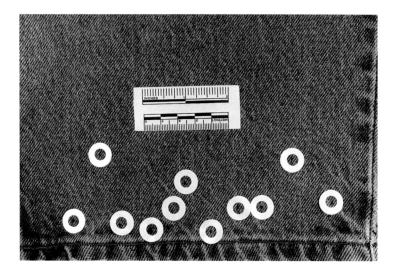

Figure 12.27 Illustrates the use of white self-adhesive ring reenforcers used to visualize the distribution of spatters on denim.

- For photographic and demonstrative purposes, the location of small blood spatters may be better visualized by encirclement with white ring-binder reinforcements (Figure 12.27) or tape triangles (Figure 12.28).
- Mirror images of bloodstains may be created when wet, bloody clothing is folded.
- Alternate clean or void areas on bloodstained clothing may indicate folds in the clothing during bloodshed. Void areas on clothing may also indicate the presence of an intermediate object.
- Search carefully for bloody weapon transfer patterns, especially in association with stabbing and blunt force injuries.
- If blood has soaked through clothing, be sure to determine on which side of the fabric the blood initially made contact (Figure 12.29 and Figure 12.30).

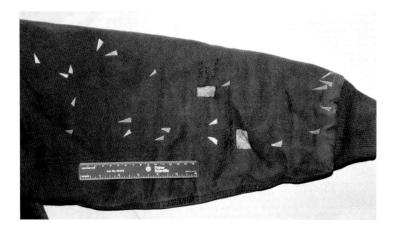

Figure 12.28 Orange tape triangles depict the distribution of spatters identified on this dark green coat sleeve. An accused was wearing this coat when inflicting blunt force trauma and cutting injuries to a victim.

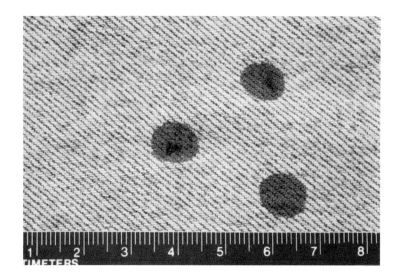

Figure 12.29 Bloodstains on the outside surface of denim jeans. Note the presence of the dark color and small crusts of blood, indicating that the stains were deposited onto that surface.

Always check the pockets of the assailant's garments for blood transferred from the assailant's hands. Spatters can also be deposited in the pockets, in shirtsleeve cuffs, or within dress shirt collars (Figure 12.31, Figure 12.32, and Figure 12.33).

- An assailant frequently steps in blood at a scene. Be sure to check the soles of the assailant's footwear for bloodstain evidence (Figure 12.34, Figure 12.35, and Figure 12.36).
- When prone victims are beaten with a blunt object or kicked, impact spatters associated with a beating are usually found in the region of the accused's lower

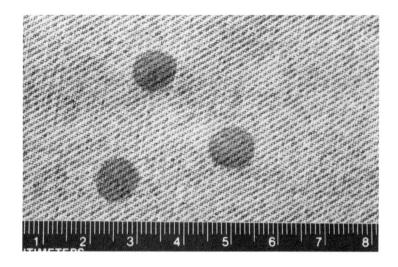

Figure 12.30 The same bloodstains as they appeared on the inside of the denim jeans that have resulted from soaking through from the outside to the inside. Note the lighter appearance of the stains.

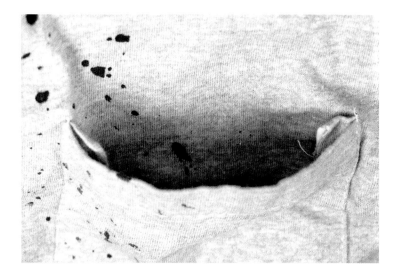

Figure 12.31 Spatters located inside of the front chest pocket of a T-shirt worn by a person accused of bludgeoning a man to death with a dumbbell.

trouser legs, shoes, and shirtsleeves. Be sure to examine the underside of the cuff areas as well as the shoes and socks of an accused in these situations.
- The rear trouser leg and rear of the shirt of an assailant may receive cast-off bloodstain patterns from a bloody weapon when it is swung over the shoulder while an assailant is on his knees administering the blows to a prone victim.
- A careful search for bloodstain evidence should be conducted on the accused's clothing in those cases where sufficient time has elapsed for the clothing to have been washed. There is always the possibility that some detectable blood may have survived the cleaning procedure. The effect of water on bloodstained clothing including

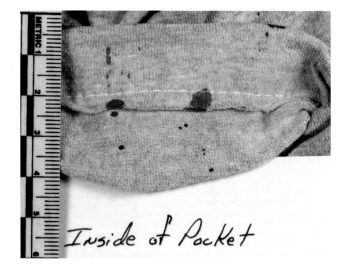

Figure 12.32 View of additional spatters identified within the shirt pocket discussed in Figure 12.31.

Figure 12.33 View of the location of a spatter under the button-down collar.

washed areas may show hemolysis and dilution of bloodstains. Evidence of diffusion of blood to areas previously not bloodstained may be recognizable.
- On apprehension, photograph the accused while they are still wearing their same clothing (Figure 12.37, Figure 12.38, and Figure 12.39).

Figure 12.34 Overall view of the sole pattern of a pair of boots recovered from the accused in a case where the victim was shot.

Figure 12.35 Closer view of the lug area of interest.

- Bloodstained clothing and footwear of the assailant may be the source of blood-transfer fabric patterns and impressions on walls, doors, floors, and other objects at the scene (Figure 12.40). When duplicating patterns of this type for purposes of comparison always use a similar material on which the original pattern was produced.
- The use of a stereomicroscope is very useful for the visualization of small bloodstains on dark fabrics and thick-knitted materials. The DAZOR corporation's electronic video microscope speckFINDER™ is an alternative to the stereoscopic microscope for examining clothing for the presence of bloodstains (Figure 12.41).

Figure 12.36 Macrophotograph of black material that was located on the edge of the boot lug. This material was collected and confirmed to be the blood of the deceased. This staining was crucial in linking the accused to the murder scene.

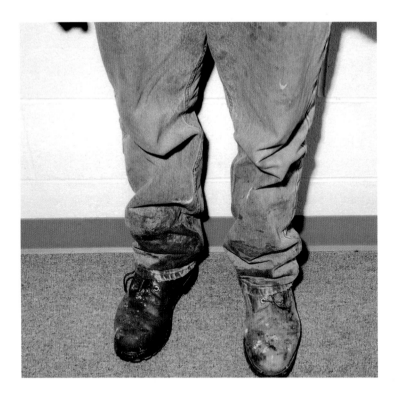

Figure 12.37 Photograph of the jeans and work boots while still being worn by a subject who was accused of beating and kicking a man to death. (Courtesy of David Williams, Rochester, New York Police Department.)

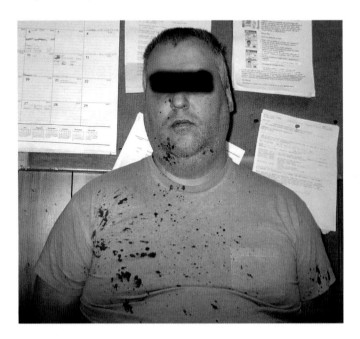

Figure 12.38 This man was the accused in a beating case. Note the blood spatter not only on his shirt but also on his face. (Courtesy of Deputy Chief Mark Flynn, Honesdale, Pennsylvania Police Department.)

Documentation and Examination of Bloodstain Evidence

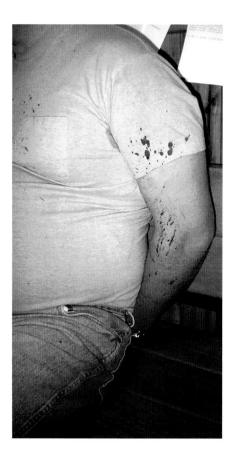

Figure 12.39 Another view of the accused depicted in Figure 12.38. Again, note the location of spatter on his arm. (Courtesy of Deputy Chief Mark Flynn, Honesdale, Pennsylvania Police Department.)

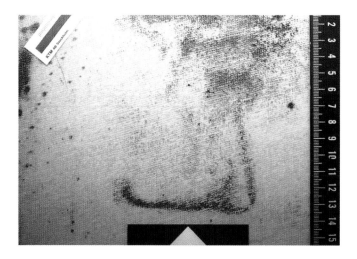

Figure 12.40 Blood transfer pattern on bed of victim that was identified as having been created by the murder weapon — a meat cleaver located in a kitchen drawer. (Courtesy of Martin Eversdijk, The Netherlands.)

Figure 12.41 Electronic video microscope speck FINDER™ with articulating arm and halogen lights. (Courtesy of Dazor Manufacturing Company, St. Louis, Missouri.)

- Leucomalachite green reagent, phenolphthalein, or other presumptive tests for blood may help distinguish blood from spots of foreign substances (e.g., paint, tar, etc.). This type of presumptive testing may prove useful during bloodstain examination of clothing, especially in the case of small spatters and areas not previously tested by a serologist that are important for the final interpretation. For purposes of further confirmatory tests for blood and DNA analysis, care should be taken not to use the entire sample for presumptive tests.

Collection and Preservation of Bloodstain Evidence

The identification of blood and forensic serologic studies for species determination, grouping, genetic marker, and DNA individualization are used in conjunction with bloodstain pattern analysis and are often mutually beneficial to the reconstruction of a crime scene. Both disciplines are important sources of physical evidence. In some cases, bloodstain pattern analysis may provide more valuable information, whereas in others the serologic studies may be more revealing and informative. The value of each should be considered when the investigator is preparing to remove bloodstained objects and individual stains from the crime scene for submission to the laboratory. It is of ultimate importance to photograph and document adequately the location, size, and shape of bloodstains prior to their collection and removal from the crime scene.

The extent of ABO grouping, testing of genetic markers, and DNA analysis in bloodstains that is possible in a given case depends on many factors including the age of the blood, the quantity of the blood available, the environmental conditions to which the blood has been exposed, and the degree of contamination that has occurred. Because of the limited availability of polyclonal antibodies, very few forensic laboratories are currently performing ABO grouping and many do not use identification of genetic markers to a great extent. Generally, with a limited sample available and with older samples, fewer systems can be tested. Exposure of the blood to extreme heat, light, and moisture is detrimental. Also, it should be recognized that laboratories differ in their abilities and techniques to individualize dried bloodstains. Hospital and clinical laboratories are better equipped to individualize fresh blood in blood banks for transfusion and in cases involving paternal disputes. Many forensic laboratories have developed sophisticated techniques for the individualization of dried bloodstains. Occasionally, the surface on which the blood has been deposited may provide a source of contamination and limit the extent of blood individualization—for example, when blood has been deposited on soiled clothing, floors, and vehicle exteriors where large deposits of crushed insects have accumulated. In all cases, it is mandatory to collect, for comparative purposes, a negative control sample from an unstained area on the same surface from which the bloodstain is to be collected.

Putrefied blood presents many problems for the serologist because of the contaminating affect of bacteria, fungi, and the enzymatic breakdown of blood components and often produces misleading or inconclusive testing results. Blood may decompose if not packaged properly. For this reason, bloodstained articles should not be packaged wet nor be sealed in airtight plastic bags, which may accumulate moisture. Individual packaging of completely dry bloodstained articles in clean paper bags is most desirable. The American Association of Crime Laboratory Directors (ASCLD) requires samples to be packaged individually in order for the sample to be accepted for analysis by an accredited laboratory. Crusts and scrapings of bloodstains may be packaged in clean folded paper or clean plastic containers. It is important during the collection process not to handle bloodstains or control areas with hands or fingers, which may contaminate the samples with DNA. All items should be properly labeled and identified for chain of custody purposes.

The following general guidelines are recommended to help ensure optimal results in the proper collection and preservation of bloodstain evidence. Investigators should consult with the specific forensic laboratory so as to follow particular protocols and to resolve the issues that may arise.

- If there is a sufficient quantity of wet, pooled blood at a scene, approximately 5 to 10 ml should be collected with a pipette or medicine dropper and placed into a glass tube containing preservative and anticoagulant for DNA analysis. Appropriate tubes containing ethylenediamine tetraacetic acid (EDTA) to preserve blood samples are available commercially or may be provided by the laboratory. Always check with the laboratory personnel for their preference of blood preservative for specific procedures. The wet blood samples should be refrigerated prior to submission to the laboratory.
- A bloodstain on a nonabsorbent surface such as glass or metal may be collected with a swab or swatch slightly moistened with a small amount of sterile physiologic saline or water. A control sample should be collected in a similar fashion. Both should be refrigerated.
- Water or other liquids to be tested for blood should be placed into clean jars or other leakproof containers and refrigerated in sterile containers.
- When feasible, the entire bloodstained article or object should be collected as evidence after the bloodstains have thoroughly dried and their locations have been photographed and properly documented.

- Do not fold wet clothing or bedding and do not package in a wet condition. When bloodstained sheets, blankets, or pillowcases are removed from a bed or other location, it is important to document top, bottom, left side, right side, and which surface is up or otherwise folded as it was at the scene. Clothing and bedding items should be suspended in a clean area for drying purposes over clean paper to collect any trace evidence that may become dislodged during the drying process. Do not use a hair dryer or heating element to hasten drying. The papers under the dried items should be folded and submitted with the individual articles. The items should be folded and packaged individually so as not to further disturb the bloodstains.
- Rugs or carpets may be lifted and removed in their entirety in many cases. If this is not feasible, the bloodstained portions and control areas should be cut and removed after photography and proper documentation and then packaged in clean paper.
- If it is not feasible or practical to submit the entire bloodstained surface of the object, the bloodstains may be scraped with a clean scalpel or razor blade onto clean paper. To avoid cross-contamination, individual disposable blades should be used. Be sure to scrape control surfaces as well. The papers may then be folded, taped, and properly identified or transferred to clean plastic containers for submission to the laboratory.
- Collection of blood samples from the victim at the time of postmortem examination is within the protocol of the medical examiner or forensic pathologist. Consultation with the pathologist will help ensure that sufficient blood is obtained for all procedures necessary including the toxicologic and serologic studies. The blood should be collected in separate tubes for each purpose with the appropriate preservatives and anticoagulants.
- Authority for collection of blood samples from suspects or defendants is obtained through appropriate legal procedures. The forensic laboratory that will test the samples should be consulted to ensure that proper samples are collected by a physician, registered nurse, or licensed medical technologist with the proper rules of chain of custody observed in each case.

Evaluation of Bloodstain Patterns at the Scene   13

Introduction

Forensic science has been called the art of observation governed by the rules of science. Nowhere is this more true than during the processing of crime scenes. The ultimate goal of any criminal investigation is to create links between the perpetrator, the victim, and the crime scene. In most cases, the crime scene is the first known common denominator — the crime scene is undeniably linked to the victim, to the perpetrator, and to the events that took place. Although no one can deny such a relationship, physical evidence is required to lock these elements together. Occasionally, the clues are readily visible and obvious. More often, however, the desired information will only be found by keen observations confirmed through the results of methodical processing.

Processing Scenes

Before one can effectively process a crime scene for bloodstain patterns or any other evidence, it is necessary to consider the scene in its totality. The fact that the living room is the primary location of bloodshed does not mean your examination should be limited to that area. To the contrary, the initial walk through should include the residence as a whole. We are searching for things that have been scattered or appear to be "out of place" either directly or indirectly as a result of the events under investigation. The quickest, most reliable means of determining what had changed is to observe how things are "normally" in any given area.

A person's home is a prime example of this method of processing crime scenes. The contents and the environment in an individual's domicile can tell you how that person lives. By establishing how a person lives — what is normal in this house — the abnormal or the unusual becomes more readily recognizable. The contents of a residence can certainly be displaced as a direct result of the violent event or as a result of associated actions such as searching for valuables. However, a home in complete disarray may provide valuable insight into the physical or mental condition of the occupant.

In a recent case, the victim was described as a member of a locally prominent family who was discovered in the blood-soaked bedroom of his home by his wife. She stated that they had been living separately for the past 2 weeks and had decided to make that a permanent arrangement. They had mutually agreed that she would come over to collect some of her belongings on that date. Having been unable to reach him by phone, she used her key to enter the residence.

The victim had sustained two gunshot wounds and was completely nude when officers responded. There were no signs of forced entry, and no gun had been found. Suspicion heightened when the wife quickly referred to this as a "tragic suicide." Several members of the victim's family insisted that he would not have committed suicide because of his deeply held religious convictions. They reported that he had been in good spirits and had recently agreed to rejoin the family business. They were certain the wife had murdered him, although her claims of having been with friends during the time in question had been confirmed.

The medical examiner was faced with a difficult ruling—homicide or suicide? Both gunshots were near contact, but neither would have necessarily been fatal. Death resulted from blood loss secondary to the gunshots. In an attempt to resolve this dilemma, the medical examiner requested bloodstain pattern analysis of the residence.

Immediately on entering the home, a problem became apparent. What had been reported as a neat, orderly home 2 weeks prior was in total chaos. One entered via a living room that was stacked with cardboard boxes (Figure 13.1). The contents of the boxes were a hodge podge of clothing and other household goods intermixed with half-empty drinking glasses, recent prescription bottles containing antidepressants, and canned food (some empty, some partially eaten, and some full). The hallway leading to the bedroom was filled with stacks of

Figure 13.1 Living room in complete disarray.

Evaluation of Bloodstain Patterns at the Scene 299

Figure 13.2 Hallway leading to bedroom with decedent's body. Access was limited to one team member per trip as a result of the piles of clothing, trash, etc. in the hallway.

boxes and arrays of refuse, limiting access to a single-file procession (Figure 13.2). After much searching, a recently fired handgun was located on a bookshelf. The scene in its totality provided the most definitive indicators of the mental and emotional status of the victim. The bloodstains examined could certainly shed light on the question of "what" had occurred, namely gunshots and copious bleeding, but could not explain the "why" (i.e., murder or suicide).

The treating psychiatrist was located from the information on the prescription bottles and confirmed that the man had a long history of mental illness and suicidal ideology. He had been described as "peculiar" or "odd," but friends, family, and his estranged wife denied any knowledge of a mental illness. Had the residence not been examined, this information might never have come to light. Although a ruling of suicide is often troubling to loved ones, a definitive ruling as to manner of death is preferential to a ruling of undetermined. In this instance, it may very well have prevented the wife from being wrongly accused of murder.

Most scenes do not reveal this sort of information so readily, but the value of using a total approach to crime scene processing cannot be denied. Every scene is different and unique in the challenges it presents. Recognizing that there is no panacea for every situation, there are some general guidelines that have proved helpful. Because documentation of bloodstain evidence is covered in a separate chapter, this discussion will be limited to the more theoretic aspects of crime scene processing.

General Processing for Bloodstain Pattern Analysis

The philosophy used by journalists of finding the who, what, when, where, and how of an event can serve as a helpful outline of objectives for processing a bloodstain pattern scene. If these basic questions can be answered, it is likely that a scene has been thoroughly processed and the necessary data has been collected.

The first few scenes one processes for bloodstain patterns can initially be overwhelming. Perhaps the most valuable advice is to remember the adage: Forensic science is the art of observation governed by the rules of science. By adopting this as a philosophy, the first action that should be taken is to observe — simply look. Put aside any and all preconceived notions or suggested scenarios and simply "see" the patterns and evidence as they exist. At this step, formulating theories as to the activity that led to the pattern's creation should not enter the equation. The overall appearance alone of the bloodstains can provide key information. For example, does the blood appear to be diluted? If so, ask yourself what is likely to have caused the dilution? If the dilution is thought to be the result of cleanup efforts, is there other evidence such as bloody sinks or rags to support this hypothesis? If the dilution appears to be the result of intermixing with other body fluids, is it the result of cerebrospinal fluid from a gunshot wound to the head or does it potentially represent a sexual assault and mixing with seminal fluid?

Perhaps the stains demonstrate the coffee-ground appearance typically associated with peptic ulcers or other causes of gastrointestinal bleeding. Because the same principles used in bloodstain pattern analysis also apply to other fluids, this is an opportune point to consider the role that other stains such as accelerant patterns at an arson scene may play. Simple observation can also provide data needed to sequence events. What stains overlie other stains or patterns? Once causative mechanisms are added to the initial observations, this information can tell the order in which events occurred. Perhaps the presence or absence of stains will yield information about positioning. Once a victim becomes sufficiently bloody, it is highly unlikely that they could have moved into areas that are devoid of blood. Or perhaps a drag mark combined with the absence of impact spatters is evidence that the victim was moved after the assault ended (Figure 13.3).

Simple observation can provide telling information about the surroundings. For example, stains on the thresholds (Figure 13.4) or the edge of doors (Figure 13.5) can prove that the door was open during the assault.

The observation of bloodstains in one area of a house can be an indicator that other areas should be more carefully scrutinized. For example, the presence of what appeared to be a carefully washed telephone in the kitchen sink at one scene was initially thought to mean nothing more than an effort to remove any fingerprints (Figure 13.6). A soda can on the kitchen counter was not congruent with what appeared to be the extremely neat habits of the homeowners. It was further noted that there was a broken pop tab lying next to the can and a diluted bloodstain on the top of the can (Figure 13.7). The realization that the perpetrator had apparently taken the time to get a soda and was possibly bleeding led crime scene investigators to theorize that there might have been more time and effort put into cleaning after the murder than had initially been thought. In the laundry closet adjacent to the kitchen, investigators had noticed a pair of household rubber gloves hanging to dry. Initially this action had been attributed to the homeowner because of the fastidious nature of the home (Figure 13.8). With suspicions of more extensive cleaning having occurred,

Evaluation of Bloodstain Patterns at the Scene

Figure 13.3 Body of shotgun victim was found lying at pool near the bottom left of the photograph. Another large pool of blood is barely visible at the top of the photo (in the hallway). Drag marks show the path along which the victim was dragged from the hallway to the living room.

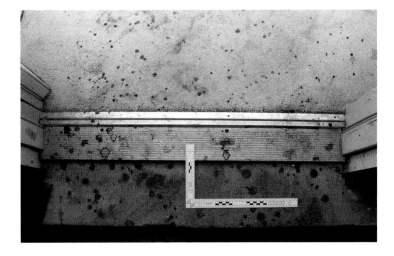

Figure 13.4 Spots of blood and skeletonized stains on threshold show the door was open at the time of the incident.

Figure 13.5 Bloodstains on the hinge side doorjamb show the door was open during the incident.

the gloves were examined more closely, and small bloodstains were noted on the interior. Analysis of the stains proved to be the only biological evidence linking the defendant to a double murder.

Initiating scene processing by limiting oneself to a focused observation of the stains and their perceptible characteristics can also prove invaluable in helping the analyst overcome the anxiety and sense of confusion that may be caused by those first glimpses at a jumbled up mess commonly referred to as a crime scene. No amount of study or completed laboratory

Figure 13.6 View of kitchen. Telephone in the sink appeared to have been washed. The doorway beside the refrigerator led to laundry area where gloves were found.

Evaluation of Bloodstain Patterns at the Scene

Figure 13.7 Pepsi soda can on kitchen counter. Pop top broken and diluted blood on top.

data sheets can fully prepare someone for those initial feelings of bewilderment precipitated by some crime scenes. Before one can possibly begin, it will be necessary to evaluate physical characteristics of the stains, and this can only be accomplished by careful examination. Making observation the first task will not only help to replace any anxiety with more directed concentration, but it will also help the analyst to break down the monumental mixture of information at a scene into simpler, more digestible pieces. The immediate benefit of this simplification process is that it becomes much easier to identify the elusive, but necessary,

Figure 13.8 Household cleaning gloves found in laundry area. Gloves yielded the defendant's DNA.

starting point. Once chaos begins to be replaced by order, the analyst will be ready to assimilate the many pieces into useful facts.

The next step in the process is to apply scientific principles to the observed patterns. This is the mechanism by which the "how" is answered. Does this pattern demonstrate characteristics of impact spatter or does it exhibit the characteristics of a transfer stain? If it is consistent with impact spatter, what mechanism caused its formation? Is there evidence to support either the presence or the absence of possible alternative mechanisms such as fly activity or the expiration of blood?

The third step is to stop and think. Think about what you now know about the case by virtue of what has been observed combined with what the application of the rules of science tells you about these observations. Is this information consistent or inconsistent with other evidence at the scene or with previous theories about the events that occurred? Does the information at hand indicate other areas that should be examined? Maybe the axe used in the murder has been stowed away in a storage shed not previously considered an integral portion of the crime scene.

In many ways, the steps proposed are a basic outline of the scientific method: observation, hypothesis, collection of data, analysis of data, and decision. This is not only good procedure, but if correctly applied, this approach will minimize the effects of any prejudice or bias. Although presented as a stepwise process, there may be considerable overlap and a virtual forest of decision trees involved in actually processing a scene. Acknowledging that every scene is unique and may require a good deal of innovative thinking or actions, the basics of the scientific method must still be followed. Every decision must be founded in light of the data collected and the application of science—not influenced by the desired outcome overtly or otherwise desired by investigators or attorneys. Once a decision is made, it must then be diligently and conscientiously scrutinized for possible alternative explanations. Only after religiously following the principles of the scientific method and exhaustive reflection can the ultimate goal be reached—sound, scientifically correct conclusions and testimony.

Processing for Data

In the many texts dealing with crime scene processing, avenues of access or egress are invariably mentioned as areas to note. How does one determine which door was used and when? Bloodstain pattern analysis may be able to provide that answer, but more often in spite of our best efforts, we are not able to answer such questions. In some instances, we will reach certain conclusions but never discover what, if anything, they mean. The simple fact remains: in the absence of effort, valuable information could be lost. The recommended philosophy in processing a scene for bloodstain pattern analysis is much like the rules governing collection of physical evidence. If you fail to collect a piece of physical evidence simply because its potential value is not congruent with the current theory, that evidence may be lost forever. This is especially painful when new facts or new theories develop and the evidence to support or refute the theory is no longer available. It is better to have volumes of valid, supportable observations that are never used in trial because their role is never established than to miss that one crucial bit of information because it did not seem to "fit" at the moment. In this vein, there are several specific areas or locations that have repeatedly been found to provide helpful, sometimes crucial, information. Following is a description of such locations for the reader's consideration.

Evaluation of Bloodstain Patterns at the Scene 305

Figure 13.9 Blood transfer on doorknob and locking mechanism.

Doors and Locks

As previously mentioned, entrances and exits are important parts of any scene investigation. The presence of bloody transfer stains on the doorknob (Figure 13.9) may show which door the assailant used to exit the scene. Similar stains on or around thumb latches (Figure 13.10) or locking mechanisms can be used to determine if a door was locked or unlocked and at what point in the sequence of events did this occur. If the lock is engaged by turning the thumb latch clockwise, the corresponding sides of the latch may be bloodstained if the door was locked after the bloodshed commenced. Conversely, transfer stains on the opposite sides of the thumb latch can confirm that the door was unlocked afterward.

Figure 13.10 Blood transfer on thumb latch for door lock.

A couple of words of caution are in order. Doors can certainly be used without any traces of blood being transferred to the surface. This is an instance when a negative finding may be meaningless. Second, doorknobs and locks are typically constructed of metal. Traces of blood can be readily removed from metal surfaces from even minimal disturbances. Such examinations may be aided with the use of magnification, additional lighting, or presumptive tests. Examination and documentation of points of entrance and egress should be accomplished early in the investigation. Inevitably, some of these doors will be necessary for investigators to enter and exit the scene during the course of the investigation.

Refrigerator

James Spencer Bell, MD, insisted that the refrigerator contents should be examined at every scene. Initially, it seemed that this was most likely nothing more than Dr. Bell's epicurean nature, but time has shown the wisdom of his advice. One such inspection recorded a plate of uneaten takeout food. The contents and the source restaurant were identical to those described as having been brought to the victim by one of the suspects. The suspect had not only voluntarily offered detailed information as to the restaurant and the items purchased but had also claimed to watch the victim consume the meal. This discrepancy gave investigators the leverage needed to effectively confront the suspect. An overabundance of bland food or dietary supplements may be the only indicator of the resident's recent state of health. Empty, obviously neglected refrigerators may confirm long-term neglect or drug and alcohol abuse. Refrigerators are also commonly used to hide valuables and drugs. If a perpetrator has searched the refrigerator for loot or possibly even prepared a snack, the refrigerator handle may bear witness to the perpetrators bloody hands. In one instance, the perpetrator actually made sandwiches and left his blood around the lid and on the top of the pickle jar (Figure 13.11).

Figure 13.11 Bloodstains around and on top of the lid of a pickle jar were matched to the suspect. Examination of the refrigerator at the crime scene yielded this and other evidence that the assailant had made a snack before he left.

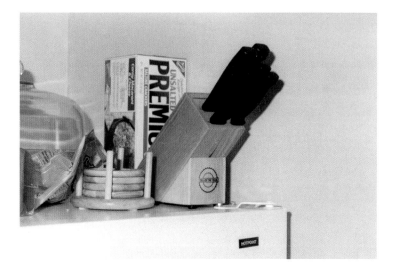

Figure 13.12 Knife is missing from the set in the knife block. This knife was found in the floor of the apartment (see Figure 13.13).

Knife Blocks

The household knife block (Figure 13.12) often serves as a convenient supply source for the assailant. Whereas the kitchen drawer requires a bit more effort, determined assailants are generally able to find a sharp knife in most kitchens. In some cases, the weapon is left at the scene and can easily be matched to the remaining knives (Figure 13.13). If the perpetrator removes the knife from the scene, it may later become necessary to compare the murder

Figure 13.13 Knife in floor was found to be the knife missing from the knife block (see Figure 13.12).

Figure 13.14 Bloodstains were found on the front and the sides of the drawer containing the kitchen cutlery, as well as on several utensils in the drawer. The assailant had rummaged in the drawer for a second weapon.

weapon to other knives in the home. For this reason, an inventory and description of the household knives, supplemented with photographs, are recommended in stabbing cases.

The knife block and the cutlery drawer should also be examined for the presence or absence of bloodstains (Figure 13.14). Such stains may be the result of bloody hands seeking an alternative weapon. They may also indicate that the murder weapon has been returned to its original location in an attempt at concealment. In one case, the killer had actually put the bloody knife in the dishwasher (Figure 13.15).

Figure 13.15 Transfer stains on the side of the dishwasher and on the racks of the dishwasher. In the back of the utensil basket (barely visible behind the table knifes in the basket) a bloody steak knife was recovered.

Figure 13.16 Photographic record of medications observed in a residence. This photographic record was supplemented with a written list.

If the murder weapon remains missing or has not been confirmed, any other knives present should be thoroughly examined for blood residue. Such an examination may be conducted at the scene with or without the aid of magnification or may necessitate laboratory submittal for stereomicroscopic examination.

Medicines/Medicine Cabinets

The presence of certain prescription drugs may provide insight into the mental state of an individual when close associates are unaware of any problems. Other prescription medications may be present because of medical conditions such as heart disease or chronic pain (Figure 13.16). Excessive quantities of commonly abused drugs may be indicators of suicidal plans or drug abuse. Drug abuse will often present with multiple dispensing physicians, overlapping fill dates, or frequent fill dates.

Medicine cabinets are also the customary location for first aid supplies. If a perpetrator has been injured, bandages or other medical supplies may be sought. Bloodstains on the edges of cabinet doors or discarded wrappers can be valuable sources of isolating the perpetrator's blood (Figure 13.17). In one case, the perpetrator never touched the

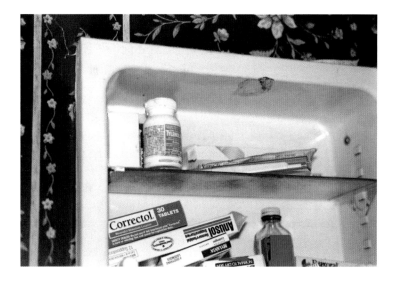

Figure 13.17 Transfer stains at the inner top of the medicine cabinet and at various other areas in the medicine cabinet at a double murder scene.

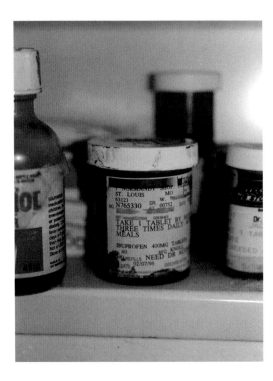

Figure 13.18 On the bottom shelf of the medicine cabinet, transfer stains indicated that this bottle had been opened. The assailant bypassed narcotics but apparently needed some ibuprofen.

narcotics in the medicine cabinet but apparently did feel the need for some ibuprofen (Figure 13.18).

Sinks and Faucets

Almost invariably anyone who seeks to apply first aid to an injury will first wash the wound. For this reason, sink traps may contain diluted blood that can be matched to the bleeding individual (Figure 13.19). The underside of the protruding faucet can also be stained with diluted blood as it splashes upward from the sink. Inspection of this area is easily accomplished using a mirror and supplemental lighting.

In some situations, the assailant or the victim may bathe. Tubs, showers, and drains should be examined for such evidence (Figure 13.20). The finding of bloodstained items such as bloody bath towels may indicate such activity. A bathroom void of towels may also be a valuable clue that cleaning has taken place—either of persons or perhaps of areas or items within the scene (Figure 13.21).

Even if visible traces of blood no longer remain in or around sinks or tubs, a second examination using an alternate light source or luminol may prove fruitful. If the examiner thinks through the process of using sinks and tubs, it will be readily apparent that water is supplied only after manipulation of the handles. Bloody transfers on control handles or knobs (Figure 13.22) are not as susceptible to being removed during washing and therefore more apt to reveal less diluted stains. Even if caution has been taken to remove obvious bloodstains, the perpetrator may overlook unanticipated stains such as bloody foot transfers (Figure 13.23).

Evaluation of Bloodstain Patterns at the Scene

Figure 13.19 Diluted blood in and around a bathroom sink where an assailant washed a wound.

Figure 13.20 Bloodstains in shower stall drain used by assailant before leaving the crime scene.

Trash Cans

Perpetrators frequently place valuable pieces of evidence in wastebaskets inside the house or in outdoor trash cans. Bloodstained paper towels have been found in kitchen trash cans that not only revealed blood that could not be tied to the victim's activities but also more immediately armed investigators with information of possible injury to the perpetrator. When the suspect was apprehended, this information was used to question him regarding a visible cut on his hand. Coupled with finding an identical roll of paper towels in his residence,

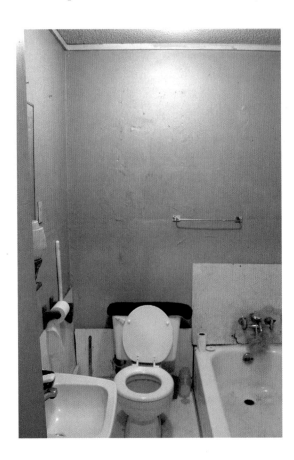

Figure 13.21 The bathroom had several indicators of recent use, but no towels could be found. It was later learned that the assailant had cleaned the sink and tub and had taken the washcloth and towels when leaving the scene.

Figure 13.22 Diluted bloodstains on handles of bathroom sink. Also note the transfer stain on the knob to engage the stopper and very small stains on the back of the sink just to the left of the faucet.

Figure 13.23 Bloody footprint left by perpetrator who showered before leaving crime scene. Towels, etc. were taken, but the footprint was apparently overlooked.

the suspect confessed to the murder months before DNA analysis of the bloodstains could be completed.

The murder and partial dismemberment of a young woman took place in the home of the man with whom she had spent the night following their first date. The roommate who reported overhearing a violent confrontation placed the 911 call that alerted police. The roommate had already fled the residence but agreed to meet officers on the street. First responders were on the scene within minutes. As they were approaching the door of the residence, officers heard a good deal of movement inside the residence. On entry, they observed the obviously deceased female and the male suspect lying on the floor. As officers approached, he began to move about and to hold his head. He immediately began to ask what had happened to his date. He claimed to have been rendered unconscious by someone who knocked at the door and he asserted that he knew nothing of the murder. Although officers were unconvinced by his story, the man was transported to the emergency room where physicians confirmed the complete absence of any head trauma or other injury.

During the usual scene processing and documentation of the bloodstains inside the residence, the sequence of the patterns was followed as they led to the back door. Obviously, remote from the site of the actual assault and attempted dismemberment, the reason for the bloodstains in this location was a source of curiosity. Stains on the doorknob and the night latch confirmed the use of this door by a bloody individual. The possibility that the bloodstains represented an attempt by the victim to flee her attacker was discounted by the fact that the stains consisted entirely of extremely diluted blood. Outside the back doorway, the bloodstains that had previously been plentiful were noted to abruptly cease. Some sort of receptacle or container would be necessary to cause such a sudden dichotomy in stain distribution. The trash can immediately to the right would certainly suffice (Figure 13.24). Examination of the contents of the trash can revealed the victim's clothing, jewelry, and one of her broken fingernails (Figure 13.25).

Figure 13.24 Dark green, rolling trash receptacle outside the kitchen door of the scene. This receptacle is used to collect trash for curbside pickup.

Figure 13.25 Contents of the rolling trash receptacle (Figure 13.24) included the victim's clothing, jewelry, and one of her broken fingernails.

The male was confronted with the unlikelihood that an unknown assailant, with no discernable motive, who had completely eluded detection, had killed a woman using a barbell weight from his closet and then attempted to dismember her in only a matter of minutes. It was even more preposterous to believe that such an assailant would exert the time or the effort to dispose of items that would link the victim to the residence. He eventually confessed to the crime and was convicted of capital murder.

Temperature

The need to record the temperature, humidity, and degree of air flow at a scene are obvious in light of the influence these factors play in issues centered on drying times. The perpetrator

Evaluation of Bloodstain Patterns at the Scene 315

may also alter thermostats and fans for reasons that are self-serving. An assailant may adjust thermostats because the noise generated by heat and air conditioning can help muffle the sound of the assault. It may also be an attempt to cool the environment and prolong the amount of time before the odors of decomposition lead to discovery of the body. Switches on thermostats that control heat, air conditioning, or fans should be inspected for possible evidence transfer bloodstains or fingerprints.

Lighting

The presence or absence of lighting can play a role in ascertaining how well a perpetrator could see. This information can be critical. In one investigation, there was testimony that the accused would have been unable to see certain evidence because the area was completely without light. Photographs that were taken during the processing of the scene clearly showed the presence of a suspended light that was illuminated.

Perpetrators may leave bloody transfers on light switches or switch plates when they turn the light on or off. In Figure 13.26 the light switch had certainly been flipped up to the "on" position. Such observations may also provide clues about the time of day the crime was committed. Did the perpetrator need to turn the lights on to search for valuables? Perhaps no lighting was needed because they were very familiar with the surroundings.

In a different turn of events, it was important to determine that the light had actually been turned off before police arrived. The homeowner claimed to have surprised and shot an unknown intruder inside his home. He asserted complete shock when he was told that the deceased was actually his former business partner (who just happened to be currently dating the shooter's estranged wife). For reasons that should be fairly obvious, investigators were suspicious of this story. When questioned as to how he could have failed to recognize the victim, the homeowner stated that it was dark when he entered his home and heard someone moving about. He said he fired three shots in complete darkness, heard a thud, and then left to call police from a nearby convenience store. He claimed to have never gone

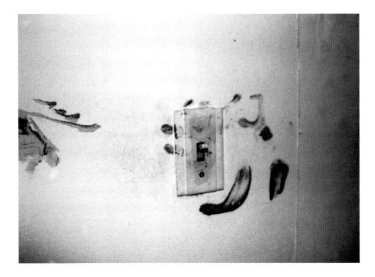

Figure 13.26 Bloody transfer stains consistent with flipping light switch up, into the "on" position.

any further inside the house and had not even turned on the lights. A very faint bloodstain on the light switch was consistent with the light having been turned off after the shooting. The shooter later admitted to having arranged the meeting, shooting his former business partner, and then turning the lights off to support his claim of not knowing the identity of the alleged intruder.

Transient Evidence

Early responders may notice odors or other evidence that could disappear or no longer be of noticeable proportions by the time the scene processing begins in earnest. Everyone admitted to a scene should be cognizant not only of what they see but also of what their other senses detect. There also needs to be an atmosphere conducive to sharing this information with other members of the processing team. Odors such as bleach or other cleaning agents may have dissipated by the time a search for bloodstains can begin. One scene had puzzling gaps in bloodstain patterns, but none of the usual indicators of cleanup could be found. A rather unpleasant, close inspection of what appeared to be simply an old, very dirty bucket of mop water prompted one officer to mention that he had also wondered if the mop water was significant. He continued by saying that, although it was no longer noticeable, the smell of bleach was almost overpowering when he first arrived. A simple presumptive test for blood proved the value of his relaying that information.

Cigarette butts still smoldering or food or drinks that remain at the usual serving temperature may be of great significance but no longer apparent by the time the full team has arrived. Bloodstains that were wet initially may have dried. Other stains may have been altered or even created by necessary medical intervention or by the transport of the victim. Even the best efforts to preserve a scene will not be perfect. If every member of the team remains vigilant in their observations and documentation and conscientious about exchanging such information, valuable evidence may be salvaged or wasted effort may be eliminated or minimized.

Sequencing

Observations related to sequencing frequently fall into the category of "we know what happened, but don't have a clue what it means." It can also be the seemingly insignificant factoid that later proves crucial. In one case, a claim of self-defense was refuted because of the observable sequence of events. According to the defendant, he shot the victim in self-defense as the victim was advancing toward him with a knife. The simple fact that the knife remained in the victim's hand after falling from a shotgun wound that destroyed his brainstem was immediately suspicious (Figure 13.27). Doubt was further increased when it was learned that a previous injury had left the victim with only minimal grasping movement in that hand. The presence of spatter in the palm underneath an unstained knife was decisive in refuting the defendant's claims (Figure 13.28).

In another case, it was noticeable that several items had been moved at the scene of a brutal stabbing. Notably, a pair of glasses was observed on the dresser that was overlying spatter (Figure 13.29). This finding led to several additional observations that indicated the perpetrator had most likely spent the night inside the residence after the murder. The victim's girlfriend had been eliminated from suspicion because several people had verified her whereabouts at the time of the murder and the evening that followed. The victim's

Evaluation of Bloodstain Patterns at the Scene

Figure 13.27 Knife remaining in the hand of an individual felled by a shotgun wound that destroyed his brainstem. The shooter claimed self-defense because the victim was advancing with a knife.

drug buddy achieved the status of primary suspect. Largely as a result of his lifestyle, he had been unable to produce any reliable alibis or other plausible explanations. An arrest seemed imminent. The question loomed, however—would he feel so certain that he would not be discovered that he would have been comfortable staying in the residence overnight? Overwhelmingly the investigation said "no." The suspect was known to be very paranoid

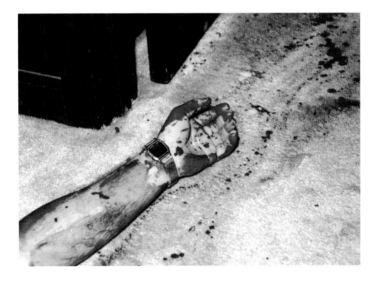

Figure 13.28 Spatters underneath the knife being held in the hand in Figure 13.27. The knife was noted to be free of bloodstains despite the spatters underneath the knife. It was later learned that the decedent had only minimal ability to grip with this hand as the result of a previous injury.

Figure 13.29 Impact spatter on the dresser scarf and the blue and silver pen. No bloodstains were found on the reading glasses, confirming their placement after the assault.

and would not even stay after their mutual drug use. They would do the drugs together, and then he would retreat to his apartment.

The investigation refocused on the girlfriend—someone who *would* feel comfortable and secure enough to stay overnight. This time the alibis collapsed, and the girlfriend confessed to the murder. The glasses turned out to be a pair of reading glasses she needed for small print. She had inadvertently placed them on top of the impact spatter and implicated herself.

Sequences as shown in Figure 13.30 could also be telling. The presence of bloodstains on the uppermost portion of the slacks was only considered unusual in that they did not "match" what was visible in that immediate vicinity. The original margin of the slacks was

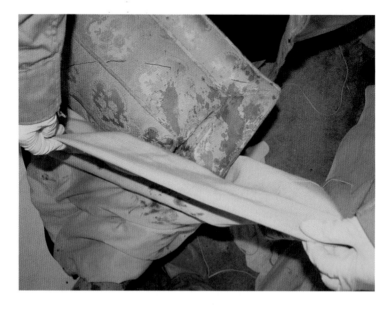

Figure 13.30 Slacks that had been placed over the bloodstains on the sofa. The black marks outline the original position of the slacks.

Evaluation of Bloodstain Patterns at the Scene 319

Figure 13.31 Coins that had been dumped on top of bloodstains. The cord is attached to a telephone that had also been moved after the bloodshed.

outlined on the sofa back (the black line visible), and the slacks were then carefully lifted. Heavy bloodstains were present underneath the slacks, but the blood had dried sufficiently that none was transferred onto the slacks. In that same crime scene, several coins were obviously dumped on top of the bloodstains on the floor and the telephone had been moved such that the cord was placed over a heavily stained area (Figure 13.31).

Findings of this nature are certainly indicative of an assailant who remained at the scene for a period of time after the assault. Blood drying time experiments may be in order if there is a need to establish the minimum length of elapsed time. Either way, this is not an instance of the perpetrator fleeing immediately. As in the previous case, it could indicate that the perpetrator feels comfortable and relatively safe from detection — characteristic of an intimate acquaintance. A second alternative is that the perpetrator stayed and moved objects because they were searching for something — perhaps valuables, drugs, or incriminating evidence. Regardless, a new avenue of investigation has been opened as a result of proper interpretation of the bloodstain patterns.

Vehicles as Crime Scenes

One prominent bloodstain pattern expert has described crime scenes contained within vehicles as closely resembling serious beatings within the confines of a very small box. No doubt the spatial limitations can be challenging. In one case, the pathologist misinterpreted the tiny satellite spatters created by an accumulating pool of blood being deposited on the interior of a door as spatter resulting from the gunshot (Figure 13.32). Apparently his conclusions failed to consider some very important things. This posture would have required the head of the victim to be forcefully jammed inside the 2- to 3-in. gap between the seat and the necessarily closed door. Had this been the case, the gun that fired the fatal shot would have to be small enough to fit inside this small space, yet powerful enough to fire the fatal bullet. At the same time, his conclusions would have required that the victim's head be trapped within this small space — no doubt resulting in a blunt trauma injury not

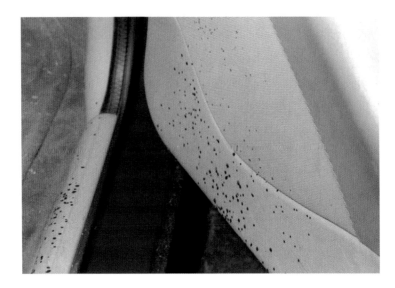

Figure 13.32 Satellite spatters in a car incorrectly identified as impact spatter created by a gunshot.

reported in the autopsy report. Also, after slamming the victim's head in the door, why would a gunshot have even been necessary? All of this over nothing more than satellite spatters that would have probably gone relatively unnoticed in more spacious areas. Such error can often be prevented by examination of the actual vehicle or a similar vehicle in conjunction with the scene photographs.

Although they can certainly present problems, spatial limitations can also be helpful to the bloodstain pattern analyst. In one case, an individual seated in the driver's seat of a vehicle received a tight contact gunshot wound to the left temple. Initially information pointed to a suicide, but questions later arose about the possibility of the shot having been fired by an individual standing in the open doorway.

Largely because of the stellate tearing produced by the tight contact gunshot wound, a large amount of back spatter had been produced. The very heavy distribution of the back spatter on the interior of the driver's door was completely uninterrupted. It was further noted that the driver's side window was also heavily stained, precluding the shot having been fired through an open window. The initial ruling of suicide was upheld.

As a note of precaution, it should be added that in the aforementioned case the possibility of the victim's head being turned and the shot having originated from the passenger side of the vehicle was eliminated by other factors. More importantly, the vehicle in question was a two-door sedan. Had the vehicle been a four-door vehicle, the possibility of a shooter being in the seat behind the driver would have to have been considered. Beware of four-door, hatchback, or rear-door access vehicles.

On a bright note, at least vehicles represent a crime scene that has finite boundaries. In reality, however, vehicles represent crime scenes with the ability to move at speeds normally quantified in miles per hour, scenes that may stop at unknown locations and be thoroughly cleaned or be moved to remote locations and burned. There is a true continuum of possibilities, all of which should be kept in mind when processing vehicles.

Once again, we need to establish some logical starting points. This exemplifies the need to observe, apply the rules of science, and then think. On initial observation, some bloodstains

Figure 13.33 Transfer stains made by opening car door.

may be readily apparent; more elusive stains will require thought. If the situation concerns movements by a possible driver, think about the areas one routinely touches when driving a car. It can even be helpful to pantomime these movements. Do not anticipate; just go through the motions. We do the same things and touch the same surfaces every day in the process of driving our cars, and perpetrators are no different. To drive a car, one must access the car. Door handles, both exterior and interior, are one of the first places that one would touch. Figure 13.33 shows an example of bloodstains created when the car door was opened. Look carefully around the ignition switch, gear shift, and steering wheels as well (Figure 13.34). Anything one might use or touch when driving a car is a potential source of evidence — radio control buttons, heater or air conditioner switches, and rear view mirror, to name a few.

The same principles apply to situations of suspected body transport. Look for the obvious stains first, and then stop and think about the activities that would have occurred. Stains in the interior of a trunk are certainly subject to notice and removal by a perpetrator. In Figure 13.35 bloodstains are visible on the lip of the truck, and in the background a portion of the carpet that had been pulled away by the suspect is also visible.

Stains on the underside or in between the two raised portions of the trunk gasket are more apt to be overlooked (Figure 13.36). The term "dead weight" bears more truth than most people realize. Moving a dead body, especially without assistance, will lead to more dragging and scraping across surfaces than actual lifting. These limited motions are prone to leaving traces not anticipated and therefore not removed, by killers.

An interesting case involved a 4-year-old child who was killed by the maternal aunt in what supposedly was an act of revenge directed at the mother/sister. Authorities verified that the aunt had cleaned the car trunk, and the body of the child has never been found. One very determined prosecutor insisted that the aunt's car be inspected although cleaning

Figure 13.34 Transfer stains on steering wheel of car.

was known to have occurred. As a result, bloodstains were located underneath the trunk gasket and between the gasket and the metal body of the car. Apparently the gasket had been inverted when the child's body was dragged into the trunk. Once the pressure was released, the gasket resumed its normal configuration and the bloodstains were not visible. DNA matched these stains to the missing child, and even in the absence of a body, the aunt was successfully prosecuted.

Figure 13.35 Bloodstains on trunk gasket and trunk lip when a body was transported in car trunk. In the background the carpet from the trunk had been pulled up from the adhesive backing, probably for disposal, but remained in the car when the suspect was arrested.

Evaluation of Bloodstain Patterns at the Scene

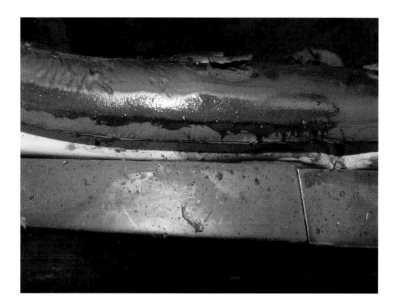

Figure 13.36 Transfer stains on lip of trunk and between the two raised portions of the trunk gasket.

Negative Checklist for Vehicles

In some situations, the significant findings when processing cars is limited to a negative visual examination or negative presumptive testing for blood. The necessary notetaking in such cases can be laborious and difficult. Negative checklists (Figure 13.37) can greatly simplify and expedite this process. A simple, quick checkmark will serve as a historical reminder that these areas were or were not examined and did or did not reveal stains.

Collisions

Seatbelts

When collisions occur, a host of events has potentially affected the bloodstains and the time is right to go back to the basics. Was the victim restrained by a seatbelt or free to move about the passenger compartment under the influence of prevailing forces? A simple inspection of the exposed lap and chest portion of the seatbelt for the presence or absence of anticipated bloodstains may provide valuable information. In Figure 13.38 the continuation of bloodstains onto the belt and the doorpost behind the belt confirms the seatbelt was not in use at the time of the bloodshed. If paramedics have extracted a victim, they may cut the belt away but the seatbelt catch will remain fastened. This is important information waiting to be discovered (Figure 13.39 and Figure 13.40).

Airbags

Once a crash has occurred, the airbags usually will be deployed. In most models, the release of the airbags is dependant on significant frontal impact. Impacts directed toward the sides of a vehicle, or even at a sharp enough frontal angle to avoid bumper contact, may result in a failure of the airbag system. There are also many late-model cars remaining in use that

```
                              PAGE ____ OF _____          LAB #: _____
NO SUSPECT STAINS CHECKLIST------VEHICLE INTERIOR
         Check areas below that were examined and did not exhibit suspect staining
LEFT FRONT--DRIVER SIDE              RIGHT FRONT--PASSENGER SIDE
_____ Seat                         _____ Seat
_____ Side of seat __ Left __ Right _____ Side of seat __ Left __ Right
_____ Front of seat                _____ Front of seat
_____ Center Armrest               _____ Center Armrest
_____ Seat Backrest                _____ Seat Backrest
_____ Side of seat back __ Left __ Right  _____ Side of seat back __ Left __ Right
_____ Back of Seat Back            _____ Back of Seat Back
_____ Floor Mat                    _____ Floor Mat
_____ Floorboard                   _____ Floorboard
_____ Running board                _____ Running board
_____ Headliner                    _____ Headliner
_____ Visor                        _____ Visor
_____ Upper portion of Door        _____ Upper portion of Door
_____ Midsection of Door           _____ Midsection of Door
_____ Armrest of Door              _____ Armrest of Door
_____ Lower portion of Door        _____ Lower portion of Door
_____ Door Knob                    _____ Door Knob
_____ Door Lock                    _____ Door Lock
_____ Window                       _____ Window
_____ Window Knob/Control          _____ Window Knob/Control
_____ Windshield                   _____ Windshield
_____ Dash                         _____ Dash
_____ Steering Wheel               _____ Front Ashtray
_____ Gear Shift                   _____ Empty
_____ Turn Signal                  _____ Contained_____
_____ Accelerator                  _____ Contents recovered
_____ Brake pedal
_____ Clutch pedal

LEFT REAR SEAT                       RIGHT REAR SEAT
_____ Seat                         _____ Seat
_____ Side of seat __ Left __ Right _____ Side of seat __ Left __ Right
_____ Front of seat                _____ Front of seat
_____ Center Armrest               _____ Center Armrest
_____ Seat Back Rest               _____ Seat Back Rest
_____ Side of seat back __ Left __ Right  _____ Side of seat back __ Left __ Right
_____ Floor Mat                    _____ Floor Mat
_____ Floorboard                   _____ Floorboard
_____ Running board                _____ Running board
_____ Headliner                    _____ Headliner
_____ Upper portion of Door        _____ Upper portion of Door
_____ Midsection of Door           _____ Midsection of Door
_____ Armrest of Door              _____ Armrest of Door
_____ Lower portion of Door        _____ Lower portion of Door
_____ Door Knob                    _____ Door Knob
_____ Door Lock                    _____ Door Lock
_____ Window                       _____ Window
_____ Window Knob/Control          _____ Window Knob/Control
_____ Back Window                  _____ Back Window
        _____ Ashtray                      _____ Ashtray
        _____ Empty                        _____ Empty
        _____ Contained_____           _____ Contained_____
        _____ Contents recovered           _____ Contents recovered
```

Figure 13.37 Negative checklist used for processing vehicles. This checklist is used to note that areas were inspected and no bloodstains were discovered.

Evaluation of Bloodstain Patterns at the Scene

Figure 13.38 The continuation of spatter on the seatbelt and on the doorpost visible behind the seatbelt confirms the belt was not being worn by the occupant.

are not equipped with airbag features; some models have only a driver's side airbag; and sometimes airbags do not deploy because they have been expended in a previous collision. Complete failures of intact systems can also occur.

The goal of an airbag safety system is to deploy quickly enough to interrupt the forward motion of the occupant and prevent more injurious contact with the steering wheel, dashboard, and windshield. The nitrogen gas produced by a sodium azide reaction inflates the nylon or polyamide bags. Manufacturers maintain that the smoke sometimes reported after a collision is actually talc that was packed with the deflated airbags to facilitate a more efficient unfurling.

To successfully halt the forward progress of the occupant within such a short distance, the airbags must be able to move quickly. Their speed had been cited as between 150 and 200 miles per hour by some manufacturers. If an airbag moving at such speeds strikes a blood source, very small impact spatters will be produced (Figure 13.41).

The occupant is driven backward in response to the force exerted by the airbag. This rearward force will direct spatter along the same vector, causing stains on the headrest, seat back, and/or rear compartment. The physical location of most airbags, in relationship

Figure 13.39 Seatbelt that had been cut from victim by paramedics.

to the occupants, will produce a rearward force that also has an upward component. This upward vector may direct bloodstains onto the headliner of the vehicle (Figure 13.42).

One additional note on this topic: if the reader requires more specific information on airbags, experience has shown it to be well worthwhile to contact the automotive manufacturer. The safety component of their engineering department can provide valuable data. Specifically request and scrutinize any videotape they may be able to provide.

Figure 13.40 Seatbelt catch remains fastened, indicating the seatbelt was fastened.

Figure 13.41 Airbag that impacted an individual injured by a gunshot prior to the car crash. Initial impact with the bag left hair transfers near the bottom of the airbag. The very small spatters (at approximately the 1 o'clock position) were produced as a result of the impact of the bag with the victim's head which was already bloody.

Effects of Motion and Impact

When the forward travel of a car is rapidly decelerated or stopped as the result of an impact, the unrestrained occupant will continue to travel forward—thus the necessity for airbags to reduce injury. Any blood or other fluid that might be present such as vomitus, will also be directed forward and may account for stains seen on dashboards or the interior of the front windshield. The effect of this sudden deceleration/stopping is readily visible in Figure 13.43. In this case, the driver of a speeding car sustained a gunshot wound to the head. When his

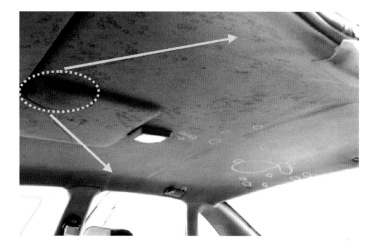

Figure 13.42 Bloodstains created on the headliner of the vehicle as a result of the airbag striking the bloody occupant.

Figure 13.43 Rear left door of a sedan showing blood flow emanating from the driver's injury. These flows are directed toward the rear of the car (from left to right) as the car continued down the roadway. The smaller lines of blood flow, directed toward the front of the car (right to left), were created when the car impacted a house.

head fell toward the open driver's window, blood was deposited on the left rear door. As the car continued forward, the flow of the blood was directed rearward along the prevailing air currents directed around the body of the car. The car subsequently crashed, and the blood rapidly reversed directionality and shifted toward the front of the car (Figure 13.44).

Figure 13.44 Closeup view of the stains described in Figure 13.43.

Figure 13.45 A relatively wide cast-off stain on the rear window of a compact, sport utility vehicle created by the whiplike motion of the victim's head when the car crashed. The victim was shot prior to the wreck but continued approximately $1\,^1\!/_2$ blocks before crashing into a wall.

The same forces affect occupants, but occupants differ from blood in their weight. People go forward initially at impact but can subsequently change direction of travel and shift toward the rear compartment. This whiplike motion will generate centrifugal forces that create cast-off patterns. The location of any resultant cast-off is dependant, as always, on the surface that lies tangential to the source. If the occupant remains primarily upright, the head swiveling on the neck is an excellent source of cast-off trails on headliners. Given enough force cast-off trails can reach the rear windshield of a vehicle.

Also characteristic of traditional cast-off patterns, the width of the trail may be influenced by the width of source object. Heads are relatively wide in comparison to a baseball bat and can create very wide trails of cast-off. The cast-off stains on the rear window of a sports vehicle shown in Figure 13.45 is a classic example of the need to consider stain characteristics when considering the types of mechanisms that could have caused the pattern. In this case, the driver was stopped on a city street and was shot in the head by a pedestrian. The driver accelerated to flee the scene, lost control of the vehicle, and struck a wall. The backlash of his blood-covered head created the wide trail of cast-off on the back window.

Processing the Body for Bloodstain Patterns

The evidentiary value of bloodstain patterns on the body is possibly one of the least exploited aspects of the field. The human body has many contours and is capable of movement along many planes, instead of standing still to be stained in only one plane. The head alone is better viewed as resembling a bobble-head toy than as the stationary model suggested by an anatomic chart. All of these factors present a myriad of challenges when studying bloodstains on bodies. They also require the judicious application of precautionary measures before any definitive conclusions are reached. Despite the challenges, a great deal of valuable information can be uncovered by studying the bloodstained body.

Documentation of Bodies at the Scene

If medical intervention is warranted, any alterations in the patterns that result are simply unavoidable situations that must be accepted. Activities that move or change the body in any way should be completely restricted to lifesaving measures until complete documentation—both written and photographic—can be completed. In some jurisdictions, there appears to be some sort of informal contest under way to see how quickly the body of the deceased can be moved from the scene. Unless a quick removal is warranted by special circumstances, wait. Adequate documentation of the bloodstains and the body takes time, just like any other evidence.

Any reliable preliminary medical information available prior to the completion of scene processing can be very helpful. Most pathologists will be willing to share at least a thumbnail sketch of major injuries on a provisional basis. This will allow the bloodstain pattern analyst to examine the scene for any patterns inconsistent with the medical history. Follow-up discussion with the pathologist, or any other specialist consulted, is appropriate as examinations continue or reach completion. This is necessitated not only for confirmatory reasons but also to quickly identify any inconsistencies that may be discovered. Potentially valuable information may be unearthed.

Bodies are another form of physical evidence and, as such, should be photographed from overall, mid-range, and closeup perspectives. Inclusive photographs must represent all views as well—from both sides, from the head down, and from the feet upward. In many scenes, it is difficult to arrange perspective directly downward, or a bird's eye view. Ladders or photographic booms may be required. It is also helpful if the photographer recognizes that a shot inclusive of the entire length of a body is apt to be too distant to provide sufficient detail. Full-length photographs are needed but should be supplemented with photos of less distance. Think of body photographs as presenting the same challenges as panoramic landscape shots. By using a quadrant approach—overlapping photos of the head, chest, abdomen, thighs, legs, feet, arm, and hands—photos can be taken from shorter distances and show greater detail.

In evaluating one's photographic efforts for completeness and accuracy, remember that it may be just as important to establish the *lack of bloodstains* as it is to establish their presence. Photographers do not usually think in terms of capturing an image of "nothing." The absence of bloodstains may not represent a good photographic subject, but such data can be pivotal to the bloodstain pattern analysis.

Be mindful that the various body parts also have many different surfaces. For example, photographs of hands should include the dorsal and ventral surfaces, lateral and medial edges, and similar angles of the fingers (Figures 13.46a–f).

Use of Scales

Because of the variety of contours, a body presents unique challenges with scale photographs. A scale is used in a photograph as a measuring tool. It needs to be placed in the same plane as the object being studied (Figure 13.47). For the bloodstain pattern analyst, this means that the scale should be placed in the same plane as the bloodstains—not in the same plane as the body part. Figure 13.48 is a good example of the scale being placed in the wrong plane. Be mindful, too, that the analyst must be able to read the scale for it to be of any value. Photos taken at excessive distances or not having the scale in focus are useless for measurement purposes (Figure 13.49).

Body Removal

When the body is ready to be moved, placing a protective sheet of plastic or even an open body bag as close as possible to the body can minimize cross-contamination of

Evaluation of Bloodstain Patterns at the Scene 331

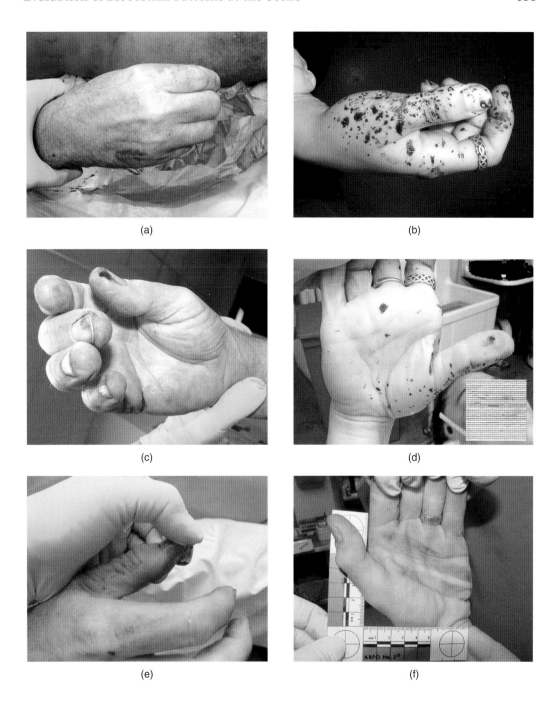

Figure 13.46 Hands have many different surfaces that should be examined, documented, and photographed. a) Back of right hand; b) Web of thumb; c) Palm of hand; d) Palm opened for photograph; e) Side of thumb; f) Rust in palm of hand from holding rifle barrel.

other areas. It is also good practice to photographically document the process of moving the body to more accurately record any changes that might result.

Once the body has been moved, the area previously covered by the body should be documented. The need to record discrete stains or other physical evidence is obvious, but

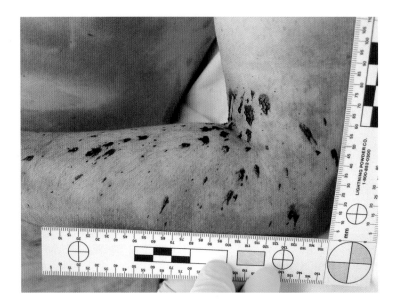

Figure 13.47 Scales are often incorrectly aligned anatomically with the body part. Scales should, instead, be placed in the same plane as the bloodstains being documented.

this is another location where the lack of stains may be telling. In one case, the area surrounding the victim's body was uniformly stained with impact spatter. The relative absence of spatter underneath the body established that the victim was rendered prone very early in the prolonged assault (Figure 13.50).

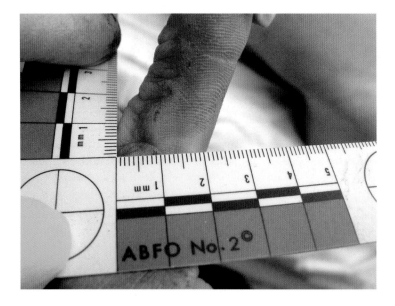

Figure 13.48 Scale not placed in the same plane as the bloodstains.

Evaluation of Bloodstain Patterns at the Scene

Figure 13.49 Larger scales should be used for mid-range or overall photographs. Scales must be readable in the photograph.

Figure 13.50 Void area underneath victim's body confirmed that the victim was rendered prone very early in the prolonged assault. Her unmoving body shielded the carpet creating the void.

Alterations in Transport

The need for thorough documentation prior to moving the body cannot be overemphasized. Once that body is moved, changes are inevitable. Good photographs and supporting documentation at the scene are essential in establishing what is real and what is artifactual.

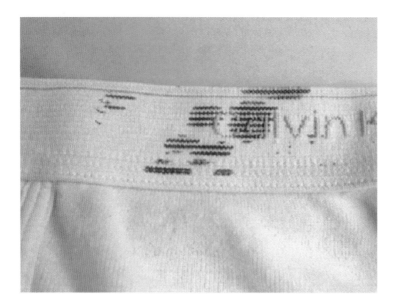

Figure 13.51 Bloodstain on the waistband of undershorts created by alteration during body transport.

Personnel from the medical examiner's office attended the scene of a suicidal gunshot wound to the head. When the body was examined the next morning, a questionable bloodstain was noticed on the waistband of the victim's briefs (Figure 13.51). A comparison with photographs taken on site quickly established the artifactual nature of this stain.

Changes seen in transport have gone from this aforementioned relatively small stain to finding the victim almost entirely covered in blood. Good scene documentation may salvage a case. Transport creates stains, but it can also cause stains to be lost. Very small stains, such as back spatter, are extremely fragile and can easily be removed during the process of body transport.

Mouth and Nose of the Decedent

Photographs clearly showing the nose and mouth of the decedent are highly recommended. In the presence of very small bloodstains, the possibility that the stains resulted from expiration of blood must be considered.

In some instances, the pathologist will find evidence of blood in the mouth and nose but be unable to definitively state whether this results from wounds or from actual expiration. Photographs taken at the scene that clearly show the mouth and nose and the presence or absence of blood can establish or refute the possibility of expiration as a mechanism.

Feet or Shoes of the Decedent

A fairly common question that confronts a bloodstain pattern analyst examining stains on bodies is whether the victim was moving about during the assault. Autopsy findings will establish if the victim had the capacity to move, but the ability to move does not equate to actual movement. The victim may be restrained from movement or incapacitated from other injuries. They may simply be rendered motionless from fear. A careful examination of the feet and lower extremities are important observations. Movement within the scene

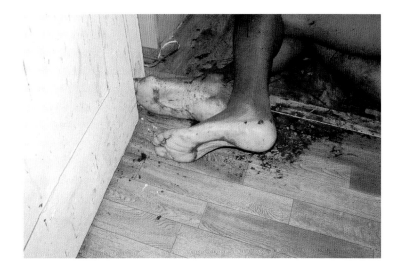

Figure 13.52 Blood on bottom of feet consistent with the victim moving about after injury and stepping into blood.

of a violent assault will result in an increased incidence of stepping into blood. Such activity is likely to be evident by the presence of blood on the bottom of feet or shoes (Figure 13.52).

There is also a very good chance that blood will be deposited on the top of the victim's feet as well because any site of injury will be above the top of a person's feet when they are ambulatory (Figure 13.53). The significance of the absence of any blood on the top or bottom of a victim's feet would have to be interpreted in context of the particular case. In some cases, it can be beneficial to establish lack of movement.

As with any other bloodstain pattern, the overall distribution is of great importance. Blood can certainly be on the bottom of a victim's feet from activities wholly unrelated

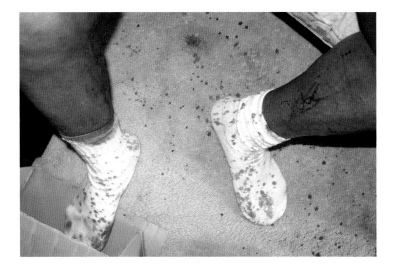

Figure 13.53 Blood on tops of feet created by drips from wounds on the upper body of the victim. During movement, blood dripped onto the tops of his feet.

to walking. For example, blood in the apex of the arch only, with no stains on the heel and fore-foot, would certainly be suspect as having resulted from walking. The possibility exists that there was very limited movement or perhaps there was a limited amount of blood in which the victim could step. Again, these situations must be reconciled in terms of the individual case circumstances. The analyst should always remember that there is no shame in saying "I don't know." Erring to the side of caution should always prevail over unfounded opinions.

Nonambulatory Movement

Victims may move any part of the anatomy, either purposely or as the result of agonal activity. In one case, a very clear, repetitive series of transfers could be seen in the bloodstains on a floor as a result of movement by the victim's arm (Figure 13.54).

There are also instances where the bloodstains will reveal that someone has moved the victim. In the accompanying photograph, a shotgun wound that struck the brainstem had instantly incapacitated the victim. Figure 13.55 shows a flow pattern visible on his forearm in complete defiance of gravity. This pattern proved that his body was moved a substantial period of time after the shot. Figure 13.56 shows several changes in flow patterns, and it also demonstrates the type of artifactual alteration caused by the removal of a ventilation tube. Changes in flow directionality are a quick and easy way to establish movement.

Posture at Time of Injury

The geometric principles used to determine point of origin are widely used to determine victim positioning. Careful examination of the bloodstain patterns on the body can provide valuable information in this regard even without the necessity of any additional calculations.

Figure 13.54 Imprint showing movement of the victim's arm.

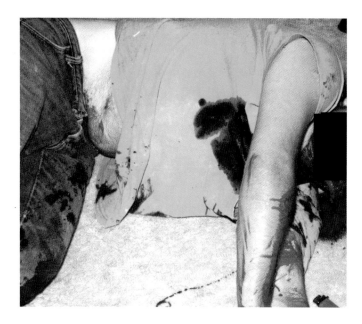

Figure 13.55 Blood flow on the top of this man's right arm is in complete defiance with the laws of gravity. This observation along with several other indicators proved this man had been moved a substantial amount of time after injury.

The young man in Figure 13.57 sustained a gunshot wound that entered the right side of the neck. The flow stains visible on his chest confirm that at least his torso was in an upright position. It is impossible that he was lying down.

In another case, relatives of the victim claimed that he had been knocked to the ground during an altercation and was then summarily "executed" by a shotgun blast to the head

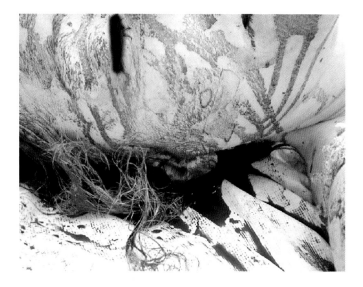

Figure 13.56 Several changes in flow patterns and artifactual alteration common to the removal of a ventilation tube that had been held in place with a rectangular strip of tape.

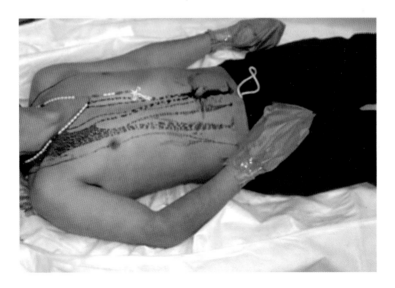

Figure 13.57 Flow from neck wound confirming the torso as being upright after injury.

while lying facedown. The vertically oriented bloodstains on his pants legs dispute this claim (Figure 13.58). The stains are evidence that he was not facedown (otherwise the front of his pants would not have been exposed to receiving stains). The stains further show him to be in a standing position, as opposed to lying down.

Negative findings on the lower extremities may very well confirm a victim as being prone at an early point in the assault. Figure 13.59 shows a woman's legs and shoes, the victim of massive, penetrating head trauma. The absence of any elongated stains on her legs or any stains on her shoe tops was combined with other findings to conclude that she was rendered supine very swiftly, and there was no evidence that she ever reassumed a standing position.

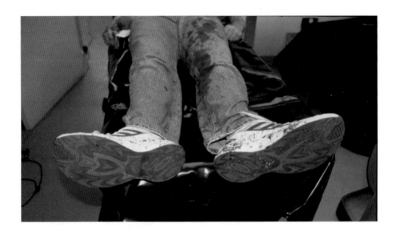

Figure 13.58 Bloodstains on the front of the pants leg disputes witnesses' claims of the deceased being facedown at the time of the shotgun wound.

Evaluation of Bloodstain Patterns at the Scene

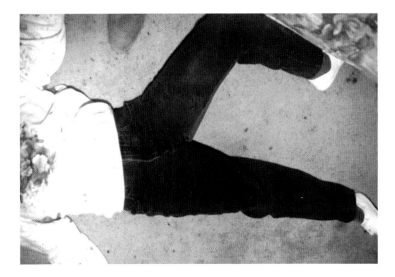

Figure 13.59 Absence of bloodstains on the pants legs and on the shoes indicates that this victim of massive, penetrating head trauma never regained an upright posture.

Claims of Self-Defense

Whether a victim managed to maintain an upright posture for any substantial period can be instrumental in supporting or refuting any claims of self-defense. It is at least difficult to be in an offense posture if one is down. Other important findings have to do with indicators of movement by the victim. Note the voided area on the front of the sweatshirt shown in Figure 13.60. Had the arms of the wearer been in motion, this area would have been exposed to receiving impact spatter. The pristine void shows that the wearer's arms were not in motion. Again, this is not indicative of offensive activity.

Time Lapse

It is crucial to correlate the injuries or other medical conditions from the final autopsy report with the observations from the crime scene or the body. Obviously a pattern considered arterial in nature should be supported by some arterial injury and exposure. There are also instances when what was assumed to be an injury at the scene is actually something very different. The bloodstain on the buttocks of the victim shown in Figure 13.61 was thought to be the result of a small-caliber gunshot wound. On closer inspection, no injury existed at that site. The stain was actually a partially formed clot that had been deposited on the buttock.

The stain on the right arm of the decedent shown in Figure 13.62 was initially thought to be blood flow resulting from a laceration. When the final autopsy report and photographs were reviewed, no injury was found. The bloodstain examination made at the scene confirmed characteristics of clotted blood at this site. Only by correlation of the two separate examinations was a correct report possible.

Sequencing Injuries

In some cases the pathologist will be unable to assign an order to injuries relying only on the autopsy data. This is common when multiple injuries occur in quick succession or

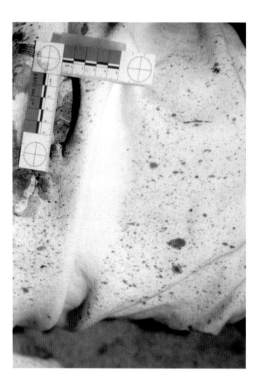

Figure 13.60 Void area in sweatshirt front is not consistent with the victim having raised her arms, either offensively or defensively, after spatter was created.

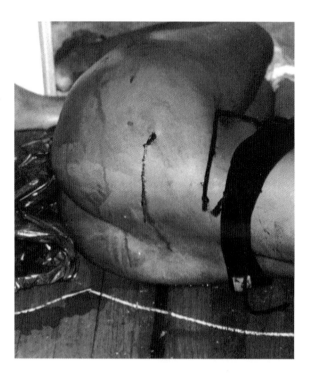

Figure 13.61 Clot on victim's buttocks thought to have been the result of a small-caliber gunshot wound. At autopsy, no gunshot wound was present.

Evaluation of Bloodstain Patterns at the Scene

Figure 13.62 Clotted blood on victim's arm thought to have been the result of injury. At autopsy, no wound was present.

injuries are remote from one another providing no sequential overlap. The solution may lie in the bloodstain patterns at the scene or on the body.

A deputy murdered in the line of duty sustained two gunshot wounds to the head, but the pathologist was not able to determine the order of the shots. Only one of the shots exited, and that shot had produced characteristic spatter at the scene. The location of this exit spatter required the deputy to have been in an upright position at the time of the shot. Because either shot would have been instantly incapacitating and would have caused the deputy to no longer be able to maintain an upright posture, the exiting shot would necessarily have been first.

In another instance, a female was discovered in a local motel with multiple stab wounds to the back and lacerations to the throat. At the time of discovery, she was lying on her back. The flow of blood from her neck wound went from the injury at the front of the neck continuously toward the back of the neck. Had the neck wound occurred first, her body position would have to have been changed to inflict the stab wounds to her back. Because the flow pattern from her neck wound did not exhibit any change in directionality, the stabs to the back were inflicted prior to the neck injury.

Victim Viability

Bloodstain evidence may also be able to establish the viability of an individual when pathology is equivocal. Although some injuries may be lethal, it is sometimes impossible to determine medically at what point in time death actually occurred. Simple observations, such as the confirmation of expirated blood, can provide a solution to this dilemma. The presence of air bubbles and air vacuoles on the shirt of one victim was accepted as mitigating evidence in the death penalty hearing of her husband (Figure 13.63).

Figure 13.63 Expirated blood — both with intact air bubbles and vacuoles — on front of victim's sweatshirt confirmed breathing occurred after the assault.

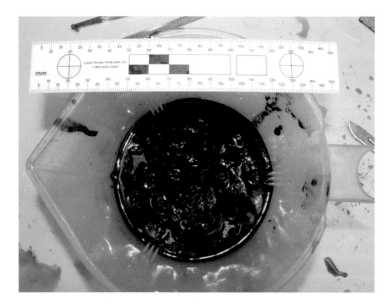

Figure 13.64 Clotted blood recovered from stomach being measured for volume at autopsy.

The presence of a significant amount of clotted blood in the stomach is also supportive evidence that the victim survived injury for some period. If the volume is measured and compared to the normal volume per swallow, an estimate of survival time will result (Figure 13.64).

Manner of Death

Many previous publications have dealt with the presence or absence of back spatter on the hands of the decedent in determining suicide versus homicide as the manner of death.

An excellent article by Yen and colleagues points out the necessity of considering the wide variety of positions that may be assumed by the victim when evaluating the location of any back spatter observed. It will be reiterated that many factors will influence the presence or absence of back spatter: ammunition, range of fire, clothing, and hair covering the site of the wound being among them.

In one situation, back spatter was not helpful in making a ruling of suicide versus homicide as the manner of death. The pivotal evidence, instead, was uncovered through drying time analysis. Her young husband stated that he had tried to persuade her otherwise, but she was so despondent that his efforts were to no avail. A gunshot wound entering between the nose and eye was instantly fatal. First responders arrived within 7 min of the 911 call the husband claimed to have placed "within minutes" of the gunshot. After taking his statement, investigating detectives were very suspicious. Despite being unconvinced by his story or his demeanor, they were more concerned about the extent of drying they observed on her face. Essentially there was a curtain of blood from her wound, continuing downward to her chin. They noted that the stain was dried and had pronounced cracks when they arrived — allegedly 7 min after the fatal shot was fired. Pertinent environmental data were collected from the scene, and the stained area was measured. Experimental results demonstrated that similar drying and cracking required a time lapse in excess of 2 h. As a result of these findings, the death was ruled a homicide. Confronted with the discrepancy in his claims and the drying time data, the husband entered a guilty plea to her murder.

Sample Collection

The effective use of bloodstain pattern analysis can lead to a beneficial reduction in the number of blood samples from a scene that will require additional characterization by laboratory analysis. Any measures that can reduce laboratory backlog are positive.

Bloodstain pattern analysis is only one tool, however. When processing any type of forensic evidence, especially crime scenes that may be available for only a limited time, one should remain mindful that theories as to stain origin might change as more information becomes available. The evolution of theories is a positive outcome of good scientific practice. For this same reason, however, the effective use of bloodstain pattern analysis does not completely obviate the need to collect samples — even if they are collected mainly as a precautionary measure. From a laboratory standpoint, collection does not necessarily mean that analysis must be performed on each and every sample. Decisions should be based on the current information available and sound reasoning. A "shotgun" approach to laboratory analysis is a waste of resources and should be avoided whenever possible.

Although it is unreasonable to expect that each stain or impact spatter is fully analyzed, representative samples can be used to confirm the presence of blood (or human blood) as opposed to staining by some other substance. The author has witnessed some rather convincing stains later found to be barbeque sauce. This can be especially beneficial when the visual appearance of a stain has been altered as a result of aging, environmental insults, or other negative factors.

Suspect Clothing

The need to confirm the source of any stains on the clothing of a suspect can be especially important. Notwithstanding these caveats, it should be remembered that the bloodstain pattern analyst's primary role is to ascertain staining mechanisms. Individual case circumstances must be considered in determining the extent to which source identification limits such conclusions.

Separating Contributors

In cases with multiple sources of blood — victim, assailant, or a combination of both — the need to identify the individual source can be significant. This is also an example of the value of effectively using bloodstain pattern analysis. By identifying patterns and then correlating those findings with the possible mechanisms of production, one will be better able to make educated decisions as to the most likely source of the blood. This information can then be applied when collecting stains that will accurately represent separate, discrete patterns. In this regard, effective bloodstain pattern analysis can make laboratory characterization efforts more fruitful by isolating patterns that represent contributors.

In other situations, a biological analysis will be necessary to complete the bloodstain pattern analysis. Frequently, multiple victims are exposed to the same type of assault within a very small space. In the absence of separation by distance or assault mechanism, biological analysis will be needed to attribute a particular pattern to the appropriate individual victim. For example, in one case two victims were beaten to death with a hammer in a hallway of a home. Numerous smears, impact patterns, and cast-off stains were observed within a space approximately 4 feet wide and 12 feet long. To place the correct victim at any given location, it was necessary to determine the biological source of several of the patterns.

Linking an Accused

Linking a perpetrator to a crime scene is a goal of crime scene processing. Bloodstain pattern analysis has proved especially valuable in locating stains that have a high degree of probability of originating from a perpetrator. Stains that are separate and distinct either in stain characteristics or geographic location may indicate stains targeted for collection from what is known about the victim's injuries, activities, or general region of assault.

Prime examples are bloodstains found in bathrooms or other sites indicative of a perpetrator seeking to attend to wounds. Detection of such stains may require special vigilance because the extent of bleeding may not necessarily be heavy. In one instance only two stains were found on the edge of a bathroom sink. The bathroom bore no other indications of postassault use. Subsequent analysis of these stains confirmed that a disgruntled employee was the source. This finding was especially valuable because other evidentiary indicators of his involvement were negated by the fact that activities required by his employment could account for fingerprints, hairs, etc. within the room where the murder took place. The presence of two stains of his blood in a bathroom restricted to use by the owners was crucial in obtaining a conviction.

In stabbing cases there is a good probability that the perpetrator will be injured and bleeding. Common household cutlery does not have the hand guard that is found on assault or hunting knives. The sudden cessation caused by unexpectedly hitting a bone or other hard surface while stabbing someone can cause the perpetrator's hand to slide down onto the blade and be cut. Injury may also result simply because their hand or the handle of the knife has become bloody and slippery and slides down onto the blade. During the struggle, assailants may actually stab themselves. As previously discussed, it is not uncommon for

Evaluation of Bloodstain Patterns at the Scene

assailants to return the murder weapon to the knife block or the cutlery drawer in the residence. This could account for passive bloodstains that are found in or around the household knife receptacles. This same type of injury may account for cast-off patterns seen in locations distant from the area of assault.

Another indicator of possible perpetrator's blood is the detection of a sustained trail of blood drops. In the absence of a replenishing blood supply, the length of a dripping trail created by carrying a bloody weapon will be of limited duration. As blood leaves the weapon, the diminished quantity will be insufficient to overcome the adherence caused by surface tension. Larger weapons and objects with rough surfaces require more blood to break surface tension and therefore release correspondingly few drops. In one case, a trail of blood drops extended in excess of 32 feet from the area in which the murder took place (Figure 13.65). Because no weapon had been located, the first personnel at the scene attributed this trail to blood dripping from the weapon as it was carried away from the scene. By applying the

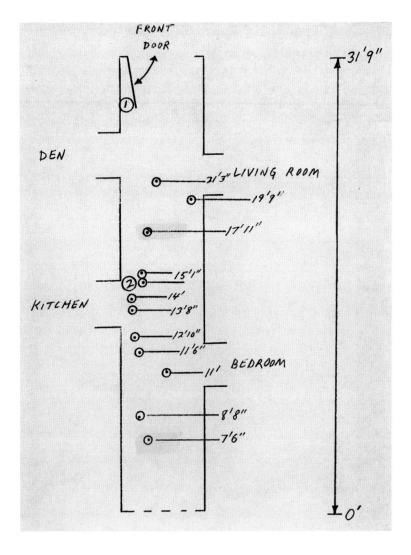

Figure 13.65 Crime scene sketch of hallway showing a trail of dripping blood that extended in excess of 32 feet from the area of the murder.

principles of bloodstain pattern analysis, it was quickly established that such a trail would not result from carrying a bloody weapon but instead represented a bleeding individual leaving the scene. Samples were collected and later confirmed to have originated from the suspect. A point of interest in this case is the absence of any visible wounds when the suspect was apprehended a short time after the murder. Obviously his blood was at the scene, but what caused the bleeding if no wound was visible? The suspect refused to elaborate about any injury he might have sustained, but the most likely source of such bleeding would be a bloody nose sustained during the struggle. Such an injury could certainly account for the blood left at the scene but leave no visible physical indicators of injury.

In another case, an even heavier trail was located outside the residence of a murder victim. Neighbors were alerted when the victim ran outside screaming for help, and they were rushing to her side when they saw a man run from the apartment, get into the victim's car, and drive away. These eyewitnesses identified her husband as being the individual they had seen running from the scene, and he was quickly located and placed under arrest. The husband denied any involvement and produced several witnesses who placed him elsewhere when the murder occurred. Officers were extremely skeptical of his alibi but did request bloodstain pattern analysis of the apartment in hopes of a solution.

A trail of blood was immediately noted that extended down the sidewalk leading away from the apartment to the parking lot (Figure 13.66). Although this was the sidewalk used to remove the body of the deceased, the trail turned in a different direction than the path taken by the ambulance crew. This was confirmed by the statements of the crew as well as photographs taken at the scene showing the ambulance on site. Despite this inconsistency, investigators had discounted the value of this trail because it led to the right side of a parking space — the side that would equate to the passenger side of a vehicle instead of the driver's side. The neighbor was interviewed again, but this time he was asked if the victim's car was

Figure 13.66 Trail of blood drops from parking lot leading to the victim's car. The husband of the victim had been identified as the killer by an eyewitness. As a result of a CODIS hit on these bloodstains, the real killer was found and the husband was exonerated.

Evaluation of Bloodstain Patterns at the Scene

Figure 13.67 Samples should be collected when stain characteristics differ drastically from the predominant stains in that area. The drop of blood on the bedpost was found to originate from another individual. Although this individual has yet to be located, this philosophy proves that, when in doubt, one should collect the stains for further analysis.

parked headfirst into the parking space or backed into the parking space. The look on his face revealed his surprise at such a question. He reiterated that he had already told the police that he had seen her unloading groceries from her car only a short time before he heard her screams. Of course she had backed her car into the parking space. Samples collected and analyzed from that trail were linked to a known sex offender through CODIS, and the husband was released immediately.

Suspicion is also warranted when stain characteristics differ drastically from the predominant patterns. In one such instance, the bloodstain patterns showed that the victim had been knocked down quickly and had remained on the floor during the assault. Impact spatter originating from near the floor would be anticipated in correlation to these findings. Several passive drops, found to overlie impact spatter, were thought to be distinct and warranted collection (Figure 13.67). Analysis confirmed a source other than the victim, but the contributor remains at large. Although the guilty party has not been apprehended, this case demonstrates a good point: When in doubt, collect the sample.

Presumptive Testing and Species Determination of Blood and Bloodstains

14

ROBERT P. SPALDING

Introduction

In recent years, a profusion of television programs devoted to the various fields of forensic science and the widespread reporting of a number of criminal trials have increased both public awareness of and interest in this area of criminal investigation. Technologies such as DNA analysis have captured our attention both for the uniqueness involved and the certainty with which an individual may be identified. A newcomer to the "scene" might even ask what did we do prior to the arrival of DNA. The answer is forensic serology.

A study of serology involves the examination and analysis of body fluids, and among those fluids, blood. In the medical setting, blood is analyzed to assess one's state of health. In the forensic realm, blood might be analyzed to determine the source of blood at a crime scene or on an item of evidence. The difference does not end here. A clinical serologist typically deals with samples that are fresh, normally liquid, and usually recently acquired from a source individual. Forensic serology deals not only with a variety of body fluids (blood, saliva, semen, and urine), but, and more frequently, with samples that are in stain form and often degraded or deteriorated, often making successful analysis more difficult.

The quantity and condition (degree of degradation or putrefaction) of evidentiary stains may depend on a number of factors, many of which are not within the power of the forensic analyst to control. Sample quantity and condition often dictate the manner or strategy an examiner might adopt in dealing with a specific piece of evidence, or indeed, whether any analysis can be done at all. Requirements of the legal system must be followed to protect the security of the evidence and thereby enable its introduction in court. The integrity of the evidence itself may easily be compromised by failure to maintain control over it at all times or by allowing it to degrade during storage under inadequate conditions. Uncontrolled exposure of evidence to heat and humidity can, in a biochemical sense, destroy much of the information contained in a stain by enhancing degradation of the chemical substances of importance to the analyst. This can occur before or after receipt of the evidence by a laboratory.

One might inquire as to the importance of forensic serology today when DNA is available in many jurisdictions across the United States. Forensic serology enjoyed a prominence in forensic laboratories throughout the world from the 1950s until the late 1980s when the ability to analyze DNA on evidence became a reality. The emergence of DNA on the forensic scene resulted in a steady growth in the number of laboratories using these methods of analysis. However, a survey of laboratories would likely show that even among those laboratories routinely using DNA analysis, a significant number still use many of the basic serologic testing procedures. Additionally, there are laboratories that still use a more complete array of serologic testing because DNA analysis may not be done as a result of limited manpower or financial restrictions.

Another and perhaps more important reason for us to consider serology focuses on the reliability and discriminatory power of DNA analysis. In recent years, the application of DNA analysis to cases in which a defendant was convicted and imprisoned has resulted in the release of a number of individuals. The original serology work was not necessarily wrong or fraudulent, but the greater power of DNA to specifically characterize an individual and the evidence that may or may not have come from that individual are emphasized. With the release of some of these individuals has come inquiry into the validity of the results originally presented in court. Finally, there are cases in which convictions have been reversed and new trial(s) have been ordered. These may involve pre-DNA serology and will normally require testimony regarding the original results. From the foregoing, it should be evident that a knowledge of serology or at least an availability of information on the subject is important for thorough forensic study. It is the intent of this chapter to review and discuss the more commonly used techniques of forensic serology regarding blood, to provide sources of information, and to present newer information as well.

The Analysis of Blood in Forensic Serology

Although it might seem that visual identification of a stain as blood would be sufficient to warrant further analysis of the material, proper scientific approach and legal requirements dictate that such an identification be established to a scientific certainty before it can be presented in court. Further, a disciplined scientific approach dictates the use of a routine protocol incorporating a logical series of testing procedures to provide the basis for such an identification. Use of a protocol is intended both to prevent inadvertent elimination of a step in the process and ensure similar treatment of all items of evidence under consideration. An example of such a protocol familiar to the author is set forth in the following. The use of such a protocol, or a variation of it, can be applied to any of the body fluids most frequently encountered in forensic serology.

- A careful visual examination of the item of evidence to locate any stains or material visibly characteristic of blood
- Application of a suitable presumptive screening test
- Application of a specific and sensitive test to confirm the presence of blood
- Determination of biological or species (animal or human) origin
- Characterization the blood using one or more genetic markers or DNA

Presumptive Tests for Blood

Identification of blood using the previously outlined approach often uses a presumptive chemical screening test that is frequently followed by a confirmatory test to clearly establish the identification. A visual observation of the untested stain, coupled with positive chemical presumptive and confirmatory tests, then provides sound data to support the identification of blood. Some of the more commonly used tests used for this are discussed here. A presumptive test is one that, when positive, allows the forensic examiner a qualified conclusion that the blood is indicated in the tested sample and confirmatory tests are needed to scientifically prove that blood is present. When negative, the test often helps to eliminate stains that need not receive further consideration. In the event of a positive test and sufficient sample remains, further action to confirm the presence of blood is usually taken because no single test is absolutely specific for blood. Presumptive tests may be recognized as those that produce a visible color reaction or those that result in a release of light. Both types rely on the catalytic properties of blood to drive the reaction.

Catalytic Color Tests

Catalytic tests use the chemical oxidation of a chromogenic substance by an oxidizing agent catalyzed by the presence of blood, or more specifically, the hemoglobin or red pigment in the blood. Those tests that produce color reactions are usually carried out by first applying a solution of the chromogen to a sample of the suspected material/stain followed by addition of the oxidizing agent (often hydrogen peroxide in a 3% solution). The catalyst is actually the peroxidase-like activity of the heme group of hemoglobin present in the red blood cells (erythrocytes). A rapidly developing color, characteristic of the chromogen used, constitutes a positive test. Some methods can be found that use the chromogen and the oxidant in a single solution. There is a potential disadvantage in this, however, because the order of addition of the specific reagents can be important. A nonblood sample capable of producing a color reaction will normally do so without the addition of the oxidizing agent so that when the color reagent is added first there is a reaction (without the addition of the oxidant). With a single solution, a reaction resulting from a nonblood material might be incorrectly interpreted. Twenty or more substances have been investigated over the years as chromogens, and those discussed in the following seem to have been used more than the rest.

It is important to recognize that these tests are presumptive — that is, they indicate, but are not specific for, the identification of blood. Further, any discussion of presumptive tests would be incomplete without mention of false-positive reactions. Sutton (1999) has correctly pointed out that a false-positive result is an apparent positive test result obtained with a substance other than blood. This eliminates many results that consist of an uncharacteristic color when nonblood materials are tested. Accordingly, it is possible that some of the tests discussed may display some reaction when testing nonblood samples, but experience, careful observation, and routine application of confirmatory testing will usually prevent errant identification of a nonblood sample as blood.

Misleading results usually can be attributed to chemical oxidants (often producing a reaction before the application of the peroxide); plant materials (vegetable peroxidases are thermolabile and can be destroyed with careful heating); or materials of animal origin (to include human), which are not blood but may contain contaminating traces of blood. Microscopic examination may give insight as to the true nature of the material. Sutton (1999) presented the results of testing a wide variety of nonblood materials with several different presumptive tests.

Despite all the work done on false-positive reactions, there is surprisingly little information available concerning false-negative reactions. Recognizing this, Ponce (1999) studied the effect of reducing agents on presumptive tests, using a representative reducing agent (ascorbic acid) to contaminate bloodstains tested with o-tolidine, phenolphthalein, leucomalachite green (LMG), and tetramethylbenzidine (TMB). They demonstrated that the ascorbic acid did, in fact, result in false-negative results, although there was variation in degree to which the individual tests were affected.

A common method of applying the presumptive tests involves sampling a questioned stain with a clean, moistened cotton swab and adding a drop of the color reagent solution followed by a similar amount of hydrogen peroxide. With this procedure, the immediate development of the color typical of the particular reagent used indicates a positive *presumptive* presence of blood in the test sample. Alternatively, the evidence could be sampled by removal of a thread or fragment of dried material and testing it with the previously mentioned reagents in a spot plate. Color development would then be observed in the spot plate. Immediate reading (within a few seconds) and recording of results are important aspects of test result interpretation. A clearly negative result may appear positive several minutes after the test is completed as a result of a slow oxidation that often occurs in air. These tests are not usually affected by stain age, and it has been the author's experience that benzidine, phenolphthalein, and TMB have given positive results with blood crusts and/or stains up to 56 years old.

Benzidine (Adler Test)

Benzidine has been used probably more extensively than any other single test for the presumptive identification of blood. Much of the earlier forensic work dealing with benzidine was done by the Adlers (1904). The reaction (Figure 14.1), normally carried out in an ethanol/acetic acid solution, results in a characteristic blue to dark blue color. The blue, in turn, may eventually turn to a brown. Benzidine, however, is recognized as a carcinogen by the Occupational Safety and Health Administration (Federal Register, 1974) and is seldom used in forensic laboratories today. Its inclusion here is mainly of historical interest. The test has largely been abandoned in favor of safer color reagents.

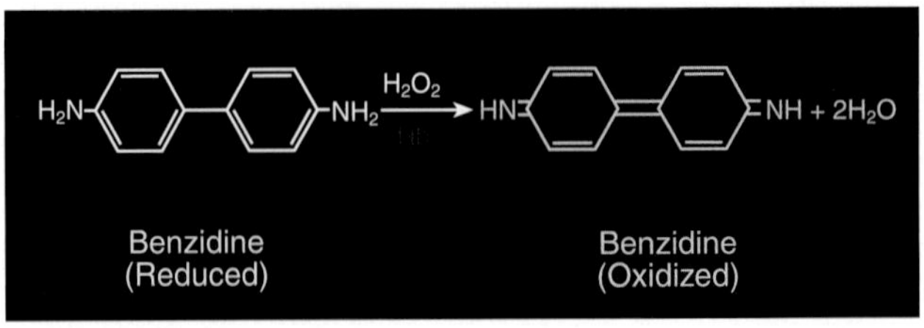

Figure 14.1 Benzidene oxidation.

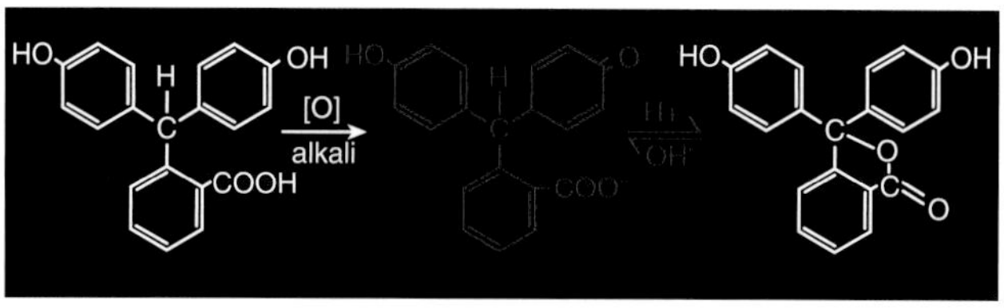

Presumptive Testing and Species Determination of Blood and Bloodstains

Figure 14.2 Phenolphthalin oxidation. Phenolphthalin (*left*, colorless) is catalyzed to phenolphthalein (pink in alkaline solution, *center*), which is colorless in acid (*right*).

Phenolphthalein (Kastle-Meyer Test)

A test procedure commonly used in many forensic laboratories today involves the simple acid–base indicator phenolphthalein. Investigated by Kastle (1901, 1906), phenolphthalein produces a bright pink result when used as explained earlier in testing suspected blood. The reagent consists of reduced phenolphthalein (phenolphthalin) in alkaline solution, which is oxidized by peroxide in the presence of hemoglobin in blood. The reaction shows phenolphthalin (colorless in alkaline solution) being oxidized to phenolphthalein (pink in an alkaline environment), as shown in Figure 14.2. Normally the reagent preparation involves reducing the phenolphthalein over zinc in potassium hydroxide solution to phenolphthalin. As with any of the catalytic tests, the result is read immediately and a positive result a minute or more after the test is performed is usually not considered reliable. False-positive results with phenolphthalein usually are not really positive in that the reaction is often not the characteristic pink but usually some other color change. Figure 14.3 shows both positive and negative results with phenolphthalein.

o-Tolidine

Ortho-tolidine is the 3,3′-dimethyl derivative of benzidine (Figure 14.4). The reaction, similar to that of benzidine, is conducted under acidic conditions and produces a blue color reaction resembling that of benzidine when testing blood. Earlier workers proposed

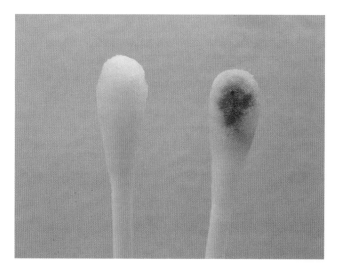

Figure 14.3 Negative and positive phenolphthalein reactions.

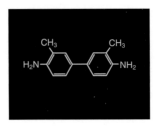

Figure 14.4 o-Tolidine.

it as a replacement for benzidine (Hunt, 1960); however, the compound was reported to be carcinogenic in rats by Holland (1974), and evidence exists (Bell, 1980) that o-tolidine based dyes may be metabolized to the parent compound in humans. Figure 14.5 shows a positive reaction with ortho-tolidine obtained when testing blood.

In 1981 o-tolidine was the active ingredient in the Hemastix[7] Test (Miles Laboratories, 1981), a diagnostic test used in a medical context for the detection of blood in urine. The test was used with positive results on the fabric of a wounded Canadian Militia Officer's uniform coat from the War of 1812 (Macey, 1979). During the 1980s, interest continued to develop in the test for forensic work; however, the carcinogenicity of o-tolidine eventually resulted in it being replaced by TMB in 1992.

Leucomalachite Green

Another compound investigated by the Adlers (1904) was the reduced or colorless form (leuco) of the dye malachite green (Figure 14.6). Often referred to as McPhail's Reagent (Hemident®), LMG oxidation is catalyzed by heme to produce a green color. Like benzidine and o-tolidine, the reaction is usually carried out in an acid (acetic acid) medium with hydrogen peroxide as the oxidizer, although there are procedures that use sodium perborate with the LMG in a single solution (RCMP, 2001). Figure 14.7 displays positive and negative LMG reactions. An interesting point is that the Adlers (1904) preferred the benzidine and LMG tests over the 30 plus reagents they tested.

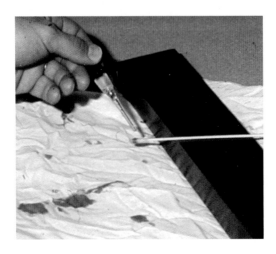

Figure 14.5 Ortho-tolidine positive reaction.

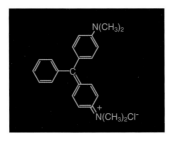

Figure 14.6 Malachite green.

Tetramethylbenzidine

With the recognition of benzidine as a carcinogen (Federal Register, 1974), the search for replacement substances was underway. A number of laboratories changed to phenolphthalein as a primary presumptive reagent, but interest also centered on TMB, synthesized by Holland (1974). TMB is the 3,3′,5,5′ tetramethyl derivative of benzidine (Figure 14.8). It had been observed that the 3,3′-dihydroxybenzidine metabolite of benzidine was more carcinogenic than benzidine itself. It was thought that orthomethylation of benzidine would

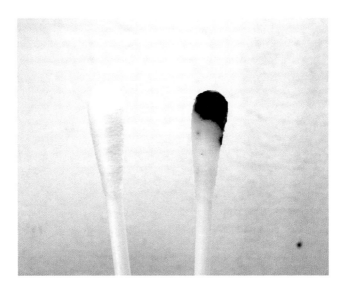

Figure 14.7 Negative and positive leucomalachite green reactions.

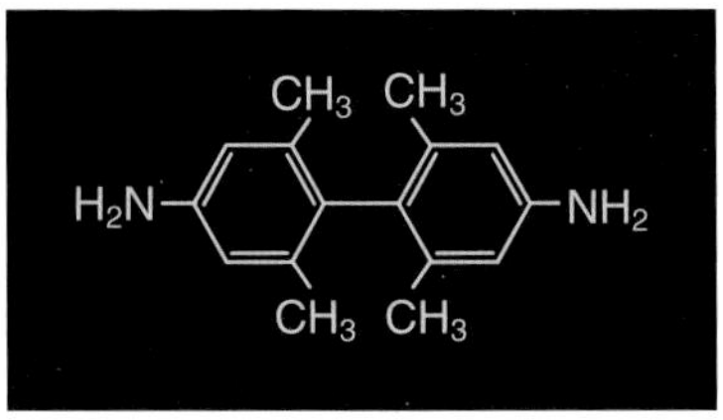

Figure 14.8 Tetramethylbenzidine.

prevent the *in vivo* orthohydroxylation. Holland did, in fact, observe that the methylated benzidine gave a negligible yield of tumors in rats as compared with benzidine and o-tolidine. As with benzidine, o-tolidine, and malachite green, TMB is used in an acid medium (acetic acid) when used as a solution and the resultant color is from green to blue-green.

The Hemastix® Test (Miles Laboratories, 1992) mentioned earlier has been adopted for field use by a number of laboratories, particularly at crime scenes when containers of solutions can be hazardous or at best inconvenient. The test itself consists of a plastic strip with a reagent-treated filter paper tab at one end. The tab contains TMB, diisopropylbenzene dihydroperoxide, buffering materials, and nonreactants. Testing a bloodstain may be accomplished by moistening a cotton swab with distilled water, sampling the stain, and touching the swab sample to the reagent tab on the strip. The reagent tab is originally yellow, and a normally immediate color change to green or blue-green indicates the presence of blood (Figure 14.9). The test kit (Figure 14.10) is often available in drug stores and from some forensic suppliers, and only swabs and distilled water are required. Experience has shown, however, that swabs pretreated with cosmetic substances may contribute false-positive results.

Comparing the Tests

A reasonable search of the literature will show that more than a single method of preparation exists in the literature for each of the presumptive test reagents discussed here. Various protocols for applying these tests can also be found as well (Sutton, 1999 and Lee, 1982). Accordingly, varying sensitivities and specificities are reported based on various authors' different experimental designs. As a means of providing some sense of how the various tests compare on a qualitative level, the results of several authors have been reviewed and are presented in Table 14.1.

Differences in specificity and sensitivity become of less consequence when it is realized that subsequent confirmatory testing and further characterization of the stains will usually disclose any irregularities in the earlier stages of testing.

Figure 14.9 Hemastix® negative and positive reactions.

Presumptive Testing and Species Determination of Blood and Bloodstains 357

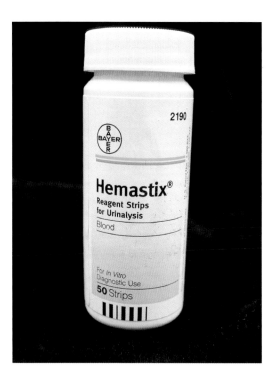

Figure 14.10 Hemastix® test kit.

Tests Using Chemiluminescence and Fluorescence

Often the presence of blood is suspected based on witness information or expected in a particular location, but under normal lighting and viewing little is to be seen. A drag pattern across a floor that has been cleaned up or a washed spatter pattern on a wall might be typical examples. At this point the luminol and fluorescein tests may come into play. These tests are presumptive and do not identify blood. They involve spraying

Table 14.1 Comparative Sensitivities and Specificities of Presumptive Test Reagents

Author	Tests Examined/Discussed	Most Sensitive	Most Specific	Overall Most Suitable
Hunt et al. (1960)	Benzidine, o-tolidine, LMG, phenolthalein, Luminol	Not stated	Not stated	Benzidine or o-tolidine
Kirk (1963)	Benzidine, phenolphthalein, LMG, Luminol	Luminol	Phenolthalein	Not stated
Higaki and Philip (1976)	Phenolphthalein, benzidine	Benzidine	Phenolphthalein	Phenolphthalein
Garner et al. (1976)	Benzidine, TMB	Equal	Equal	TMB (due to carcinogenic nature of benzidine)
Cox (1991)	Phenolphthalein, TMB, LMG, o-tolidine	o-Tolidine and TMB	Phenolphthalein and LMG	Phenolphthalein

a chemical mixture on a suspected bloodstained area, usually *in situ*, and observing (sketching, photographing, etc.) the result, either in darkness or in reduced light with the aid of an alternate light source (ALS). The observed result is a production of light that often enables the observer to determine the limits, shape, and some degree of detail in the original bloodstained area, often including an enhancement of blood patterns already present.

The presence of blood and patterns displayed by blood can provide information of value, and one should be alert to both. Further, the nature of these tests makes them potential sources of contamination of the blood. So, if the stain can be seen and collected, these tests probably should not be used. Because these tests do add material to the tested area, their use should be carefully considered, often as a last resort. These tests are of more value in locating and defining areas of possible blood than in specifically identifying it.

A word should be said about these substances regarding safety. Both luminol and fluorescein are classified as irritants, and Material Safety Data Sheets (MSDS) (Mallinckrodt Baker, 1999 and Fluka Chemical Corporation, 2001) toxicologic data do not indicate the substances to be carcinogenic at this time. However, although neither compound is classified as a carcinogen, safety recommendations are described for each and they should not be treated carelessly.

Luminol and fluorescein each produce light but in different ways. It should be understood that the terms chemiluminescent and fluorescence are two entirely different phenomena that result from the chemical mechanisms for each of these tests. Chemiluminescence is the process by which light is emitted as a product of a chemical reaction. No additional light is required for the reaction to take place. Luminol relies on this process. Fluorescence occurs when a chemical substance is exposed to a particular wavelength of light (usually short wavelengths, such as ultraviolet) and light energy is emitted at longer wavelengths. Light produced by fluorescein is a result of this irradiation, usually by ultraviolet light in the range of 425 to 485 nm.

Luminol

Since the early 1900s (Curtius and Semper, 1913 and Huntress, et al., 1934), it has been known that luminol or 3-aminophthalhydrazide (Figure 14.11) would luminesce after oxidation in acid or alkaline solution. It was found that various oxidizing agents could cause the luminescence in alkaline solution, and positive results were observed with old as well as fresh bloodstains. It became apparent that the compound, which came to be called luminol, could be useful in forensic investigations involving blood, and the years have confirmed its suitability in this regard. Luminol reacts in a similar fashion as the color tests discussed earlier, wherein luminol and an oxidizer are applied to a bloodstain. The catalytic activity

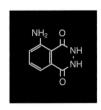

Figure 14.11 Luminol (3-aminophthalhydrazide).

of the heme group then accelerates the oxidation of the luminol, producing a blue-white to yellow-green light (depending on the reagent preparation) where blood is present.

The forensic application of luminol at crime scenes involves spraying a mixture of luminol and a suitable oxidant in aqueous solution over the area thought to have traces of blood present. A resultant blue-white to yellow-green glow (depending on reagent composition) will indicate the presence of blood. Outlines and details are often visible for up to 30 seconds before additional spraying is required. Excessive sprayings will usually result in stain pattern diffusion.

The effects of luminol on stains destined for future analysis have been a concern. The more contemporary studies of Laux (1991) and Grispino (1990) found that pretreatment of bloodstains with luminol failed to significantly affect the presumptive, confirmatory, species origin and ABO tests but interfered with several of the enzyme and protein genetic marker systems, notably, erythrocyte acid phosphatase (EAP), esterase D, (EsD), peptidase A (PEP A), and adenylate kinase (AK). Even more recently, Gross and colleagues (1999) determined that the effects of pretreating bloodstains with luminol did not adversely affect polymerase chain reaction analysis of DNA. Hochmeister and colleagues (1991) found that in bloodstains pretreated with ethanolic benzidine, phenolphthalein, or luminol, the recovery of high-molecular–weight DNA for restriction fragment length polymorphism analysis was unaffected, with the exception of the phenolphthalein where the yield was only slightly lower.

Perhaps the greatest advantage of luminol is its sensitivity. Proescher and Moody (1939) found that luminol would detect hematin from hemoglobin in dilutions up to 1 in 10,000,000, and Kirk (1963) reported that blood in dilutions of 1 in 5,000,000 could be detected. It is more than capable of detecting blood not present in sufficient amounts to be seen with the naked eye. It is also important to record the luminol reaction with photography (Zweidinger, 1973, Lytle, 1978, and Gimeno, 1989a, 1989b), as demonstrated in Figures 14.12 and 14.13.

Figure 14.12 Enhancement of area of room subjected to prior cleanup of blood with luminol reagent photographed with digital camera with flash after 27 sec. (Courtesy of Martin Eversdijk, Politie LSOP Institute for Criminal Investigation and Crime Science, Zutphen, The Netherlands.)

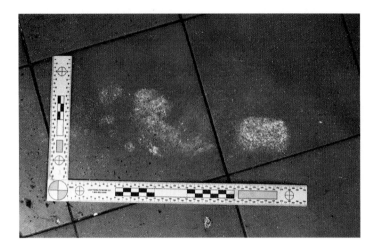

Figure 14.13 Enhancement of latent bloody footprint with Bluestar™ luminol reagent photographed with digital camera (30-sec exposure at f/2.8). (Courtesy of Philippe Esperanca, French Gendarmerie Forensic Laboratory.)

Fluorescein

The use of fluorescein to detect blood was recognized as early as 1910 by Fleig. It appears that Cheeseman (1995, 1999, 2001) is the first to have considered the fluorescin/fluorescein system for larger area application at crime scenes. As with luminol, careful consideration of the ultimate aim of the investigation should guide fluorescein application at a scene. If there is blood that can be seen and collected without the aid of fluorescein, then its use may be unnecessary. However, if the purpose is to define and/or enhance patterns not visible but thought to be present, then any visible blood (desired for subsequent analysis) can be covered to shield it from chemical contamination or can be secured before the area is tested.

Fluorescein is prepared for use much like phenolphthalein. Fluorescein is reduced in alkaline solution over zinc to fluorescin, which is then applied to the suspected bloodstained area. The catalytic activity of the heme then accelerates the oxidation by hydrogen peroxide of the fluorescin to fluorescein that will then fluoresce when treated with ultraviolet light, as shown in Figure 14.14. Fluorescein and luminol are similar in that they produce light to indicate the presence of blood; however, practical differences exist in the

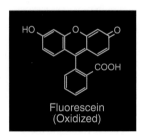

Figure 14.14 Fluorescein.

Figure 14.15 Fluorescein reaction. (Courtesy of L. Allyn DiMeo, Arcana Forensics, La Mesa, California.)

use of the two. Where an aqueous solution of the luminol reactants is simply sprayed on a surface bearing suspected bloodstain residues, a common fluorescein system includes a commercial thickener, which effectively causes the mixture to adhere to the surface and is thus more effective on vertical surfaces. Once the mixture is sprayed, visualization of fluorescence requires the use of an ALS, typically set at 450 nm (Figure 14.15). It also may be noted that fluorescein does not fluoresce with household bleach where luminol will react (Cheeseman, personal communication). Studies regarding the interference of fluorescein with subsequent short tandem repeat (STR) testing of blood for DNA were conducted by Budowle (Budowle et al., 2000). They observed that the results from STR typing showed no evidence of DNA degradation.

Other Methods

While the bulk of forensic testing to identify blood has been limited to the methods outlined in this chapter or similar ones, more novel or innovative approaches have been used. Microscopic examination of reconstituted residues identified erythrocytes on archeologic materials determined with radiocarbon dating to have originated between 7400 to 6900 B.C. (Loy, 1989). Procedures involving crystallography, chromatography, spectrophotometry, immunology, and electrophoresis have all been used. The scope of this writing precludes a detailed discussion of each of them; the reader is directed to works by Lee (1982) and Gaensslen (1983).

Confirmatory Tests for Blood

Crystal tests are regarded by many as confirmatory to the presumptive tests (Hunt (1960), Dilling (1926), and Kerr (1926), to name several). These tests involve the nonprotein heme group of hemoglobin, the oxygen-carrying protein of erythrocytes, which belongs to a class of compounds called porphyrins. The heme structure contains a hexavalent iron atom. Nitrogen atoms within the ring structure bind four of the iron coordination positions, and one is bound to histidine nitrogen in the globin protein. In hemoglobin, the remaining coordination position is normally bound by water or, in oxygenated hemoglobin, oxygen. In dried bloodstains, these last two positions are used in the formation of crystals that are the basis for confirming the presence of blood.

Teichmann Test

First described by Teichmann (1853), the Teichmann test consists of heating dried blood in the presence of glacial acetic acid and a halide (usually chloride) to form the hematin derivative (Figure 14.16a). The crystals (Figure 14.16b) formed are observed microscopically, usually rhombic in shape and brownish in color. Stain age is not an obstacle with the test in experienced hands, and the test has even appeared in basic biochemistry student laboratory manuals (Harrow et al., 1960). The crystals are formed by placing a sample of suspected blood on a microscope slide and adding a small amount of chloride-containing glacial acetic acid, followed by heating. In the experience of this author, the greatest difficulty in performing the test is controlling the heating of the sample because it is easy

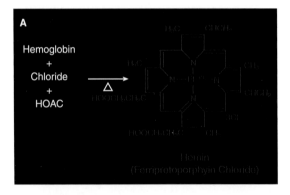

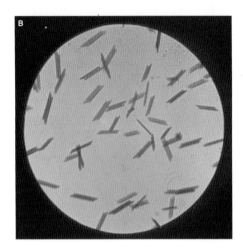

Figure 14.16 (a) Teichmann reactions. (b) Teichmann crystals.

to overheat or underheat the preparation, resulting in no crystal formation even when blood is known to be present.

Takayama Test

If heme is gently heated with pyridine under alkaline conditions in the presence of a reducing sugar such as glucose, crystals of pyridine ferriprotoporphyrin or hemochromogen (Figure 14.17a) are formed. The reaction was examined by Takayama (1912), who examined several mixtures and found best results with a reagent containing water, saturated glucose solution, sodium hydroxide (10%), and pyridine in a ratio of 2:1:1:1 by volume. The normal procedure is to place a small stain sample under a coverslip and allow the reagent to flow under and saturate the sample. After a brief heating period, the crystals are viewed microscopically (Figure 14.17b). A very small (2 mm square or smaller) coverslip allows the test to be carried out on a small stain quantity. The test has been effective on aged stains and crusts and produced crystals from blood on the uniform coat of a wounded Canadian Militia officer from the War of 1812 (Macey, 1979). Further, the test has given positive results on bloodstains that failed to test positively with the Teichmann test.

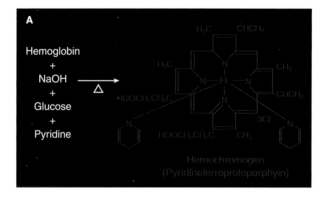

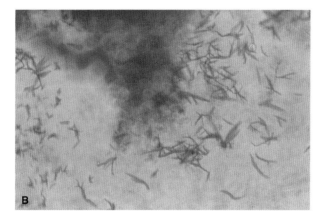

Figure 14.17 (a) Takayama reactions. (b) Takayama crystals.

Probably the most recent improvement in the test procedure is that offered by Hatch (1993). It is known that oxygen and pyridine compete for the same binding site on the heme molecule. Hatch used Cleland's reagent (dithiothreitol) in the reagent to reduce this competition and shift the reaction in favor of pyridine, increasing the rate at which the hemochromogen crystals are formed.

Species Origin Determination for Blood

Serum Protein Analysis

The serum proteins are a large collection of proteins in serum and consist of five classes: albumin, α-1 globulins, α-2 globulins, β globulins, and γ globulins. Albumin is present by far in the highest concentration, and the globulin fraction contains antibodies that are important in much of the following discussion.

General Methods

The following methods discussion provides a foundation for considering both origin determination and the analysis of genetic markers to be covered in the following. A wide variety of tests are available for the determination of species origin of an identified bloodstain, and most use immunoprecipitation to affect a result. If a host animal (i.e., a rabbit) is inoculated with a serum protein (i.e., human), the immune system of the animal will normally recognize the protein as foreign and produce antibodies (gamma globulins) against it. Harvesting the antibodies provides an antiserum to the protein (antigen), and when a sample of the antiserum and the antigen are brought in contact, a precipitin reaction normally occurs. The precipitate may form in a gel or in a solution, but its formation signals a reaction has occurred. The antibody and antigen are then said to be homologous.

Antibodies are also proteins but differ in structure and characteristics from the antigen proteins (mainly albumin). Key to the use of this in a forensic context is the fact that the antigen proteins bear species-specific characteristics that enable the serologist to identify the animal source of the stain extract protein. Several common test systems using this reaction are discussed in the following paragraphs, and additional test systems are discussed elsewhere (Lee, 1981 and Gaensslen, 1983).

Electrophoresis is a technique in which charged molecules (i.e., proteins) are caused to migrate in an electric field in a suitable support medium under controlled conditions that include temperature, pH, voltage, and time. Support media often used include starch gels, starch-agarose gels, cellulose acetate, membranes and polyacrylamide gels. Positively charged molecules will migrate toward the cathode, and negatively charged ones migrate toward the anode. The individual charges of the molecules will play a role in the rate of migration. If two molecules of differing charge are placed in a field at the same point, they will migrate at different rates and will be separated. Once the separation is accomplished, the bands are visualized with staining or through a series of chemical reactions. Forensic electrophoresis is largely done in a gel support medium on a glass plate, and samples as blood-soaked threads or stain extracts are placed in the gel. The characteristics of the sample, the gel itself, and those conditions cited earlier all play a role in how well separated and resolved the final result will be.

Isoelectric focusing (IEF) is an electrophoretic method that takes advantage of the fact that at a certain pH protein in aqueous medium will exhibit a point of no net charge (the isoelectric point, IEP). In electrophoresis, the buffer in the gel controls the pH of the system uniformly throughout the gel. However with IEF, there is a pH gradient set up in the gel with the low end being at the anode and the high end at the cathode. When current is applied, proteins will then migrate to the point where they encounter their IEP and stop. The result is that the migrating molecules focus at their IEPs and form sharp, well-defined bands, which can be much more easily observed than results sometimes seen with electrophoresis. As with electrophoresis, different band patterns are detected through immunologic or chemical techniques to define the phenotypes present.

Ring Precipitin Test

The ring precipitin test uses simple diffusion between two liquids in contact with one another in a test tube. The two liquids are the antiserum and an extract of the bloodstain in question. If the antiserum (antihuman) is placed in a small tube and a portion of the bloodstain extract (human) is carefully layered over the denser antiserum, dissolved antigens and antibodies from the respective layers will begin to diffuse into the other layer. The result will be a fine line of precipitate at the interface of the two solutions (Figure 14.18). In the case where the bloodstain extract is not human, there should be no reaction. Occasionally, there will be more than one precipitin band, indicating different reacting components with different diffusion coefficients. A standard operating procedure would be to include necessary positive and negative controls with a questioned sample.

Ouchterlony Double Diffusion Test

Named after the man who devised it, Ouchterlony (1949, 1968), this test is carried out in a gel on a glass plate. A hot agar solution is plated on small glass plates or a petri dish and allowed to cool, producing a gel layer usually 2 to 3 mm thick. Holes (wells) are punched in the gel layer at specified locations, and the gel is removed. Commonly used patterns are a square of four wells rosette of six wells surrounding a center well or two rows of parallel wells. When the antiserum and stain extracts (antigen) are prepared, the antiserum is placed in the central well and the different stain extracts are placed in the surrounding

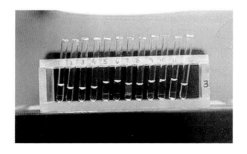

Figure 14.18 Ring precipitin test.

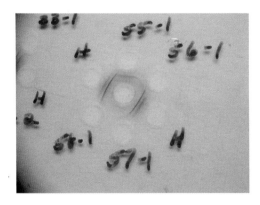

Figure 14.19 The Ouchterlony double diffusion test.

wells. The system is then incubated at a constant temperature (4°C, 18°C, or 37°C are commonly used). After a given time, the diffusion of the reactants through the gel results in the formation of immunoprecipitate lines in the gel between the wells (Figure 14.19). This precipitate is a stable antigen–antibody complex that has grown beyond the limits of its solubility.

Antihuman serum from the center well forming an immunoprecipitate with samples from one or more of the surrounding wells indicates the presence of human protein in the respective outer well(s). The precipitate may be stained with a suitable protein stain. Although the point at which the immunoprecipitate forms is controlled by the diffusion coefficients of the reactants, its formation is dependent on the concentration of reactants and the identities of the reactants. When antigen and antibody are present in relatively equivalent amounts, the line grows in an arc between the two wells. Incubation at cooler temperatures is a slower process but has the advantage of producing sharper lines and sometimes additional precipitin lines as a result of reduced solubility at lower temperatures.

Crossed-Over Electrophoresis

So far, methods discussed have relied on simple diffusion to move antigens and antibodies toward one another. The crossed-over method uses electrophoresis to achieve this. Rows of opposing wells are cut in an agarose gel plate so that well pairs can be arranged with one well toward a cathode source and the other nearer the anode source. The stain extract (antigen) is placed in the cathodic well so that it can migrate toward the anode (because of the pH of the gel buffer used). The antiserum is placed in the anodic well to migrate toward the cathode. When the current is applied, extract proteins move toward the anode.

Antibodies are carried toward the cathode by a phenomenon known as electroendosmosis, which involves anions integral to the gel itself (such as sulfate and pyruvate), the associated cations (sodium), and their water of hydration. When electrophoresis is applied, the cations and water of hydration tend to move in a cathodic direction. This flow carries the antibodies along with it to meet the stain extract proteins between the two rows of wells and results in the formation of immunoprecipitate bands at a site between the wells. This technique has the advantage of forcing the reactants to move in directions optimal for maximum reaction, as opposed to diffusion in 360° around a diffusion plate well.

Nonserum Protein Analysis

Antihuman Hemoglobin

Hemoglobin is a protein that has species-specific immunologic characteristics and is unique to blood. If both characteristics could be addressed in a single test, the work of the serologist would be considerably simplified. Lee and DeForest (1977) presented their work on this, offering a procedure capable of producing a high-titer antihuman hemoglobin serum. Such an antiserum would potentially be very effective in any of the systems described earlier. An obstacle with such a test is the cross-reactivity exhibited by proteins from other higher primates, which, of course, is balanced by the relative expectation of encountering nonhuman primate blood in an investigation.

The efforts of the commercial world are not without merit in this regard. Hochmeister and colleagues (1999) evaluated an immunochromatographic system, Hexagon OBT1®, and found the one-step test to be suitable for laboratory or field use, sensitive, and primate specific. They concluded the test was useful with aged and degraded material, produced easy-to-interpret results, and was unaffected by environmental insults except for detergents, household bleaches, and prolonged exposure to luminol. The test could be used in conjunction with current forensic DNA extraction techniques without loss of DNA. Johnston and colleagues (2003) examined the ABAcard® HemaTrace®, Abacus Diagnostics, Inc. (Figure 14.20), another immunochromatographic test device for potential use in forensic

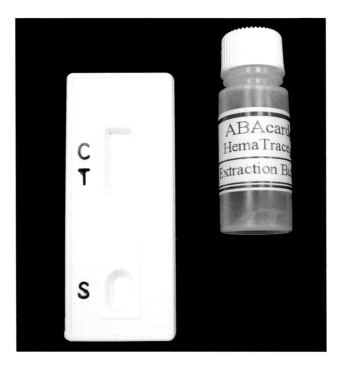

Figure 14.20 ABAcard® Hematrace® kit components.

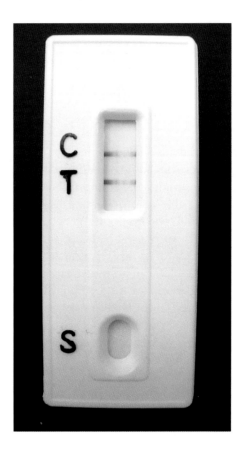

Figure 14.21 ABAcard® Hematrace® positive reaction.

investigations, and found it sensitive and specific for human, higher primate, and ferret bloods (Figure 14.21). Other investigators (Vandenberg and McMahon, 2004 and Silenieks et al., 2004) have found the ABAcard® HemaTrace® to be rapid, sensitive, and accurate in the identification of human (higher primate) blood. Reynolds (2004) evaluated the ABAcard® HemaTrace® as well and concluded the test to be reliable for the identification of human blood.

Other Techniques

Although additional origin determination methods are available, those presented in this chapter are frequently encountered in forensic serology. Such techniques as immunoelectrophoresis, rocket (Laurel) electrophoresis, variations on single and double diffusion, and tests involving agglutination of red blood cells (hemagglutination) have been investigated and provide interesting and valuable approaches to the problem of origin determination.

The Detection of Blood Using Luminol

15

DALE L. LAUX

> Fausak had never experienced anything like it. He felt as if he were in a sci-fi movie. Initially, there was the same pale green light. It got greener and brighter. It began to glow. And through its luminosity he could see the trail of blood. The trail was solid, but with streaks in it, as though someone had taken a big wet mop and wrung it out and dragged it along the floor. The length of the bloody trail measured some 55 feet.
>
> The shimmering glow hung in the air, above Fausak's knees. It had become so bright that he could see the faces of the forensic men and the chemists.

Introduction

The fascinating chemical luminol, which produces a bright luminescence when in the presence of the most minute amounts of blood, is without a doubt an asset to the investigator's repertoire. Chemiluminescence occurs when a molecule capable of fluorescing is raised to an excited level during a chemical reaction. On its return to the ground state, energy in the form of light is emitted. Only a few molecules are known to emit appreciable amounts of light, and of those, luminol is one of the most outstanding. Much of the early work with luminol occurred in Germany, and consequently, many of the articles require translation. The following text, however, is an attempt to unravel the history of luminol and its use in the detection of blood.

Discovery

A very good historical treatise of luminol's discovery and mechanism of action appears in Gaensslen's *Sourcebook* (1983). Most reports cite A. J. Schmitz as the first to synthesize luminol in 1902; however, a paper by Gill states luminol's discovery to be around 1853. What is certain is that in the early 1900s a chemical having chemiluminescent properties was available and being studied. A better understanding of the chemical's nature came about in 1928 with Albrecht's investigations. This study demonstrated that hypochlorites, plant peroxidases, and blood greatly enhanced the luminescence of the compound with hydrogen peroxide. Albrecht also speculated on a mechanism for the reaction.

In 1934 Huntress and coworkers published an article describing a method of synthesis for the compound and coined the term luminol. Gleu and Pfannstiel (1936) prepared luminol and found that the alkaline solution, with the addition of hydrogen peroxide or sodium peroxide, produced an intense luminescence with hematin. This was independently confirmed by Tamamushi and Akiyama (1938). Luminol caught the attention of the forensic science world after the extensive studies by Specht (1937). On the suggestion of Gleu and Pfannstiel, Specht studied fresh and old bloodstains as well as milk, coffee, sperm, saliva, urine, feces, and other body fluids. He tested numerous materials such as wallpaper, fabrics, leather, oils, varnish, wood, grass, leaves, and soil. It is interesting that he tested metals (copper, steel, brass, lead, and zinc) and obtained negative results. In the presence of hypochlorite, a weak luminescence was produced. The test was said not to interfere with subsequent crystal or spectral tests or with serologic tests for species or blood groups. Specht demonstrated the ability to photograph the positive luminol reaction and obtained quality photographs of blood on various substrates including an outdoor setting. Two preparations of luminol were used and were applied as an aerosol with a glass sprayer. Specht found that dried stains produced a more intense chemiluminescence than fresh blood, which produced only a "weak glow." According to Specht, the older the bloodstain, the more intense the reaction, with blood traces a year old or more yielding luminescence. Specht commented on the ability to cover large areas of a crime scene or large pieces of evidence quickly with luminol and recommended that luminol be used for medicolegal examinations, believing that it was quite specific for blood.

A very nice paper by Proescher and Moody (1939) confirmed many of Specht's findings and made three important observations:

1. Although the test is presumptive, large areas of suspected material can be examined rapidly.
2. Dried and decomposed blood gave a stronger and more lasting reaction than fresh blood (3-year-old stains gave a brilliant luminescence).
3. If the luminescence disappears, it may be reproduced by the application of a fresh luminol–hydrogen peroxide solution; dried bloodstains may thus be made luminescent many times over.

These authors were also among the first to describe parameters for obtaining photographs of the luminescence produced.

McGrath (1942) found the luminol reaction to be rather specific for blood, testing a wide variety of substrates including serum, bile, sputum, pus, seminal stains, and feces and obtaining no reaction. However, he recommended that the test should not be used as a specific test for blood. He recommended that after its application, which should be kept to a minimum, the stain be photographed and allowed to dry. The stain is then available for all the usual serologic analyses. McGrath was unable to detect any differences between stained and unstained samples. Nevertheless, he used a sprayed area with no luminescence as a control for typing purposes.

Luminol's preparation has changed over the years. Specht (1937) used a mixture of luminol, calcium carbonate, and hydrogen peroxide or luminol with sodium peroxide. Proescher and Moody (1939) used sodium carbonate with hydrogen peroxide or sodium peroxide. McGrath (1942) used the methods of both Specht and Proescher and Moody and noted a trace of inherent luminescence of the solution on preparation, which could be removed with the addition of indazolon-4-carboxylic acid. Essentially, the preparation requires luminol, the addition of a base (to make the solution alkaline), and an oxidant. Grodsky and colleagues (1951) described the preparation of a solution composed of 0.07 g sodium perborate in 10 mL water, to which is added 0.01 g luminol and 0.5 g sodium carbonate.

Weber (1966) published a preparation of luminol consisting of the following stock solutions:

- 8 g sodium hydroxide in 500 mL water
- 10 mL 30% hydrogen peroxide in 490 mL water
- 0.354 g luminol in 62.5 mL of 0.4 N sodium hydroxide diluted to 500 mL with water

The test reagent is prepared by mixing 10 mL of each of the previously listed stock solutions with 70 mL of water. This preparation was stated to be an improvement over earlier ones in that it reacted well with fresh bloodstains and was more sensitive to dilute blood samples.

The Luminol Reaction

Some proteins possess a nonpeptide portion called a prosthetic group. The prosthetic group is intimately involved in the biological function of the protein. Such is the case for hemoglobin. The prosthetic group for hemoglobin is heme, and it is involved in the transport of O_2 and CO_2 in the body. Heme as shown in Figure 15.1 contains iron bound to the pyrrole system known as porphin. In the normal function of hemoglobin in the body, the iron atom does not undergo change in valence as oxygen is bound and lost; it is in the Fe(II) state. On aging and subsequent oxidation, the iron atom changes from heme to hemin or hematin; with this transformation comes a change in the color of the stain from red to brown. It is this conversion that is crucial to the subsequent oxidation/reduction reactions that are ultimately used to test for the detection of blood.

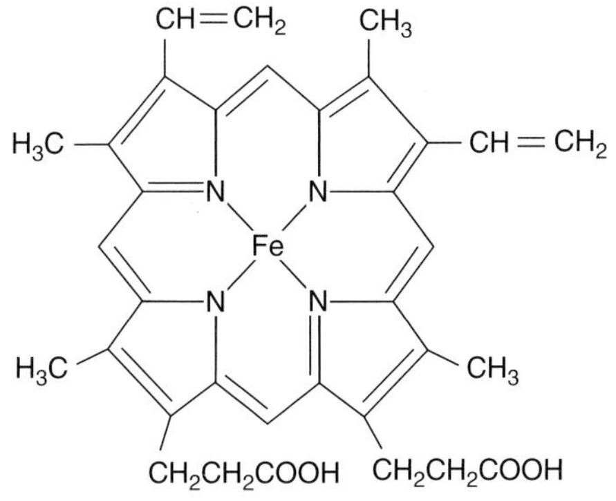

Figure 15.1 Chemical structure of Heme.

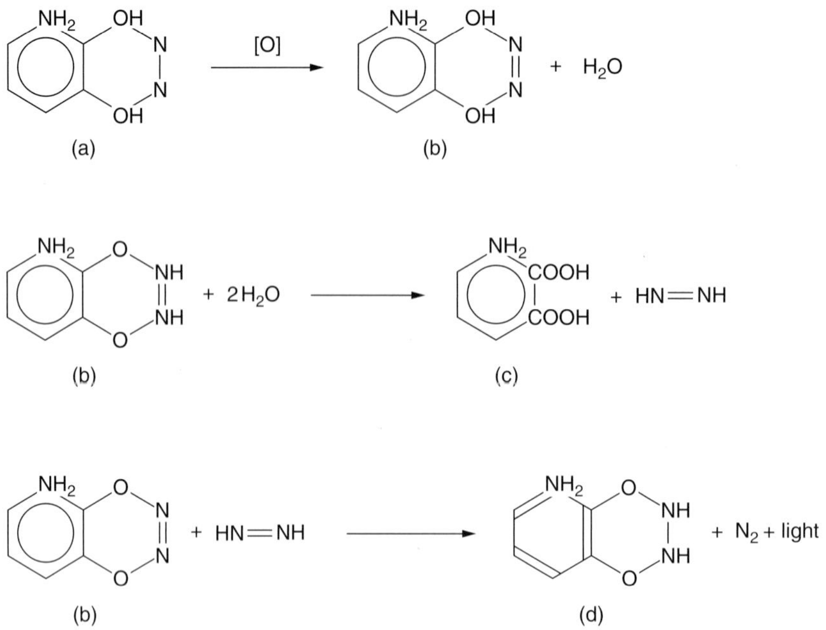

Figure 15.2 The Albrecht Mechanism. (From Gaensslen, R.E., *Sourcebook in Forensic Serology, Immunology, and Biochemistry,* Superintendent of Documents, U.S. Government Printing Office, Washington, DC, 1983.)

The reactions of luminol, or 3-aminophthalhydrazide, with blood producing what is termed chemiluminescence are complex. Many authors have made attempts to describe the reaction. Early studies by Albrecht (1928) indicated that luminol (Figure 15.2a) was oxidized to form intermediate compounds (Figure 15.2b), which were hydrolyzed in the basic solution to yield a dianion, phthalic acid (Figure 15.2c), and N_2H_2. The dianion would further react with N_2H_2 and form nitrogen, light, and luminol (Figure 15.2d). Because Albrecht could only isolate unchanged luminol from the reaction mixture after chemiluminescence had ceased, he thought that luminol itself was the light-emitting species.

Proescher and Moody (1939) correctly predicted the structure of luminol in acid (Figure 15.3a) and the tautomeric structures (Figures 15.3b and c) in alkaline solutions. Form (a) gave a strong blue luminescence. It was believed that the chemiluminescence was produced by oxidation in an alkaline solution of either form (b) or (c). The authors speculated that the reaction proceeded as in the Albrecht mechanism.

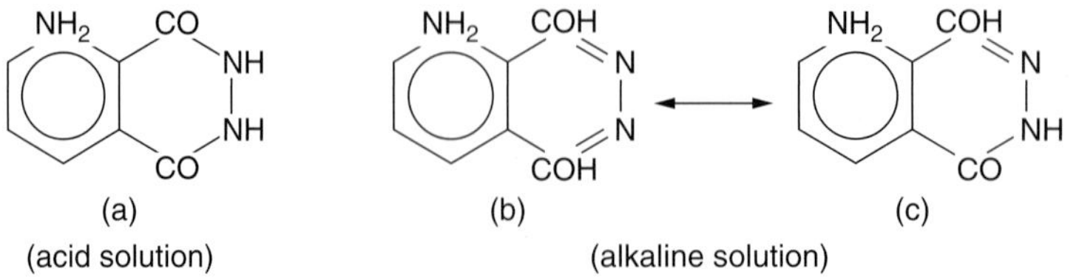

Figure 15.3 Luminol structures in solution. (From Proescher, F., and Moody, A.M., *J. Lab. Clin. Med.,* 24:1183–1189, 1939. With permission.)

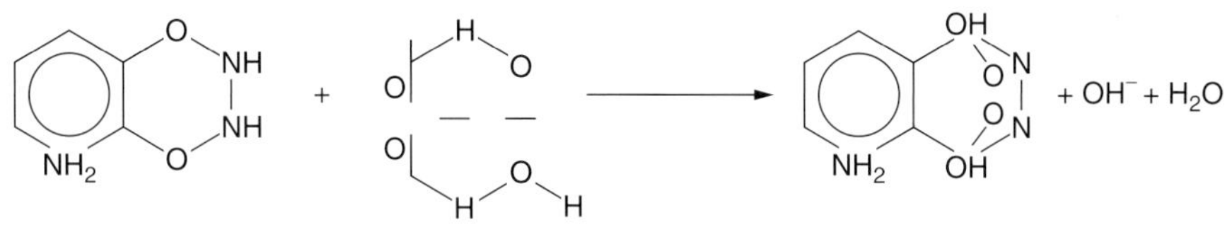

The Detection of Blood Using Luminol

Figure 15.4 Formation of a luminol endoperoxide.

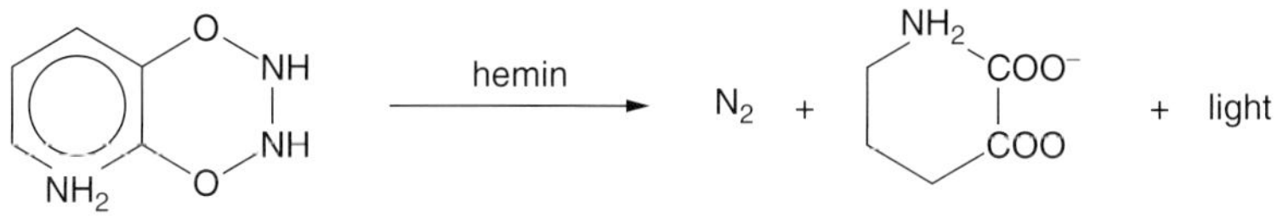

Figure 15.5 Irreversible oxidation of luminol.

The Albrecht reaction was seriously challenged in the early 1960s. Erdey and colleagues (1962) speculated that depending on the pH, hydrogen peroxide exists in different forms. At a pH greater than 12, hydrogen peroxide forms a transannular peroxide and oxidizes luminol to form a luminol endoperoxide compound, as shown in Figure 15.4. Then in the presence of catalysts such as hemin, luminol is irreversibly oxidized, as shown in Figure 15.5. White and colleagues (1961, 1964) speculated that the intermediate dianion, 3-aminophthalate (Figure 15.6), is the light-emitting species. The fluorescence spectrum of this compound matches the chemiluminescence spectrum of luminol. Gundermann (1965) restated White's findings that the dianion is both the product and the light-emitting particle of the luminol reaction.

Finally, and most recently, Thornton and Maloney (1985) have presented an excellent overview of the study of the luminol reaction and have proposed a reaction of their own. In this mechanism, the hemin acts as a catalyst, triggering the oxidation of luminol by hydrogen peroxide in an alkaline solution. As was stated earlier, the iron in heme can exist in different states. The normal valence for iron in hemoglobin is in the Fe(II) ferrous state. On aging of the bloodstain, hemoglobin is converted to methemoglobin and the iron is oxidized to the Fe(III) ferric state. On the addition of hydrogen peroxide, the iron is oxidized to a transition state Fe(IV). At the time of the oxidation of luminol, the Fe(IV) is reduced to Fe(III), thus allowing the iron in heme to participate in another reaction. This phenomenon may explain the finding that stains can be dried and sprayed with luminol repeatedly, often resulting in brighter luminescence on latter applications as a result of the further conversion of heme to hemin in the stain. The exact mechanism of luminol's reaction with blood may not be completely understood, but the final outcome can be dramatic and provide the investigator with important information. It is for this reason, after all, that luminol is being used. It is luminol's preparation, application, and interpretation that will now be considered.

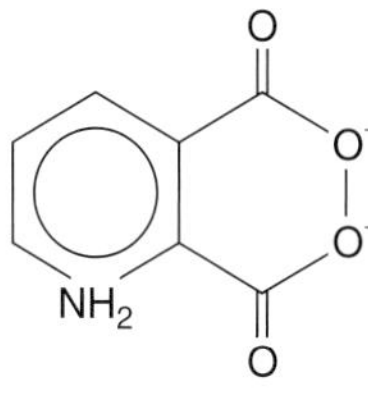

Figure 15.6 Structure of 3-aminophthalate.

Preparation of Luminol

A number of preparations for luminol exist. Two of the most commonly used preparations are detailed in Table 15.1 and Table 15.2. The preparation described by Grodsky and colleagues (1951) has an advantage in that the components can be placed in a luminol "kit" in the trunk of a car or crime scene vehicle and remain there indefinitely. The dry ingredients can be placed in discarded plastic 35-mm film containers, which conveniently hold the amounts of sodium perborate and sodium carbonate required to make 500 mL of solution. The 0.5 g of luminol required can be stored in a 1.5-mL microfuge tube. It is important to dissolve the sodium perborate first in 500 mL of distilled water before adding the luminol and sodium carbonate. Sodium perborate does not dissolve completely if all the reagents are mixed at once. The preparation of the solution should be performed in a 500-mL volume (or greater) plastic container. Never use water from the crime scene to prepare the reagent because "hard" water or well water often contains chemicals that may cause a severe background luminescence.

This author prefers to use the luminol preparation described by Weber (1966). This preparation has two advantages over the preparation of Grodsky and colleagues. First, it is reported to be more sensitive than other luminol preparations because of the lower concentrations of reagents, which tend to be inhibitory. Second, the lower concentrations of reagents make it safer to use. There are some disadvantages, however. First, the preparation of this luminol solution is more difficult and probably could not be accomplished by an investigator in the field from scratch. Second, it has been observed that the luminescence produced by this preparation does not last as long as others and may require repeated sprayings during photography. Nevertheless, this author prefers to use the Weber formulation. Stock solutions are kept refrigerated. Preparation of the working solution is easy and can be accomplished in the field very quickly from previously prepared stock solutions.

A word of caution here: although luminol is not a known carcinogen, the mutagenic effects are not known at this time. According to Material Safety Data Sheets available at the time of this writing (2004), luminol is an eye, skin, and respiratory irritant and may irritate eyes causing redness, watering, and itching. Sodium perborate is harmful if inhaled or swallowed, can cause eye and skin irritation, and is irritating to mucous membranes and the respiratory tract. Sodium carbonate is toxic and a severe eye, skin, and mucous membrane irritant. Dusts or vapors may cause mucous membrane irritation with coughing, shortness of breath, and gastrointestinal changes. Sodium hydroxide

Table 15.1 Preparation of Luminol According to Grodsky, Wright, and Kirk (1951)

1. Weigh out 3.5 g sodium perborate and add to 500 mL distilled water.
2. Stir until dissolved.
3. Add 0.5 g luminol and 25 g sodium carbonate.
4. Stir until dissolved.
5. Let stand 5 min.
6. Decant solution into plastic spray bottle.
7. Use immediately. The reagent does not keep well once prepared.

Note: The chemicals may be stored dry and in separate plastic film containers until needed.

Table 15.2 Preparation of Luminol According to Weber (1966)

Stock Solutions

8 g sodium hydroxide in 500 mL distilled water (0.4 N)
10 mL 30% hydrogen peroxide in 490 mL distilled water (0.176 M)
0.354 g luminol in 62.5 mL 0.4 N sodium hydroxide to final volume of 500 mL (0.004 M)
Store stock solutions in a refrigerator.

Test Solution

10 mL solution (A) + 10 mL solution (B) + 10 mL solution (C) + 70 mL distilled water
Final volume = 100 mL working solution
After thoroughly shaking the mixture, the solution should be allowed to stand for several minutes allowing any undissolved ingredients to settle. The solution is then decanted into a plastic spray bottle. If mixing is performed in the spray bottle, the end of the nozzle will most likely clog.

is corrosive and an irritant and may irritate the eyes and skin on contact. Hydrogen peroxide can cause irritation of the eyes and is a severe irritant for lungs and the respiratory tract.

Because the luminol preparation is applied as an aerosol, protective equipment is recommended. Dust and mist respirators, splash-proof safety goggles, gloves, and protective clothing are suggested (Figure 15.7). Contact lens wearers may notice that they are more sensitive to the luminol mist than others and should wear eye protection without air vents. It is recommended that contacts not be worn during the luminol application. The luminol application should not be conducted in the presence of an audience. Limit the number of people in the room for their own safety.

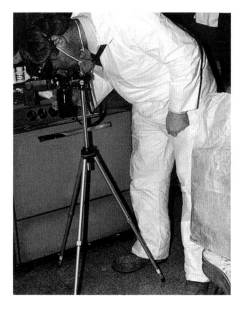

Figure 15.7 Protective clothing including eye goggles, dust and mist respirators, and gloves should be worn during luminol application.

Interpretation of Luminescence

Luminol is a presumptive test and has been found to react with plant peroxidases (fresh potato juice), metals (Figure 15.8), and some cleaners (especially hypochlorites). This author has found it to react with soil, fresh bleach solution, and cigarette smoke inside cars. It does not react with semen, saliva, or urine. Frequently, on reaction with metals, there will be a "twinkling" or "rippling" of the luminescence instead of an even, long-lasting glow as with blood. It is possible to obtain an even, long-lasting, moderate glow from an old porcelain sink or bathtub that has been exposed to cleaners. The reaction of luminol with a true bloodstain produces an intense, long-lasting, even glow frequently in patterns such as spatters, smears, wipes, drag marks, or even footwear impressions (Figure 15.9).

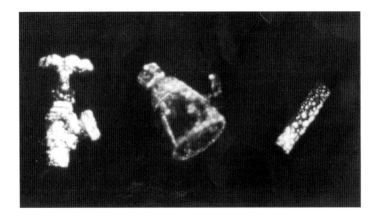

Figure 15.8 Luminescence given off by metal objects. **From left:** water valve, shower head and copper pipe.

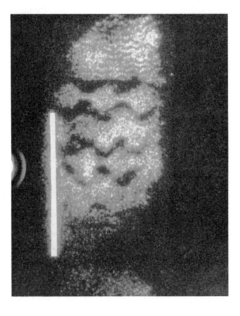

Figure 15.9 Footwear impression in blood as developed with luminol. Note the luminescent tape in the photograph as a reference.

Caution should be exercised when an entire object luminesces such as a sink, carpet, automobile seat, etc., because this is most likely the result of background luminescence. In the case of carpeting, a sofa or chair should be moved and the area should be sprayed to check for luminescence. If luminescence appears, it is probably the result of a cleaner or something in the substrate.

Use of Additional Presumptive Tests

After any positive reaction with luminol, the stained area should be circled with a grease pencil or other suitable marker. The stained area should be checked again with another reagent for blood such as tetramethylbenzidine (TMB), phenolphthalein, or o-tolidine. A clean, dry, cotton-tipped swab should be wiped lightly over the area of luminescence. The reagents can be applied directly to this swab, and the swab can be examined for its reaction. The luminol preparations do not induce a false-positive reaction with either TMB, o-tolidine, or phenolphthalein (Laux, 1991, and Gross, 1999).

Interpretation of Patterns

Blood can be deposited on an object in many ways. Blood can be transferred by contact, splashed, projected, and cast-off by many different means on a vast array of surfaces. Compound this with the ways and means of cleaning up the bloodstains, and the investigator can be faced with difficulties in locating stains. Luminol is a valuable tool for the investigator because it not only discloses the presence of blood but its distribution. Frequently, in "cleaning up," blood is distributed over a larger area, making the detection with luminol easier. The fact that blood is present and was cleaned up is often more important than the typing of the bloodstains. Experience is the best educator in locating and interpreting bloodstains with luminol, and it is recommended that the new investigator witness several demonstrations of luminol at crime scenes before interpreting any themselves. No mock crime scenes can duplicate all the variables in the real world. Finally, the investigator should be aware of and recognize the following types of blood distribution possible at crime scenes that have been cleaned up:

- Impact spatters
- Cast-offs from weapons, tools, etc.
- Pooled blood that has soaked into carpet padding, into cracks in wood floors, between tiles, etc.
- Wipe/smear marks associated with cleaning
- Footwear impressions
- Drag marks along walls, floor, etc.
- Partial or entire handprints on walls, furniture, etc.
- Outlines of weapons, cleaning tools, etc.

One should look for clearcut demarcations of luminescence. An excellent example of the different types of blood distribution one may encounter is in Lytle and Hedgecock (1978).

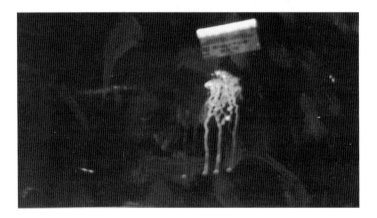

Figure 15.10 Luminol reagent "running" down the side of a vinyl car seat during application.

Effect of Substrates

The luminol test, like most analyses, can be quite easy at times and difficult at times. One of the most important considerations facing the investigator when using the reagent is the nature of the substrate possessing the stain. Absorbent materials such as fabric, carpeting, etc., are fairly easy to analyze, and stains on materials such as these can be resprayed and photographed successfully. Nonabsorbent substrates such as glass, vinyl, porcelain, tile, and linoleum are more difficult to examine. Stains on materials such as these will quickly "run" when sprayed (Figure 15.10). The investigator should use a minimum amount of reagent (use fine-mist spray bottle) and be ready to photograph the stain quickly. Reapplication of the reagent on these types of surfaces is usually not successful.

When one is asked to analyze a large room with linoleum, ceramic tile, etc., luminol can be applied and large areas can be covered quickly. However, should luminescence appear, spraying should be immediately halted. If photographs are desired, a camera should be set up with a tripod over areas not yet sprayed. These areas can be photographed with a flash, the room can be darkened, luminol can be applied, and photographs of the luminescence can be obtained. Gradually, the entire area can be covered and photographed. As stated earlier, reapplication of the reagent on these types of surfaces is usually not successful and the investigator often gets only one chance to obtain photographs.

Collection of Samples

After bloodstains are located, it is customary to collect the stains and take them to the crime laboratory. Obviously, if the object with the bloodstain can be moved, the entire object should be taken to the crime laboratory. This allows for the analyst to sample and collect stains to their own specifications. If the bloodstained object cannot be moved (e.g., walls, floors, ceilings, etc.), the stains can be collected on cotton threads or swabs. A very diffuse stain—on a wall for example—may be best collected on the end of a dry, cotton-tipped swab by rubbing the swab tip over the entire stain. Allow the swab to dry and place it inside a paper evidence envelope. Label the envelope accordingly. Controls should be collected by swabbing an area of the substrate that has been sprayed but does not exhibit luminescence.

Various sources list luminol's sensitivity ranging from 1 in 1 million to 1 in 1 billion dilution (Proescher and Moody, 1939; Grodsky and colleagues, 1951; Weber, 1966; Kirk, Lytle, and Hedgecock, 1978). However, these results were often obtained photographically with long exposures. A more practical dilution one may expect to see in the field is 1 in 1000.

Tests conducted by this author on bloodstains ranging from fresh to several years old revealed that the luminol preparations exhibited strong luminescence on all samples equally well except the very fresh samples, which exhibited little or no luminescence. Stains older than 17 years stored at room temperature produced luminescence. It is recommended that luminol not be used on very fresh stains (e.g., the day of the bloodshed) but rather wait a few days to allow for ample degradation of the heme. Pretreatment of the stains with hydrochloric acid (Ballou, 1995) is not recommended because it may have deleterious effects on any subsequent typing attempts and appears to decrease sensitivity and increase the level of background luminescence (Lytle and Hedgecock, 1978).

Effects of Luminol on the Subsequent Analysis of Bloodstains

Generally, when luminol is used, it is because a visual search with the naked eye has not revealed any bloodstains. Keeping this in mind, if bloodstains are discovered with luminol, chances are that there is not a great deal of blood present. Further tests on bloodstains that have been sprayed with luminol have been performed with varying degrees of success. A brief mention of luminol's effect on the typing of bloodstains appeared in Lytle and Hedgecock (1978). They mentioned that luminol does not prevent subsequent identification tests or ABO grouping but does interfere with erythrocyte acid phosphatase and phosphoglucomutase typing. Most studies with luminol have shown that the reagent does not interfere with subsequent confirmatory tests (Specht, 1937; Proescher and Moody, 1939; McGrath, 1942). Duncan and colleagues (1986) studied the effects of superglue, ninhydrin, black powder, and luminol on the subsequent analysis of bloodstains and found that luminol did not interfere with catalytic examinations, crystal tests for hemoglobin, species tests, or ABO blood grouping by absorption elution. They reported that enzyme typing was affected with a weakening or destruction of the entire band pattern.

Grispino (1990) tested Specht's original luminol preparation and a modified formulation on many subsequent blood tests. He found, as before, that the reagents did not affect presumptive tests, hemochromogen crystal formation, or species testing by the precipitin method. He found that luminol treatment did diminish the ability to detect blood group antigens by absorption elution. Luminol also routinely denatured most blood enzymes he studied after a short exposure.

Laux (1991) studied the effects of two preparations of luminol, those of Grodsky and colleagues and Weber and colleagues, on the subsequent analysis of bloodstains. The results were consistent with those observed by others. Luminol had no effect on further presumptive chemical tests, the Takayama confirmatory test, species identification by rocket electrophoresis, and ABO typing by absorption elution. Depending on the amount of blood detected by luminol and the manner in which it is collected, one may be able to determine the species origin and the ABO grouping of a sprayed bloodstain with the limiting factor being sample size. Extracts from bloodstains are added to a test chamber that contains a monoclonal antibody specific to human hemoglobin. Migration of the complex down an absorbent membrane results in a visible band if the source is human blood (apparent cross-reaction with ferret blood). Luminol had no apparent deleterious effects on this type of assay (Hochmeister, 1999). New immunochromatographic tests for the presence of human hemoglobin have become widely used throughout the forensic community.

Laux found that protein markers were affected by both preparations of luminol with a general decrease in band intensity. Phenotyping of bloodstains with the enzymes esterase D (EsD), phosphoglucomutase (PGM), PGM subtyping, glyoxalase (GLO), adenosine deaminase (ADA), and adenylate kinase (AK) was possible. The most deleterious effects of luminol were on the Group IV system with distortion of hemoglobin migration and decrease peptidase A (PEP A) and carbonic anhydrase II (CA II) bands to the point that they could not be typed.

Hochmeister and colleagues (1991) tested the effects of presumptive reagents on DNA typing of bloodstains and semen stains. Their results indicated that the amount of DNA yielded and its quality as typed by restriction fragment length polymorphism (RFLP) was the same from luminol-treated bloodstains as uncontaminated controls.

Cresap and colleagues (1995) studied the effects of luminol and Coomassie Blue on DNA typing by polymerase chain reaction (PCR). Based on their study, luminol had no effect on the ability to obtain PCR results, and "field investigators should feel free to use both luminol and Coomassie Blue as appropriate when searching a crime scene for bloodstains."

In the 7 years since the second edition of the book *Interpretation of Bloodstain Evidence at Crime Scenes* where this chapter first appeared, individualization of bloodstains by DNA analysis, specifically short tandem repeat (STR) analysis, has become standard operating procedure in crime laboratories throughout the world. Studies have shown that STR results can be obtained from bloodstains after treatment with luminol (Gross, 1999; Fregeau, 2000; and Della Manna, 2000).

Keep in mind that luminol is generally used as an investigative tool in cases where bloodstains are not readily observed and is often used as a last resort. In such cases, there is probably not a sufficient amount of blood for more than a confirmatory test and perhaps species identification. Should luminol be used and lead one to pooled blood, spraying of these stains should be avoided. However, should large bloodstains become inadvertently sprayed with luminol, they should be allowed to dry completely and retained because research has shown that it may be possible to generate some useful serologic information from them.

Luminol Photography

Shortly after the discovery of luminol and its recommended use in the detection of blood at crime scenes, it was apparent that photography of the distribution of blood was important. One of the first to obtain photographs of bloodstains with luminol was Specht (1937). Proescher and Moody (1939) briefly described photographing the luminol reaction. Zweidinger and colleagues (1973) described the photographic parameters required to obtain both 35-mm photographs and Polaroid photographs using Polaroid Type 410 Trace Recording film (ASA 10,000). They noted that drawbacks existed in using Polaroid film to record luminescence, including the fact that there is no permanent negative and no adjustments can be made in the length of exposure.

Lytle and Hedgecock (1978) photographed bloodstains they prepared on a variety of surfaces and cleaned up with soap and water. They obtained best results with Kodak Tri-X film developed in HC 110 developer. They used Nikon and Pentax 35-mm single-lens reflex cameras using 35- and 50-mm f/1.4 lenses. A 30-sec exposure often sufficed in capturing the luminescence. Longer exposures often resulted in overexposure of the luminol and in the swamping of the luminescence by ambient light under field conditions.

Thornton and Murdock (1977) made brief mention of making double exposures by placing a piece of cloth over the flash unit and exposing the negative again after recording the luminescence. Gimeno and Rini (1989) and Gimeno (1989) expanded on this technique, describing detailed procedures for obtaining "fill-flash time exposures," which resulted in both the luminescence and the object being clearly visible in the resulting photograph. The photographs are dramatic when successful, but practice is recommended before this method is tried in the field.

Niebauer and colleagues (1990) describe a more conventional way of obtaining luminol photographs, a method they refer to as a "film overlay negative system." In this method, the crime scene is recorded using ambient light (or flash) on one negative. The luminescence is recorded in total darkness on the next negative. Critical to this technique is the use of a tripod and a luminescent ruler or target of known dimensions. This ensures that the two photographs are of the same area and can be directly compared.

Recently, rather good photographs have been obtained by many investigators by simply "bouncing" a flash off a nearby wall during the luminol exposure. This method adds a sufficient amount of light to visualize the object without washing out the luminescence (Figure 15.11). Trial and error with your own equipment is simply the best way to succeed.

Figure 15.11 Luminol photograph using a bounce flash during the luminol exposure. Note the spatters of blood, the left handprint on the wall, and the smearing on the carpet.

The following procedures are to assist you in the setup and actual photography of the bloodstain. They should be taken as starting points and used for guidance only.

1. Setup
   - A 35-mm camera with "bulb" setting is recommended. A fast 50-mm fixed lens is suggested. The f-stop should be at the lowest setting (i.e., the aperture is fully open) for luminol photographs. A shutter release cable should be attached to the camera.
   - A tripod is necessary for luminol photography. Fluorescent tape placed on the ends of the legs and near the top of the tripod helps to locate it in the dark. A flash unit should be attached to the camera for flash photos and should be shut off or disconnected for the luminol photography.
   - Kodak T-Max 400 or Tri-Max Pan 400 black-and-white print films or color 400 ASA print films are acceptable.
2. Flash/dark sequence
   - A photograph of the object with a flash should be taken prior to the luminol photographs. The camera and flash should be prepared for a flash photograph. The camera lens should be parallel with the object being photographed. A fluorescent ruler or tape of known length must be in both the flash and luminol photographs.
   - After the flash photograph is taken, the flash should be shut off or disconnected. The f-stop should be at the lowest setting. The camera should be set for "B" or bulb. The room is darkened, and the reagent is applied. Using the shutter release cable, the luminol photograph is obtained. The stain can be resprayed during the luminol photograph to enhance the reaction.
3. Exposure
   - Research by the author has shown that a 30- to 45-second exposure with one of the recommended films captured the luminescent image very well.

Since this chapter appeared in *Interpretation of Bloodstain Evidence at Crime Scenes* 7 years ago, crime scene technicians and police detectives have used digital photography more and more. The luminol reaction can be captured successfully with some digital cameras that feature manual settings and allow extended exposure times. Figure 15.12 shows the results of the luminol reaction on a pair of tennis shoes worn by the subject that were cleaned shortly after an assault. The shoes showed no apparent bloodstains under visible light. Luminol was applied to the shoes, and a weak luminescence was observed visually. Digital photography was performed using a Nikon D100 camera with a 20-mm lens. The camera was set to manual, the ISO setting was at 1600, and the exposure was for 30 sec with an aperture setting of 2.8. The camera was mounted on an MP4 copy stand.

As mentioned earlier in this section, trial and error is simply the best way to test your camera equipment. Practice taking photographs of the luminol reaction at different settings. A tripod or copy stand is still required. Do not forget to include a luminescent ruler to ensure that the photographs are one to one. After obtaining the photographs, practice with various software programs that produce the best rendition of the luminol reaction — most importantly, the correct color and intensity of the reaction. Of critical importance is that the photographs

Figure 15.12 Luminol reaction of a pair of tennis shoes using digital photography.

are not enhanced or distorted in any way using software programs and truly represent the distribution of blood at the scene.

Report Writing

The fact that the blood was found (verified by a confirmatory test) and the distribution of that blood should be stated in a precise and clear manner. Unusual features of the stains such as spatters, smears, pooling, footwear impressions, direction of travel, etc., should appear in the report. The exact way in which the facts are put to written word is obviously up to the individual. The following examples of reports are to be used for guidance only.

- Application of luminol to the section of carpet submitted to the laboratory revealed the presence of luminescence in several areas. Chemical tests on extracts made from these areas were positive; however, species tests for human blood were negative.
- The examination of the interior of the vehicle with luminol revealed a luminescent pattern on the inside passenger window. Analysis of the pattern revealed it to be consistent with a high-velocity spatter. Chemical and immunologic analysis of the stain revealed the presence of type "A" human blood.
- Examination of the floor in the kitchen with luminol revealed the presence of stains extending from the stove to the doorway. These stains exhibited characteristics of a right tennis shoe with a suction-type pattern and lead from the stove toward the doorway. Chemical and immunologic analysis of the stains revealed the presence of human blood.

Testimony

The key here is to remember that luminol is a presumptive test for blood—nothing more. It is useful in that it not only indicates where blood is located but also patterns of blood distribution. Further analysis of the stain must be made before it can be called a bloodstain. Luminol has been accepted in court as a recognized tool in the detection of bloodstains. With knowledge and experience, it should be recognized by the investigator that luminol has an important role in the detection of bloodstains.

Case Studies

Case 1

Two female friends made arrangements for an evening out, and one of the women went home to change. This was the last time the victim was seen. The next day, the daughter of the victim reported her mother's disappearance to the police.

The victim was married, and authorities questioned her husband regarding his wife's disappearance. They were known to have had a stormy marital relationship. The husband stated that he had seen his wife leave their trailer home with a person in a red pickup truck and that was the last time that he had seen her.

Examination of the inside of the victim's trailer commenced approximately 1 week after her disappearance (Figure 15.13, Figure 15.14, Figure 15.15, and Figure 15.16). The inside of the trailer was in disarray, with clothes piled about and dirty dishes in the sink and on the table. No bloodstains were found during routine examination in daylight. When luminol was applied to the darkened living room, several areas of luminescence developed, including marks on the surface of the couch cushions (Figure 15.14) and on the couch with the cushions removed (Figure 15.16). Enough blood was found on the couch to type the ABO group and several protein markers. Blood from the parents of the victim was

Figure 15.13 Photograph of the couch taken from the crime scene in Case 1.

The Detection of Blood Using Luminol

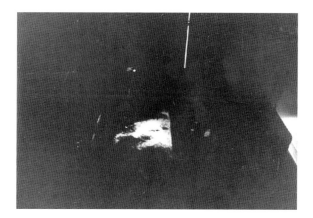

Figure 15.14 Luminol photograph of the couch in Figure 15.13 showing the distribution of blood over the surface of the cushions. Note the smeared blood on the center cushion and the partial handprint on the right cushion. The circle at the top of the couch is a penny that was used as a reference. This is not recommended; a luminescent ruler or tape should be used as a reference instead.

typed, and it was determined that the blood found on the couch could have originated from a child of the couple and was found in only 5% of the general population. The husband could not have been the source of the blood.

The husband was found guilty of murder. The body of the victim was never found. The finding of blood at the home was important in establishing that a crime had occurred. The fact that luminol demonstrated that the bloodstains had been cleaned up was important in linking the husband to the crime.

Figure 15.15 Photograph of the couch from the crime scene in Case 1 with the cushions removed.

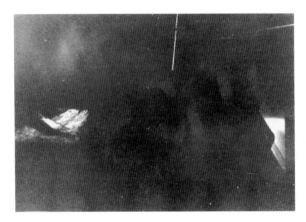

Figure 15.16 Luminol photograph of the couch in Figure 15.15 with the cushions removed. Note the attempts at cleaning and the smeared blood.

Case 2[1]

Alaska State Crime Lab personnel responded to a homicide in which the victim had died from numerous stab wounds. Examination of the victim's residence revealed evidence of a prolonged struggle in which a large amount of blood had been spattered throughout the living room. No visible bloody shoe impressions were located during the course of a detailed scene search. After completion of physical evidence collection, luminol was applied to the carpeted floors. Vivid, luminol-reactive shoe impressions were found leading from the bloodstained living room, down a hallway, and back to where a missing VCR had been located.

Closeup photographs of the individual shoe impressions were taken (Figure 15.17) along with overall views of the trail of shoe impressions (Figure 15.18). These shoe impressions were relatively unique in that they appeared to have been made by a cowboy boot with a deep "V" notch on the heel. The scene search also located the empty box from which the missing VCR had come, with the respective serial number. This information was crucial in developing a suspect. Two days later, an Alaska State Trooper responded to a residential disturbance call at which the missing VCR was located in the suspect's bedroom. A female and male suspect were arrested. During booking, the male suspect tried to hide his cowboy boots, which were found to have a hand-cut "V" on the heel (Figure 15.19). It was later discovered the male suspect had recently been released from a Colorado State penitentiary where the notches had been cut.

Laboratory comparison of the luminol impressions at the scene to the suspect's boots confirmed the scene impressions were made by a boot of the same sole design with the same type of notch cut in the heel. Because the unknown impressions were all located on carpet, insufficient detail was present for a positive identification. However, graphic displays of the impressions and the boots at trial provided powerful circumstantial evidence not only linking the suspect to the scene but also displaying his movements after the stabbing. The male suspect was convicted of first-degree murder.

[1] Contributed by James R. Wolfe, Alaska State Crime Lab.

The Detection of Blood Using Luminol

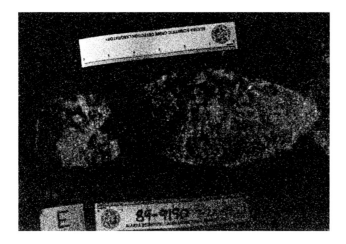

Figure 15.17 Closeup luminol photograph of the shoe impression found in Case 2. Note the "V" notch cut out of the heel.

Figure 15.18 Overall view of the trail of shoe impressions in Case 2.

Case 3[2]

Alaska State Crime Laboratory personnel responded to the scene of a brutal double homicide in Nome, Alaska. The victims, a mother and her daughter, were found murdered in the bedroom of their residence. Both had been stabbed numerous times. A knife handle

---

[2] Contributed by Chris W. Beheim, Alaska Dept. of Public Safety, Scientific Crime Detection Laboratory

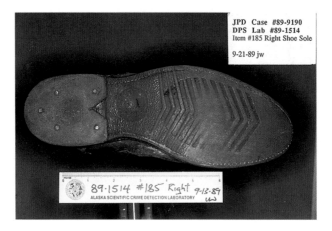

Figure 15.19 Photograph of the right shoe from the male suspect in Case 2. Note the "V" notch in the heel that corresponds to the shoe impression in Figure 15.17.

with a broken-off blade was found on the floor next to the bodies. Autopsy results revealed that two different knives had been used during the assault.

The floor of the residence was processed with luminol, and a trail of latent footwear impressions was developed (Figure 15.20). These impressions led from the bedroom into an area of the kitchen where knives were stored. The impressions then led back into the bedroom. Impressions on the kitchen floor showed the best detail, and closeup photographs were taken (Figure 15.21 and Figure 15.22).

A suspect was later apprehended, and his shoes were submitted to the laboratory for comparison to the luminol photographs. Test impressions made from the suspect's shoes matched the luminol impressions with respect to size and outsole tread design. One of the luminol photographs exhibited sufficient detail to show accidental characteristics from a

Figure 15.20 Trail of luminescent latent footwear impressions developed with luminol at the crime scene from Case 3.

The Detection of Blood Using Luminol

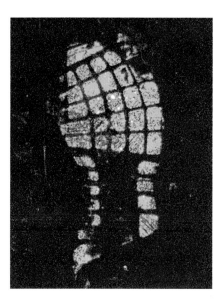

Figure 15.21 Closeup luminol photograph of latent footwear impression from Case 3 without fill-in flash.

Figure 15.22 Closeup luminol photograph of latent footwear impression from Case 3 with fill-in flash.

small cut located on the edge of the suspect's right shoe. The footwear impressions visualized by luminol proved to be the crucial piece of evidence in this case because the suspect later admitted to being in the residence but adamantly denied ever entering the kitchen. He later pled guilty to the crimes.

Chemical Enhancement of Latent Bloodstain Impressions

16

MARTIN EVERSDIJK

Introduction

Visible bloodstains and impressions present at crime scenes are often accompanied by the less obvious partial and latent impressions of footwear, palm and fingerprints, and other objects such as fabrics and weapons that may prove crucial to the investigation and scene reconstruction. Partial or latent impressions in blood may prove to be of superior quality for identification of class and individual characteristics after chemical enhancement than those impressions that are more heavily bloodstained. Therefore, a working knowledge of blood-enhancement techniques and their preparation, in conjunction with experience in their application in specific situations, is essential. There are numerous chemical enhancement procedures available for selective application, and the analyst should consider the following factors before choosing any enhancement procedure:

- The suitability of the technique for the specific crime scene with respect to the potential health hazards posed by the chemicals involved in the procedure that may dictate that the procedure be carried out under laboratory conditions
- The color of the enhancement relative to the contrast between the blood impression and the surface or substrate on which it exists
- The effectiveness of the technique on a porous versus nonporous surface
- The possible chemical damage to the surface
- The feasibility of lifting the enhanced impression from the surface or substrate
- The feasibility of optimizing the visibility of the impression using an alternative light source

The criteria by which the blood-enhancement techniques were selected for inclusion in this chapter are safety, availability of chemicals and equipment, ease of reagent preparation, and their stability and ease of application to the surface textures where blood impressions are commonly located. The following chemical enhancement techniques will be discussed in detail:

- Amido Black Water-based
- Amido Black Methanol-based
- Aqueous Leucocrystal Violet (ALCV)
- Crowle's Double Stain
- Coomassie Brilliant Blue
- Hungarian Red
- Titanium dioxide

Preliminary Procedures

Prior to beginning a chemical enhancement technique at a crime scene or with an item of physical evidence in the laboratory there are recommended preliminary procedures that should be performed. The blood impression should be examined visually with and without the use of an alternative light source. After this visual examination the analyst should photograph the visible area of the impression. Initial photographic documentation prior to using an enhancement technique may serve as a record should the enhancement technique fail and destroy the original impression. The impression should be documented as to its exact position at the crime scene or on the object.

The collection of a serologic and DNA sample must be done prior to the application of any blood-enhancement reagent. One of the best ways to take such a sample is by swabbing. The swab should either be taken from the more heavily bloodstained portion of the impression where there is not much visible detail or from smeared areas of the impression, which also lack detail. There is mention in the scientific literature of successful DNA profiles obtained from chemically enhanced blood impressions. However, it is wise to secure an uncontaminated blood sample for serologic and DNA examination. After having taken swabs of the blood impression the analyst should perform a presumptive test for the presence of blood. It is wise to do this presumptive test because impressions may appear to be blood even after chemical enhancement because the reagents used may react with, for instance, the protein albumin present in eggs, thus giving a false-positive result. For this reason, the chemical enhancement procedures should not be considered a presumptive test for blood because they are protein stains and do not react with heme in the hemoglobin molecule.

After these steps have been taken the analyst can initiate the chemical enhancement procedures. It is necessary to test the surface with the reagents prior to the enhancement in an area away from the visible impression, keeping in mind the possible presence of latent blood impressions around the visible area. The solvent can damage some surfaces and blood impressions.

The best results are obtained when the blood impression has had sufficient time to dry. The best method is to allow the blood on a surface or object to dry at a room temperature at a maximum of 30°C. Be aware that higher temperatures may cause unnecessary flaking of the blood, which can damage the impression. Hair dryers, ovens, and lamps should not be used.

Fixing Blood Impressions

One of the unknown variables the investigator has to deal with is the adhesiveness of the blood to the particular surface or substrate and the age of the blood. To ensure that the blood will not be dissolved, washed away, or otherwise destroyed when coming in contact with the enhancement reagents, the blood impression must be chemically fixed to the surface prior to applying any enhancement chemical.

There are many blood/protein fixatives, some of which are toxic, carcinogenic, flammable, or capable of causing other health-related problems. Examples of such fixatives are formalin, methanol, picric acid, and osmium tetroxide. Instead of these, a 2% aqueous solution of 5-sulphosalicylic acid has proved to be a safe and extremely effective fixative agent for water-based dyes such as Amido Black, ALCV, Crowle's Double Stain and Hungarian Red (Hussain and Pounds, 1988). The 5-sulphosalicylic acid forms a thin fixative layer across the surface of the blood. As effective as it is for water-based stains, the results are poor when used prior to methanol-based Amido Black and Coomassie Brilliant Blue. In those cases methanol is still the most effective fixing solution of choice.

Fixative with 2% Aqueous Solution of 5-Sulphosalicylic Acid

Fixing times vary based on the age of the blood and the adhesive nature of the blood to the surface, and they cannot be standardized. It is important to remember, however, that fixing should not take place unless the blood impression and the surface are totally dry. Faint blood impressions or areas where latent blood impressions are suspected should be fixed by spraying with a fine mist of 2% aqueous solution of 5-sulphosalicylic acid above the impression or by applying with a wash bottle.

The procedure of either spraying or wash bottle application should be carried out for approximately 3 min, but not constantly. A visual examination should be done every 30 sec to observe how the fixing process is progressing to prevent overfixing. The 2% aqueous of 5-sulphosalicylic acid can cause a change in color in the blood impression from its dark red to dark brown. This change of color may be an indication that the surface of the blood is fixed, but not in all cases.

Longer application time or immersion of the object into fixing solution may be necessary on larger impressions or heavier deposits of blood. The risk of damaging the thin layer over the blood created by the fixative with the 2% aqueous solution of 5-sulphosalicylic acid is always present. This risk is greater with heavily stained blood impressions on soft or flexible surfaces. Always perform a test on similar material prior to application.

Prior treatment with cyanoacrylate to develop latent fingerprint impressions does not fix the blood, and fixing must still succeed this process. Cyanoacrylate treatment will likely reduce the quality of the blood-enhancement results.

After the fixing is complete, the excess fixing solution should be removed by the following:

- Shaking it from the object
- Removing the solution with a water vacuum cleaner without contacting the surface
- Air drying with a cold blow dryer
- Using a distilled water rinse

Preparation of the Fixing Solution 2% Aqueous 5-Sulphosalicylic Acid

Fixing Solution
20 g 5-Sulphosalicylic acid
1000 mL Distilled water

- Weigh out 20 g of 5-sulphosalicylic acid on a balance paper, and place it in a 2000-mL glass beaker.
- Add 1000 mL distilled water. Stir the solution on a stirring device. A clear water-based solution will result.
- Using a clean funnel pour the solution into a 1000-mL dark glass bottle.
- Label and date the bottle.

Fixing Impressions with Methanol

When doing chemical enhancement of a bloodstained surface, methanol-based dyes are generally preferred over water-based dyes because they produce sharper detail in the boundary lines of the blood print. Methanol is still the most effective fixing solution of blood prior to using methanol based dyes. The methanol-based dyes described in this chapter are Amido Black and Coomassie Brilliant Blue. A standard time for fixing blood impressions with methanol cannot be given. Like other fixing methods it depends on the age of the blood and the adhesiveness of the blood to the surface.

Faint blood impressions or areas where latent blood impressions are suspected should also be fixed by spraying or washing with methanol. Latent or faint impressions, as well as thick deposits of blood, can be fixed in the same manner as the application of the aqueous solution of 5-sulphosalicylic acid solution previously described. Methanol can cause a change in color in the blood impression from dark red to dark brown. This change of color may be an indication that the surface of the blood is fixed, but it is not a guarantee.

Chemical enhancement processes involving methanol should not be performed at the crime scene. An exception would be if the blood impression were at risk of damage or destruction if the evidence item on which it was situated was relocated to the laboratory. When a methanol-based dye is used, both the surface and blood impression may be damaged. If the content of the surface material is unknown, it is recommended that a control test on similar material be performed with the reagents prior to the enhancement or on the article in an area away from the suspected impression. Reagent grade methanol (99+%) must be used. Less pure methanol can cause unnecessary dilution of the blood impression. Methanol will evaporate during use, leaving no excess on the surface.

General Methods for Application of the Fixing Solutions

Method 1

- Pour sufficient fixing solution into a clean, dry, metal dish to cover the bloodstained surface.
- With the stained side up, immerse the impression in the fixing solution for a few minutes. Check the impression regularly. The degree of fixing required depends on factors such as the surface texture, adhesion of the stain to the surface, and the thickness and age of the bloodstain. If the blood impression is thick, it may be necessary to replace the fixing solution.

Method 2

- Apply the fixing solution using a wash bottle above the blood impression and allow the solution to run over the print for 30 seconds. This procedure may need to be repeated.

Method 3

- Use a spraying device capable of a fine mist, and lightly spray above the blood impression for 30 seconds. Repeat until the blood impression is well fixed to the surface, but avoid overfixing.

General Methods for the Application of the Chemical Dye Reagents

Method 1

- Pour sufficient dye solution into a clean, dry, metal dish.
- Immerse the surface or bloodstained object in the dye solution and rock the dish gently for approximately 30 to 90 seconds.

Method 2

- Apply the dye solution for approximately 30 to 90 seconds by using a wash bottle above the blood impression.

Method 3

- Use a spraying device that can create a fine mist and lightly spray above the blood impression for 30 to 90 seconds.

Rinsing Procedure

Rinsing removes excess dye solution, which clears the impression of unnecessary background color and creates clearer and sharper boundary lines in the impression. For the water-based chemical enhancement techniques discussed in this chapter a Citric acid and distilled water in a 3:1000 solution will give excellent results and should therefore be used unless otherwise described in the procedure. After treatment with the rinse solution, the final rinse and chemical discharge of the substrate background of the enhanced blood impression can be accomplished with distilled water after using rinse solution methods 1, 2, or 3 described in the following section.

Preparation of the Rinse Solution for Water-Based Chemical Dyes

Rinse solution
3 mL Citric acid (99+%)
1000 mL Distilled water

- Measure out 1000 mL of distilled water and pour it into a 2000-mL glass beaker with a stirring bar.
- Add 3 mL of citric acid.
- Place the solution on a magnetic stirring device for approximately 10 minutes. A clear, water-based solution will be produced.
- Using a clean funnel pour the solution into a 1000-mL dark glass bottle.
- Label and date the bottle.

Application of the Rinse Solution for Methanol-based Chemical Dyes

For the methanol-based chemical enhancement techniques discussed in this chapter glacial acetic acid and (99+%) methanol in a 1:9 solution followed by a 1:19 glacial acetic acid (99+%) and distilled water in a solution will give excellent rinsing results. After the methanol from the dye solution has evaporated, the first rinse solution should be applied. Apply the rinse solution until maximum contrast is achieved and no excess dye is present. Follow with the second rinse.

Preparation of the First and Second Rinse Solutions for Methanol-Based Chemical Dyes

These rinse solutions are for use on methanol-based chemical blood-enhancement techniques such as Amido Black methanol-based, Coomassie Brilliant Blue, and titanium dioxide. They should be applied in sequence.

Rinse solution 1	Rinse solution 2
100 mL Glacial acetic acid 900 mL Methanol	50 mL Glacial acetic acid 950 mL Distilled water

Rinse Solution 1
- Measure out 100 mL of glacial acetic acid into a 2000-mL glass beaker.
- Add 900 mL of methanol.
- Stir for 5 minutes or until a clear methanol-based solution is produced.
- Using a clean funnel pour the solution into a 1000-mL dark glass bottle.
- Label and date the bottle.

Rinse Solution 2
- Measure out 50 mL of glacial acetic acid into a 2000-mL glass beaker.
- Add 950 mL of distilled water.
- Stir for 5 minutes or until a clear, water-based solution is produced.
- Using a clean funnel pour the solution into a 1000-mL dark glass bottle.
- Label and date the bottle.

General Methods for the Application of the Rinse Solutions

Method 1
- Pour sufficient rinse solution into a clean, dry, metal dish.
- Immerse the surface in the rinse solution and rock the dish gently until all excess chemical dye has been removed from the background and the highest level of contrast is achieved. It may be necessary to change the solution once or twice.

Method 2

- Apply the rinse solution using a wash bottle, and allow the solution to run gently over the background until all excess dye has been removed.

Method 3

- Use a spraying device that can create a fine mist, and lightly spray above the blood impression on the surface for 30 to 90 seconds or until all excess chemical dye has been removed and the highest level of contrast is achieved between the impression and the background.
- Follow the rinse methods above with a final rinse of distilled water.

Caveats

Chemical enhancement of blood impressions can interfere with or totally damage evidence, which could be useful for other forensic examination, such as the following:

- Handwriting examination and further examination of material such as ink and paper can be destroyed.
- Indented impressions may be affected or destroyed by applying enhancement chemicals, which can change the structure of the surface. As a result these impressions can fade or even become completely useless.
- Paint, fibers, gunshot residue, and other types of trace evidence may be altered or lost.
- The collection of a serologic and DNA sample must be done prior to the application of any blood-enhancement reagent.
- Be aware of cross-contamination of DNA evidence if one or more objects are processed in the same solution. However, this cross-contamination can also occur with the other described methods of applying fixative, dye, and rinse solutions.
- If after methanol-based chemical enhancement a DNA sample had not yet been secured, scrape a sample from the blood impression instead of swabbing it. Because the methanol creates a thin layer over the impression, the swab will be ineffective.
- Never blot dry the enhanced surface or object with a paper tissue or towel. In addition to the danger of destroying the blood impression, this may also cause small cracks in the impressions.
- The contrast between the blood impression and the surface optimized by chemical enhancement can fade when the print is left exposed to light for a few days. In some cases this may not necessarily be a disadvantage.
- Photographing by means of a forensic light source may increase contrast by fluorescence or absorption. There is no standard excitation wavelength for the surface or chemical technique used; therefore, it is advisable to investigate the possibility of this enhancement technique by searching the entire available light spectrum.
- When chemical enhancement is performed on a porous multiple-layered item of evidence (e.g., clothing), take precaution to avoid cross-contamination of the blood impression from one layer to the other (e.g., from the front panel of the shirt to the back panel). This cross-contamination can lead to false interpretation, damage, or total destruction of the evidence. The best way to prevent this cross-contamination

from occurring is by enhancing each layer separately or by placing water-resistant paper between the (individual) layers before enhancement.

A well-known enhancement technique in fingerprint technology is the use of 1,8-diazafluoren-9-One (DFO) and Ninhydrin. DFO and Ninhydrin are very effective chemical reagents for latent fingerprints on porous surfaces such as paper. These technologies react with the amino acids, also present in blood, and possibly other constituents of the print. One of the advantages of using DFO and Ninhydrin prior to other blood-enhancement techniques is that they will provide some enhancement of light deposits of blood in combination with any latent fingerprint deposits. DFO and Ninhydrin followed by the use of zinc toning is a useful enhancement technique because the dye will fluoresce under a forensic light source. This improves the contrast on dark surfaces.

Amido Black Water-Based

Amido Black, also known as Amido Black 10B, Acid Black 1, or Naphthol Blue Black, is a commonly used chemical dye during forensic investigations. Amido Black stains the proteins present in blood — rather than the hemoglobin — or other body fluids a blue-black color. The dye ALCV, as opposed to Amido Black water-based, can be used both on porous and nonporous surfaces. In all likelihood, ALCV will diminish the use of Amido Black as a primary chemical enhancement solution. Amido Black can be used secondarily after the application of ALCV on a nonporous surface to establish more contrast.

When using water-based Amido Black for blood-enhancement, the forensic examiner has a choice between two possible fixative solutions. The first is without 5-sulphosalicylic acid, which means the blood impression has to be fixed with 5-sulphosalicylic acid before enhancement. The second solution does contain 5-sulphosalicylic and can therefore be applied onto the impression without prior fixing. However, the second solution carries the risk of blurring the boundary lines of the impression, especially where large deposits of blood are concerned.

Advantages

- With the proper safety requirements water-based Amido Black can be used both at the crime scene and in the laboratory.
- The preparation of Amido Black is a simple and inexpensive process.
- Amido Black will not detect the normal constituents of latent fingerprints.
- Careful application of aluminium fingerprint powders before application of Amido Black should not interfere with the dye to stain proteins present in blood.

Disadvantages

- The water-based Amido Black solution is not recommended for use on porous surfaces and dark blue or black surfaces.
- Amido Black is not suitable for the detection of the normal constituents of latent finger and footprints and therefore must be used in sequence with other forensic latent print techniques when presumed blood-containing surfaces are examined.
- Amido Black may react with proteins from sources other than body fluids.

Chemical Enhancement of Latent Bloodstain Impressions

RECOMMENDED PROCESS DIAGRAM

Photograph
↓
Visual examination
↓
Forensic examination and collection
of forensic material*
↓
Photograph*
↓
Allow to air dry
↓
Forensic light source examination
↓
Photograph*
↓
DNA sampling*
↓
Presumptive test for blood
↓
DFO†
Ninhydrin + zinc toning†
Cyanoacrylate fuming†
↓
Test the enhancement technique on the surface†
↔

Heavily bloodstained impression		Lighly bloodstained Impression
↓		↓
Fix by immersion*		Fix by spraying, immersion, or wash*
↓		↓
Chemical enhancement by immersion*		Chemical enhancement by spraying, immersion, or wash*
↓		↓
Rinse by immersion*		Rinse by spraying, immersion, or wash*
→	Reenhancement*	←

↓
Allow to air dry
↓
Photograph
↓
Lift with gelatin lifter*
↓
Forensic light source examination
↓
Photograph
↓
Physical developer
↓
Photograph

---

*If possible or if necessary
†May interfere with some chemical blood-enhancement techniques

Preparation of Amido Black Water-Based Without Fixative

Working Solution Without Fixative	Rinse Solution
21 g Citric acid 2 g Amido Black 10B, or Acid Black 1, or Naphthol Blue Black 85% 1000 mL Distilled water	3 mL Glacial acetic acid 1000 mL Distilled water

- Weigh out 2 g of Amido Black 10B on a balance paper and place in clean, dry, 2000-mL glass beaker.
- Add 21 g of citric acid.
- Add 1000 mL of distilled water and stir the solution with a stirring device for at least 30 minutes. A black water-based solution will be produced.
- Using a clean funnel pour the solution into a 1000-mL dark glass bottle.
- Label and date the bottle.
- Store in refrigerator or at room temperature.
- Shelf life is indefinite.

Although the freshly prepared dye is ready for use, it is better to allow the mixture to stand for several days.

Preparation of Amido Black Water-Based with Fixative

Working Solution	Rinse Solution
20 g 5-Sulfosalicylic acid 3 g Sodium carbonate 50 mL Formic acid 50 mL Glacial acetic acid 125 mL Kodak Photo-Flo TM 600 solution 3 g Amido Black 10B, or Acid Black 1, or Naphthol Blue Black 85% 500 mL Distilled water	3 mL Glacial acetic acid 1000 mL Distilled water

- Weigh out 3 g of Amido Black 10B, or Acid Black 1, or Naphthol Blue Black 85% on a balance paper and place into a 2000-mL glass beaker.
- Add 20 g of 5-sulfosalicylic acid.
- Add 3 g of sodium carbonate.
- Add 50 mL formic acid, 50 mL glacial acetic acid, and 125 mL Kodak Photo-Flo TM 600 solution.
- Add 500 mL of distilled water.
- Stir the solution with a stirring device for at least 30 minutes. A black water-based solution will be produced.
- Add 500 mL of distilled water to the black water-based solution and stir the solution with a stirring device for at least 5 minutes.
- Using a clean funnel pour the solution into the 1000-mL dark glass bottle.
- Label and date the bottle.
- Store in refrigerator or at room temperature.
- Shelf life is indefinite.

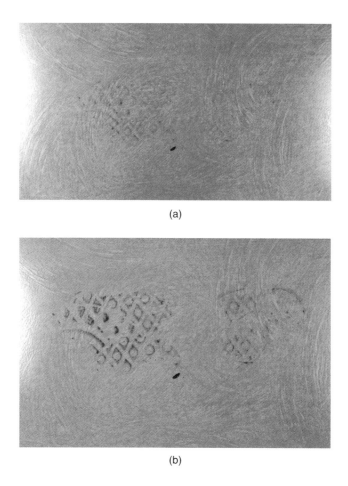

Figure 16.1 (a) A bloody footwear impression on a section of vinyl flooring prior to enhancement. (b) Footwear impression enhanced with Amido Black water-based. Amido Black water-based was chosen to chemically enhance the blood in this case because it does not alter the surface or cause unnecessary background coloring.

Although the freshly prepared dye is ready for use, it is better to allow the mixture to stand for several days.

Rinse solution: Glacial acetic acid 3 mL in 1000 mL of distilled water

Method of Use

Refer to the recommended process diagram for possible prior investigations.

Using Methods 1, 2, or 3 as previously described:

1. Apply the fixing solution.
2. Apply the working solution.
3. Apply the rinse solution.
4. Apply final rinse of distilled water.

Caveats
- Blood impressions made visible by Amido Black can be lifted with fingerprint tape or white gelatin lifter after being photographed.
- Amido Black can be used after ALCV on nonporous surfaces. This can be helpful for enhancing the contrast. Refixing is not necessary.
- Be aware that cyanoacrylate fuming can be detrimental to this dye process.
- If Amido Black is used prior to the collection of DNA samples, sampling for DNA profiling by swabbing may be ineffective. In these circumstances samples should be collected by scraping.
- If not satisfied with the achieved contrast it is possible to retreat the bloodstain with dye and rinse. It is not necessary to apply fixing solution between the first and the second enhancement. Rinse after desired contrast is achieved.

Amido Black Methanol-Based

Amido Black methanol-based stains the protein present in blood and other body fluids a blue-black color. A great advantage of Amido Black methanol-based is that it can be used on both porous and nonporous surfaces. One of the disadvantages, however, is that methanol is toxic, highly flammable, and explosive. Another disadvantage is that it can damage or destroy the substrate background. Various paints, varnishes, and plastics can be damaged, or boundary line details of the impression can be obliterated.

Nevertheless, as opposed to a water-based enhancement chemical, this methanol-based Amido Black stains more intensely and marks boundary lines more sharply.

Advantages

- Preparation of Amido Black is a simple and inexpensive process.
- The blue-black coloring is intensive, and the boundary line details are sharp.
- It is suitable for use on porous and nonporous surfaces.

Disadvantages

- Amido black methanol-based reacts with protein and is not specific for blood.
- It is flammable and explosive.
- It is a hazard to health.
- It should only be used under laboratory conditions.
- It is not suitable for use on dark blue and black surfaces.
- It can damage certain surfaces.
- Cyanoacrylate fuming may be detrimental to this process.
- Amido Black will not detect the normal constituents of latent fingerprints, so it should be used in sequence with other techniques when blood-contaminated surfaces are examined.
- Fingerprint powders containing metal, such as aluminium, can strongly react with methanol, creating the highly explosive hydrogen gas.

Fixing solution: methanol +99%.

Preparation of Methanol-Based Amido Black

Working Solution	Rinse Solution 1	Rinse Solution 2
2 g Amido Black 10B, or Acid Black 1, or Naphthol Blue Black 85%	100 mL Glacial acetic acid	50 mL Glacial acetic acid
100 mL Glacial acetic acid	900 mL Methanol	950 mL Distilled water
900 mL Methanol 99%+		

- Weigh out 2 g of Amido Black 10B, or Acid Black 1, or Naphthol Blue Black on a balance with a balance paper and place into a 2000-mL glass beaker.
- Add 100 mL glacial acetic acid.
- Add 900 mL of methanol.
- Stir with a stirring device for 30 minutes. A black methanol-based working solution will be produced.
- Using a clean funnel pour the solution into the 1000-mL dark glass bottle.
- Label and date the bottle.
- Store in refrigerator or at room temperature.
- Shelf life is indefinite.

Method of Use

Refer to the recommended process diagram for possible prior investigations.

Using Methods 1, 2, or 3 as previously described:

1. Apply the fixing solution.
2. Apply the working solution.
3. Apply the first rinse solution.
4. Apply the second rinse solution.
5. Apply final rinse of distilled water.

Amido Black Water-Ethanol Based

A relatively new and more surface-friendly dye is Amido Black water-ethanol based. The fixing and rinse solutions are methanol free. Because this dye is nonflammable it is suitable for use at a crime scene for chemical enhancement of blood impressions.

Fixing solution: 2% aqueous solution of 5-sulphosalicylic acid.

Preparation of Amido Black Water-Ethanol Based

Working Solution	Rinse Solution 1
1 g Amido Black 10B, or Acid Black 1, or Naphthol Blue Black 85%	250 mL Ethanol
50 mL Glacial acetic acid	50 mL Glacial acetic acid
250 mL Ethanol	700 mL Distilled water
700 mL Distilled water	

- Weigh out 1 g of Amido Black 10B, or Acid Black 1, or Naphthol Blue Black on a balance with a balance paper and place in a 2000-mL glass beaker.
- Add 50 mL glacial acetic acid.

- Add 250 mL of ethanol.
- Stir with a stirring device for 5 minutes. The black ethanol-based working solution will be produced.
- Add 700 mL of distilled water.
- Stir for an additional 5 minutes. The black, ethanol-water based working solution will be produced.
- Using a clean funnel pour the solution into the 1000-mL dark glass bottle.
- Label and date the bottle.
- Store in refrigerator or at room temperature.
- Shelf life is indefinite

Although the dye is ready for use, it is better to allow the mixture to stand for several days.

Preparation of the Rinse Solution for Amido Black Water-Ethanol Based

- Measure out 50 mL of glacial acetic acid and place into a 2000-mL glass beaker.
- Add 250 mL of ethanol.
- Add 700 mL of distilled water.
- Stir with a stirring device for 5 minutes. A clear ethanol-based rinse solution will be produced.
- Using a clean funnel pour the solution into the 1000-mL dark glass bottle.
- Label the bottle.
- Store in refrigerator or at room temperature.
- Shelf life is indefinite.

Method of Use

Refer to the recommended process diagram for possible prior investigations.
Using Methods 1, 2, or 3 as previously described:

- Apply the fixing solution.
- Apply the working solution.
- Apply the first rinse solution.
- Apply the final rinse of distilled water.

Caveats

- Blood impressions made visible with Amido Black methanol- or ethanol-based solutions can be lifted with fingerprint tape or white gelatin lifter after being photographed.
- If Amido Black methanol-based is used prior to the collection of DNA samples, sampling for DNA profiling by swabbing may be ineffective. In these circumstances samples should be collected by scraping.
- Be aware that cyanoacrylate fuming can be detrimental to this process.
- Amido Black methanol- or ethanol-based can be used after ALCV for enhancing the contrast. Refixing is not necessary.

Chemical Enhancement of Latent Bloodstain Impressions

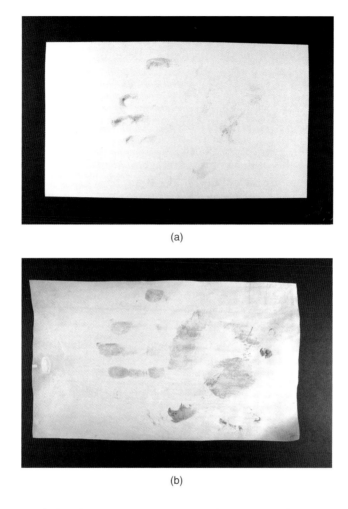

(a)

(b)

Figure 16.2 (a) Bloody handprint on paper prior to enhancement. (b) Handprint after enhancement with Amido Black methanol-based. Amido Black methanol-based is very suitable for porous surfaces as opposed to Amido Black water-based. However, this visual enhancement can only be used under laboratory conditions because of the flammability and health hazard of methanol.

- Some surfaces will be damaged by the use of Amido Black ethanol-based. It is recommended that a small part of the surfaces be tested away from the latent impression prior to starting enhancement on the impression.
- If not satisfied with the achieved contrast it is possible to retreat the bloodstain with dye and rinse. It is not necessary to apply fixing solution between the first and the second enhancement. Rinse after desired contrast is achieved.

Hungarian Red Water-Based

Hungarian Red is a dye that stains the protein present in blood and other body fluids a red color. In comparison to Crowle's Double Stain, Hungarian Red has the ability to fluoresce under a forensic light source after being lifted with a white gelatin lifter, which is an advantage when dealing with multicolor or dark surfaces. Hungarian Red also enables the forensic examiner to see more fine or barely visible details in the impression under a forensic light source after it has been lifted from the surface or article.

Advantages

- With the proper safety requirements it can be used at crime scenes and in the laboratory.
- The use of Hungarian Red is a simple and rather safe process.
- Hungarian Red will not detect the normal constituents of latent fingerprints.
- It has the ability to lift the enhanced blood impression with a white gelatin lifter
- It can fluoresce.
- It is suitable for use on nonporous surfaces.

Disadvantages

- Hungarian Red water-based is not recommended for the use on porous surfaces because this solution may be absorbed by porous surfaces and produce high background interference.
- It does not react with the normal constituents of sweat in uncontaminated latent fingerprints so it should be used in sequence with other techniques when blood-contaminated surfaces are examined.

Fixing solution: 2% aqueous solution of 5-sulphosalicylic acid.

Method of Use

Refer to the recommended process diagram for possible prior investigations.
  Using Methods 1, 2, or 3 as previously described:

- Apply the fixing solution.
- Apply the working solution.
- Apply the first rinse solution.
- Apply the final rinse of distilled water.

Rinse solution: Glacial acetic acid 30 mL:1000 mL distilled water.

Caveats

- Photograph the impression before lifting.
- For best results, the white gelatin lifter has to be at least at room temperature, and the foil should not be too old and/or discolored.
- The enhanced blood impression and the surface must be completely dry.
- Remove the protective transparent layer from the gelatin lifter and place the lifter surface over the blood impression. Be very careful not to let air bubbles develop under the gelatin lifter, otherwise there is the possibility of losing detail on the lifted blood impression.
- Leave the lifter on the blood impression for 15 to 30 minutes. (The optimal time is dependent on the level of the coloration of the enhanced bloodstain. In some cases very faint bloodstains were lifted after leaving the gelatin lifter on the trace for more than 2 hours.)

Chemical Enhancement of Latent Bloodstain Impressions

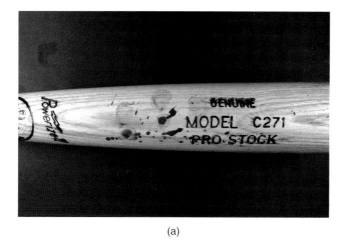

(a)

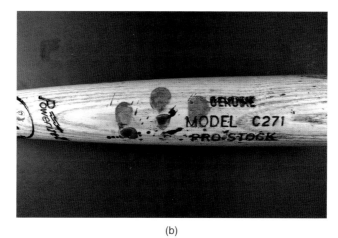

(b)

(c)

Figure 16.3 (a) A wooden baseball bat with a blood spatter pattern and several bloody fingerprints prior to enhancement. (b) Baseball bat after treatment with Hungarian Red water-based. (c) Closer detail of the enhanced fingerprints. This enhancement chemical was used because the baseball bat was varnished, making the surface nonporous. Additionally, the purple color provides sufficient contrast, even on the grains of the wood.

- After lifting the gelatin lifter from the blood impression, the protective transparent layer can be placed over the lifter surface for protection during transportation.
- Lifting the Hungarian Red dye can be done more than once with or without restaining. (Heavily blood-stained impressions may require new fixative before restaining.)
- The blood impression on the gelatin lifter must be photographed within several hours after lifting because the lift is not a permanent record. The dye will slowly spread into and over the gelatin surface of the lifter, causing loss of clarity.
- Blood impressions on a white gelatin lifter enhanced with Hungarian Red can fluoresce when excited with green light (515 to 560 nm) and viewed with red goggles.
- The fluorescence can be photographed when a red filter, such as Kodak Wratten 25, is used on the camera lens. If the surface is dark or multicolored, the fluorescence can be very helpful in examining the impression and enhancing the photographic contrast.
- Be aware that a blood print lifted with a gelatin lifter is a mirror image.
- Be aware that cyanoacrylate fuming can be detrimental to this process.
- If not satisfied with the achieved contrast, it is possible to redye and rinse the bloodstain. It is not necessary to apply fixing solution between the first and the second enhancement (except in case of heavily stained blood impressions). After the desired contrast is achieved, the final rinse can take place.

Aqueous Leucocrystal Violet

ALCV is the completely reduced form of Crystal Violet and is therefore colorless. When ALCV solution, which contains hydrogen peroxide, comes into contact with the heme portion of the hemoglobin in blood, a catalytic reaction occurs, which generates an almost instantaneous purple-blue coloration.

The use of ALCV for the enhancement of bloodstain evidence has been popular since its introduction in late 1993 because it is a quick, uncomplicated, and rather safe method of enhancing the visualization of faint or latent blood impressions and it is suitable for both porous and nonporous surfaces. It is uncomplicated because many components of the solution already contain a fixing agent, therefore making prior fixing unnecessary. Moreover, it is possible to apply Amido Black water-based after the use of ALCV on a nonporous surface because ALCV reacts with the heme portion of the hemoglobin and Amido Black reacts with protein present in blood.

Advantages

- With the proper safety requirements it can be used at the crime scene and in laboratories.
- Preparation and use of ALCV is a simple and inexpensive process.
- There is minimal background discoloration on porous surfaces and is therefore highly suitable for examination of clothing and paper.
- It will not detect the normal constituents of latent fingerprints, and therefore it is recommended to be used in sequence with other techniques when blood-contaminated latent fingerprints are examined.

Disadvantages

- Boundary lines may be outlined less sharply because of the water bases of ALCV (as compared to methanol-based chemicals).
- It is not suitable for use on blue-purple-colored surfaces and some materials after cyanoacrylate fuming process.

Fixing solution: 2% 5-sulphosalicylic acid.

Preparation of ALCV

Working Solution	Rinse Solution
20 g 5-Sulphosalicylic acid	Ethanol (lab grade)
1000 mL 3% Hydrogen peroxide solution	or
7.4 g Sodium acetate	30 mL Glacial acetic acid
2 g Leucocrystal violet	1000 mL Distilled water

- Weigh out 20 g of 5-sulphosalicylic acid on a balance paper and place in a 2000 mL glass beaker.
- Measure out 1000 mL of 3% hydrogen peroxide solution.
- Stir with a stirring device. A clear water-based solution will be produced.
- Add 7.4 g of sodium acetate while stirring. A clear water-based solution will be produced.
- Add 2 g of Leucocrystal violet and stir for 30 minutes. A clear, very light blue, water-based solution will be produced.
- Using a clean funnel pour the solution into the 1000-mL dark glass bottle.
- Label the bottle and add the expiration date (30 days).
- Although the dye is ready for use, it is better to allow the mixture to stand for 60 minutes prior to use.

Rinse solution: Glacial acetic acid 3 mL:1000 mL distilled water.

Method of Use

Refer to the recommended process diagram for possible prior investigations.
Using Methods 1, 2, or 3 as previously described:

- Apply the fixing solution.
- Apply the working solution (Method 1 or 2).
- Apply the first rinse solution.
- Apply the final rinse of distilled water.

Caveats

- Because coloring of the impression on the surface after application of ALCV occurs instantaneously, spraying or pouring the ALCV solution is recommended. This method enables the forensic examiner to control the coloration and achieve maximum contrast.

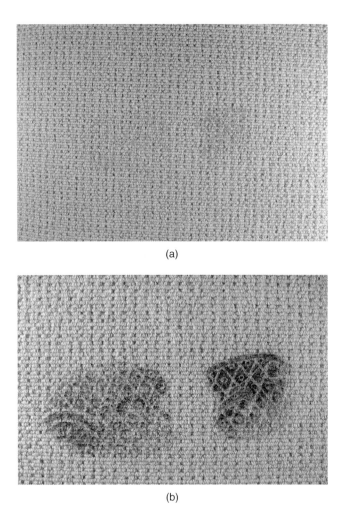

Figure 16.4 (a) A bloody footwear impression on a carpet prior to enhancement. (b) Footwear impression enhanced with Aqueous Leucocrystal Violet (ALCV). ALCV is very suitable as a visual blood-enhancement technique on the porous carpet. Although this ALCV already contains a fixing component, extra prefixing with a sulphosalicylic acid solution was used in this case. This creates additional fixing of the blood on the top of the carpet fibers to prevent the blood from being absorbed in the carpet, which would diminish the level of detail.

- Although ALCV contains 2% 5-sulphosalicylic acid, there is evidence that impressions fixed before the use of ALCV showed sharper boundary lines than traces that were not fixed prior to ALCV enhancement. Prefixing with 2% 5-sulphosalicylic acid is recommended in the case of porous surfaces.
- After the use of ALCV, Amido Black can be applied to create a higher contrast, without the necessity to refix.
- The shelf life of the working solution is a maximum of 30 days. It is therefore advisable to use the solution as fresh as possible because the detection rate diminishes as a result of chemical changes in the solution.
- Unwanted background development can occur when the ALCV process is used in direct sunlight because of photoionization. A blood impression, made visible by

Chemical Enhancement of Latent Bloodstain Impressions

Figure 16.5 (a) Bloody fingerprints on a dollar bill prior to enhancement. (b) Fingerprints after treatment with Aqueous Leucocrystal Violet (ALCV). Subsequent to this chemical enhancement Amido black methanol-based may be applied as well. Because ALCV reacts with the heme group of the hemoglobin and Amido Black with proteins present in blood, treatment with the latter may achieve an even better visual enhancement.

ALCV, needs to be photographed as soon as possible. The contrast of the impression against its background is not permanently visible. Discoloration of the surface and impression will slowly occur with exposure to light and oxygen. Over time, the whole environment will turn purple.
- The contrast of blood impressions, which are enhanced with ALCV, will fluoresce and luminesce under a forensic light source under a variety of wavelengths.
- Impressions should be photographed. The contrast of impressions can differ as time passes because of the long-lasting effect of ALCV, so it is recommended to take several photographs at different intervals.
- New Leucocrystal violet should be obtained if the solution turns from clear to yellow.

Coomassie Brilliant Blue

Coomassie Brilliant Blue is a commonly used dye during forensic investigations, which stains the protein present in blood and other body fluids a blue-violet color. There are two different preparations of Coomassie Brilliant Blue solutions. One is a methanol-based solution, and the other is a methanol-water based solution. The methanol-based solution is comparable in use with the Amido Black methanol-based solution. However, on some surfaces such as porcelain or shiny plastics it will produce a better enhancement result than the Amido Black. The Coomassie Brilliant Blue methanol-water based solution can be used instead of Amido Black methanol-based solution on varnished or painted surfaces.

Advantages

- Preparing and using Coomassie Brilliant Blue is a simple and inexpensive process.
- Coomassie Brilliant Blue will not detect the normal constituents of latent fingerprints.
- It is suitable for use on porous and nonporous surfaces.

Disadvantages

- Coomassie Brilliant Blue colors protein present in blood and also that of other body fluids.
- It is flammable.
- It is a hazard to health.
- It can only be used under laboratory conditions.
- It can damage certain surfaces.
- Cyanoacrylate fuming may be detrimental to this process.
- Coomassie Brilliant Blue will not detect the normal constituents of latent fingerprints.
- It is not suitable for use on blue-violet-colored surfaces.

Fixing solution: Methanol +99%.

Preparation of Coomassie Brilliant Blue Methanol-Based

Working Solution	Rinse Solution 1	Rinse Solution 2
2 g Coomassie Brilliant Blue 900 mL Methanol 99% + 100 mL Glacial acetic acid	100 mL Glacial acetic acid 900 mL Methanol	50 mL Glacial acetic acid 950 mL Distilled water

- Weigh out 2 g of Coomassie Brilliant Blue on a balance with a balance paper and place in a 2000-mL glass beaker.
- Add 100 mL Glacial acetic acid.
- Place the solution on a stirring device.
- Add 900 mL of methanol.
- Allow to stir for 30 minutes. The blue methanol-based working solution will be produced.

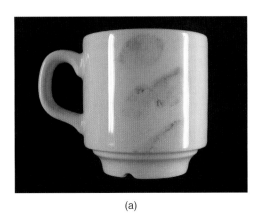 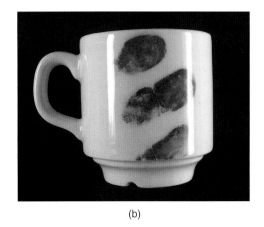

Figure 16.6 (a) A coffee cup with bloody fingerprints prior to enhancement. (b) Coffee cup after treatment with Coomassie Brilliant Blue. Coomassie Brilliant Blue is very suitable for visual enhancement of blood on surfaces made of porcelain or shiny plastics. It will produce a better enhancement result than Amido black methanol-based.

- Using a clean funnel pour the solution with the into the 1000-mL dark glass bottle.
- Label and date the bottle.

Rinse solution 1: Glacial acetic acid:Methanol = 1:9.
Rinse solution 2: Glacial acetic acid:Distilled water = 1:19.

Coomassie Brilliant Blue Water-Methanol Based

A relatively new and more surface friendly dye is a water-methanol based solution. The 2% aqueous solution of 5-sulphosalicylic acid fixing solution is recommended on varnished or painted surfaces. The chance of damage to the surface is less compared to fixative by methanol. Still, fixative with methanol prior to this dye will give the best results and should be used unless there is no chance of surface damage. This water-methanol based solution is not suitable for the use at a crime scene and must take place in a laboratory.

Fixing solution: 2% aqueous solution of 5-sulphosalicylic acid.

Preparation of Coomassie Brilliant Blue Water-Methanol Based

Working Solution	Rinse Solution 1
410 mL Methanol 99% +	450 mL Methanol
24 g Coomassie Brilliant Blue	100 mL Glacial acetic acid
84 mL Glacial acetic acid	450 mL Distilled water
410 mL Distilled water	

- Weigh out 24 g of Coomassie Brilliant Blue on a balance with a balance paper and place in a 2000-mL glass beaker.
- Add 84 mL of glacial acetic acid and place the solution on a stirring device.

- Add 410 mL of methanol and stir for 20 minutes.
- Add 410 mL of distilled water.
- Allow to stir for another 10 minutes. The blue methanol-water based working solution will be produced.
- Using a clean funnel pour the solution into the 1000-mL dark glass bottle.
- Label and date the bottle.

Preparation of the Rinse Solution for Water-Methanol Based Coomassie Brilliant Blue

- Measure out 100 mL glacial acetic acid and add to the 2000-mL glass beaker.
- Add 450 mL of methanol.
- Add 450 mL of distilled water.
- Stir for 15 minutes. A clear water-methanol based rinse solution will be produced.
- Using a clean funnel pour the solution into the 1000-mL dark glass bottle.
- Label and date the bottle.
- Store refrigerated or at room temperature.
- Shelf life is indefinite.

Method of Use

Refer to the recommended process diagram for possible prior investigations.
Using Methods 1, 2, or 3 as previously described:

- Apply the working solution.
- Apply the first rinse solution.
- Apply the final rinse of distilled water.

Caveats

- Some surfaces will be damaged by the use of Coomassie Brilliant Blue. It is recommended that a small part of the surface be tested prior to commencing with the enhancement of the impression or surrounding surface.
- If not satisfied with the achieved contrast, it is possible to redye and rinse the bloodstain. It is not necessary to apply fixing solution between the first and the second enhancement. After the desired contrast is achieved the final rinse can take place.
- Blood impressions enhanced with Coomassie Brilliant Blue can be lifted with a white gelatin lifter.

Crowle's Double Stain

Crowle's Double Stain is a dye that stains protein present in blood and other body fluids red. Since the release of Hungarian Red, which also stains proteins red, the use of Crowle's Double Stain as a red-coloring enhancement technique has declined. Hungarian Red has a great advantage over Crowle's Double Stain in that it enables enhanced impressions to be lifted with a white gelatin lifter and will fluoresce under a forensic light source. This and the often more intensive and clear print cast by Hungarian Red has caused the popularity of Crowle's Double Stain to decrease.

Advantages

- With the right safety requirements it can be used both at the scene of a crime and in a laboratory.
- Using Crowle's Double Stain is a simple process.
- Crowle's Double Stain will not detect the normal constituents of latent fingerprints.
- It is suitable for use on nonporous surfaces.

Disadvantage

- Crowle's Double Stain is not recommended for the use on porous surfaces.

Fixing solution: 2% 5-sulphosalicylic acid.

Preparation of Crowle's Double Stain

Working Solution	Rinse Solution
2.5 Crocein scarlet 7B	30 mL Glacial acetic acid
150 mg Coomassie Brilliant Blue	1000 mL Distilled water
50 mL Glacial acetic acid 99.5 %	
30 mL Trichloroacetic acid	
1000 mL Distilled water	

- Weigh out 2.5 g Crocein scarlet 7B on a balance paper and place in a 2000-mL glass beaker.
- Add 150 mg of Coomassie Brilliant Blue.
- Add 50 mL of glacial acetic acid.
- Add 30 mL of trichloroacetic acid. Add to the 2.5 g.
- Add distilled water.
- Stir the solution on a stirring device for approximately 30 minutes until all the dye is dissolved. A red-violet water-based solution will be produced.
- Using a clean funnel pour the solution into the 1000-mL dark glass bottle.
- Label and date the bottle.
- Keep refrigerated (preferably) or at room temperature.
- Shelf life is indefinite.

Method of Use

Refer to the recommended process diagram for possible prior investigations.
   Using Methods 1, 2, or 3 as previously described:

- Apply the fixing solution.
- Apply the working solution.
- Apply the first rinse solution.
- Apply the final rinse with distilled water.

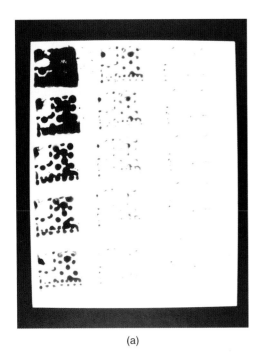 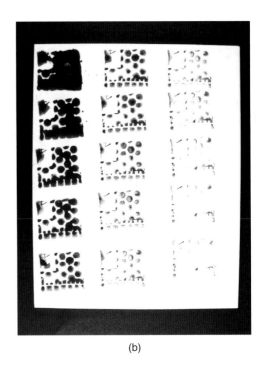

(a) (b)

Figure 16.7 (a) Blood impressions on a whiter ceramic tile prior to enhancement. (b) Impressions after treatment with the dye Crowle's Double Stain. The bloodstains on the tile are created with a stamp made of a part of a shoe sole. The print made in the upper left corner was the first print made on the tile after being in contact with blood. The print made under the left corner was the 15th print made after the first contact with blood. Clearly the thinned prints, after treatment with Crowle's Double Stain, contain more information than the first print.

Caveats

- Impressions made visible by Crowle's Double Stain can be lifted with fingerprint tape or white gelatin lifter after being photographed.
- Crowle's Double Stain can be used after ALCV has been used on nonporous surfaces. This can be helpful for enhancing the contrast. Refixing is not necessary.
- Be aware that cyanoacrylate fuming can be detrimental to this dye process.
- If Crowle's Double Stain is used prior to the collection of DNA samples, sampling for DNA profiling by swabbing may be ineffective. In these circumstances samples should be collected by scraping.
- If not satisfied with the achieved contrast it is possible to restain and rinse the blood impression. It is not necessary to apply fixing solution between the first and the second enhancement. After the desired contrast is achieved the final rinse can take place.

Titanium Dioxide Suspension

Titanium dioxide is well known as a color dye for paint and other products that must be white. If titanium dioxide is dissolved in methanol it can be used as a solution to enhance blood prints on dark surfaces, resulting in a visible white-colored impression. Because titanium dioxide is suspended in methanol, there is no need for fixing prior to the application of the dye solution. Titanium dioxide, as a blood-enhancement technique, is a simple two-step process: spray and rinse. The methanol in this dye solution makes it unsuitable for use at a crime scene. It is unsuitable for use on porous surfaces or objects. This dye may damage painted, varnished, and plastic surfaces and can obliterate boundary line details. The white coloring can be an advantage over other methanol-based dyes considering the visible enhancement on dark surfaces or objects. Titanium dioxide–enhanced blood prints showed clear white boundary lines that could be seen under normal lighting conditions.

Advantages

- Preparing titanium dioxide dye is an inexpensive process.
- Using titanium dioxide dye is a simple, two-step, spray and rinse method.
- Titanium dioxide dye reacts with protein and will not detect the normal constituents of latent fingerprints.

Disadvantages

- Titanium dioxide dye is not recommended for use on porous or white surfaces.
- It is flammable.
- It is a hazard to health.
- It can only be used under laboratory conditions.
- It can damage certain surfaces.
- Cyanoacrylate fuming may be detrimental to this process.

Preparation of Titanium Dioxide Suspension

Working Solution	Rinse Solution
100 g Titanium dioxide 1000 mL Methanol 99%+	Ethanol Lab grade

- Weigh out 100 g titanium dioxide on a balance paper and place it in a 2000-mL glass beaker.
- Add 1000 mL of methanol. Stir the solution on a stirring device for approximately 10 minutes until all the dye is dissolved. A white methanol-based solution will be produced.
- Using a clean funnel, pour the solution into the 1000-mL dark glass bottle.
- Label and date the bottle.

Method of Use

Shake the solution before use. Spray the titanium dioxide suspension solution with continuous agitation onto the blood print. Use a spraying device that can create a fine mist and do not spray directly onto, but rather lightly above, the impression. Spray for 30 seconds, and allow the methanol to evaporate for approximately 30 seconds. Rinse. The rinse solution can be applied in the same manner as the working solution. Repeat the method if necessary until the desired level of contrast is reached, but prevent overstaining. The developed blood prints should be allowed to dry completely before photographing.

(a)

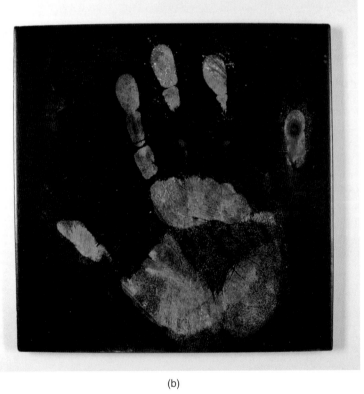

(b)

Figure 16.8 (a) A bloody handprint on a black kitchen tile prior to enhancement. (b) Handprint after treatment with titanium dioxide.

Caveats

- Titanium dioxide suspension solution can be used after other blood-enhancement chemicals. It can still develop some boundary line details in the blood print.
- If used in combination with other enhancement chemicals, titanium dioxide suspension solution will not create the same clarity or brilliance it creates when used as a single-enhancement technique.
- If titanium dioxide suspension solution is used prior to other blood-enhancement chemicals, the titanium dioxide, once dried, will create an effective shell over the ridges and stains. This shell reduces the reactions and coloration of other blood-enhanced chemicals to a minimum.
- Impressions made visible by titanium dioxide suspension solution can be lifted with fingerprint tape or black gelatin lifter after being photographed.

General Notes of Precaution

The reader is advised to follow safe work practices when handling the chemicals used in the described blood enhancement. Safe work practices include the use of appropriate personal protective equipment (e.g., gloves, laboratory coat, eye protection), engineering controls (e.g., fume hoods), and hygiene practices (e.g., washing hands, avoidance of eating or drinking and smoking during the procedures). The reader assumes the responsibility of obtaining the necessary knowledge concerning each chemical used, the hazard(s) it may pose, and the procedures and work practices necessary to prevent unhealthful exposure. This information is available from the Material Safety Data Sheets (MSDS) and the labels affixed to the chemicals purchased from the manufacturers.

Approaching the Bloodstain Pattern Case   17

Introduction

The manner in which an individual sets out to solve a problem will often determine whether the problem will be resolved. This is also true with establishing the geometric significance of bloodstain patterns. To be successful, the bloodstain pattern analyst should approach every case as a new problem to be solved, with a "new" set of facts and circumstances to be considered.

The bloodstain analyst should have the following characteristics:

- Open-minded, without prejudice, and possess a desire to learn
- Objective and not be swayed by the opinions or theories offered by the individual who has retained their services
- Never render opinions beyond what the evidence will allow no matter the amount of pressure being placed on them by their superiors, the individual who requested their assistance, or any attorney.

One must remember bloodstain pattern analysis is merely a forensic *tool* that may be used to better understand what may or may not have occurred during a bloodshed event. As with any forensic *tool*, the individual in control of it must have a solid foundation and background within their respective discipline prior to applying their knowledge to a situation that may inevitably deprive another human being of their liberty or life.

The stages of a bloodstain pattern analysis from the initial request for assistance through the generation of a bloodstain pattern report will be discussed within this chapter. These stages are meant to serve as an organized methodology for those who are asked to analyze the geometric significance of bloodstain patterns. These stages are in no way an alternative for the scientific method; rather, they are meant as a method for organizing data specific to bloodstain pattern analysis.

The people requesting bloodstain pattern analysis examinations are generally from law enforcement agencies, district attorneys, criminal defense attorneys, civil litigation attorneys, insurance companies, etc. A trend has been for criminal and civil defense attorneys to employ outside bloodstain pattern experts to reconstruct their clients' cases prior to the trial.

The bloodstain pattern analyst should realize that regardless of who is requesting an analysis of the bloodstain patterns, the person requesting the analysis will have their own theory of the case as well as a predisposition to one side. The analyst must maintain a totally objective thought process beginning with their first conversation with the individual requesting services.

Prior to the commencement of the examination of evidence, photographs, or the crime scene, one must determine whether the request is within the realm of knowledge and expertise of the bloodstain pattern analyst. Routinely, we are asked questions that are ultimately being referred to a forensic pathologist or a serologist. The bloodstain pattern analyst should not render opinions outside of the bloodstain pattern analysis discipline, unless the analyst has a previously acquired expertise in an additional discipline.

The following should be requested prior to the reconstruction of a bloodstain pattern case:

1. Brief background information
2. Examination of the actual crime scene
3. Examination of physical evidence
4. Photographs of the crime scene
5. Photographs of the physical evidence
6. Photographs of the autopsy
7. Autopsy report
8. Serology report
9. Crime scene diagram(s)
10. Crime scene investigator's notes or reports
11. Responding officer's notes
12. Emergency medical technicians' (EMTs) notes or depositions
13. Additional forensic reports

Fingerprint (bloody prints)
Shoe print (bloody prints)
Firearms (blood on weapon)
Other forensic evidence

Obviously, every case will not have each of the items listed. Although, if the information provided is too scant, vague, or poor quality, there is absolutely nothing wrong with the bloodstain analyst discontinuing his or her investigation by stating insufficient information exists to render a meaningful opinion in this matter.

Once this information has been received, a sequence for extracting the pertinent information needs to be established. Often the bloodstain analyst is only retained after all other investigative leads have been exhausted.

Sequence for Approaching Bloodstain Cases

Obtain a Brief Background of the Case

Background information should be collected at the initial contact when the request is made for a bloodstain pattern analysis.

- Name of victim
- Name of suspect
- Time of incident
- Date of incident
- Geographic location of the incident
- Who contacted you
- When you were first contacted
- Determine why the individual contacting you believes a bloodstain pattern analysis is necessary in their case

In certain circumstances, the analyst can establish early on whether the case is worthy of an examination. For instance, many attorneys have the misconception that if their client has no bloodstains on their person, they could not have been involved in the bloodshed event. The absence of evidence is very rarely an issue for a bloodstain pattern analyst who has been trained to analyze what is present and not to speculate on why something is not present. Kish and MacDonell discuss the significance, or rather insignificance, of the lack of bloodstains in considerable detail in their article entitled "Absence of Evidence is Not Evidence of Absence." Within this article, Kish and MacDonell cite several of the classic works (Piotrowski, Ziemeke, Walcher, etc.). The lack of bloodstains on an accused person should be viewed as an issue of further inquiry. Often the apparent lack of bloodstains on an accused will relate to types and location of the deceased's injuries or the actions taken by the accused during or after the violent act.

The analyst should limit the amount of superfluous information the contact person volunteers. Routinely, attorneys will attempt to provide an entire life history of their case or client. Generally, this type of information will have little if any bearing on the bloodstain evidence or the analyst's participation. To maintain a completely objective mindset when reviewing a case, the contact person should be requested to refrain from reiterating any type of scenario in regard to what they believe occurred.

The attorney retaining a bloodstain expert should be somewhat skeptical of the bloodstain analyst who asks for the attorney's opinion as to what they believed occurred prior to conducting their own independent examination of the case.

Examination of the Crime Scene

The basics of examining a crime scene for bloodstain pattern evidence has been discussed in previous chapters. Hence, in this chapter we will only be discussing various crime scene issues that apply directly to the analyst's approach as it would pertain to the bloodstain patterns.

The bloodstain analyst will be dealing with essentially three types of crime scenes:

1. Fresh crime scene: A crime scene where no scene processing has commenced prior to the arrival of the bloodstain analyst on the scene.
2. Recent crime scene: A crime scene that has been previously processed for physical evidence but remains in control of law enforcement.
3. Old crime scene: A crime scene that has previously been released from the custody of law enforcement.

A primary concern of a bloodstain pattern analyst with any crime scene, whether visited by the analyst or not, is whether the bloodstain patterns identified are a direct result of the incident rather than the result of interventions made by EMTs, law enforcement, or others trying to be of assistance.

The examination of the crime scene, whether a fresh, recent, or old scene, should be viewed as the single greatest opportunity the bloodstain pattern analyst will have to obtain information. The actual crime scene holds a wealth of useful information for the analyst.

The earlier the bloodstain pattern analyst can be brought to the crime scene the better. More often than not, the bloodstain pattern analyst works on what we would refer to as a recent crime scene. With the recent scene, the analyst will have to initially differentiate between bloodstain patterns that were the direct result of the incident and bloodstain patterns that were artifactually created by EMT intervention, body removal, and scene processing efforts made by earlier investigators. To assist with these determinations, the analyst should use a set of scene photographs taken at the time the scene was originally processed. If the bloodstain pattern analyst should ever question whether a bloodstain pattern is artifactual, ALWAYS DOCUMENT the pattern and treat it as if it were authentic until further information is obtained to either include or exclude it as being created as a direct result of the incident. Cases have been reviewed by the authors in which the bloodstain patterns located at a crime scene actually resulted from a previous altercation within the same dwelling.

The true value of examining an old crime scene may not be fully understood until the bloodstain pattern analyst discovers a key piece of information from visiting an old crime scene. You will only have to find one piece of essential evidence from one old crime scene visit to make your past and future efforts worthwhile. Once again, the bloodstain analyst should use the original crime scene photographs when examining the old crime scenes.

Initial Review of Crime Scene and Evidence Photographs

An initial review of the crime scene and evidence photographs should be one of the first tasks. This review is not meant for actual analysis of the bloodstain patterns; it is for acquainting oneself with the crime scene, victim location, layout of the scene, bloodstain locations, physical evidence, etc. One may assimilate this review with the initial walk-through done prior to processing the actual crime scene. It is important to obtain some idea of the task ahead, and the best method for accomplishing this is to first make a preliminary review of the scene and evidence photographs. Remember this is an initial review. Creating a hypothesis or case theory would be premature at this point.

Examination of Autopsy or Hospital Photographs

When examining the autopsy or hospital photographs, the bloodstain analyst should be concentrating on extracting data about the following:

- Injuries of the victim(s)
- Injuries of the suspect(s)
- Sources of bleeding
- Locations of wounds
- Types of wounds
- Clothing worn by victim(s)

- Clothing being worn by the suspect(s)
- Types of medical interventions

Examination of Autopsy Protocol or Hospital Records

From the autopsy protocol or hospital records the bloodstain analyst should extract the following information:

- Injuries resulting in blood loss (external versus internal bleeding)
- Damage to any major artery or vein
- Whether blood was located within nasal passages, mouth, or anywhere within the air passages, which may result in the blood being forcefully expelled from the victim
- Height and weight of the victim(s)
- Types of injuries
  - Gunshot wounds
  - Stab wounds
  - Blunt-force trauma
  - Lacerations
- Location of all injuries

Autopsy protocol or hospital records will indicate the location and type of blood source, which, in turn, will be useful in determining the type of mechanism that could have created the bloodstain patterns at the scene. The hypothesis of a bloodstain pattern being the result of arterial spurting would be supported by identifying information about a severed artery within the autopsy report. Likewise, to support a hypothesis of a pattern being the result of blood being expirated from the injured party, it must be established that blood was present in the nose, mouth, or both. A victim's height and weight may assist in locating a model for reconstruction efforts if needed. The type and location of an injury will be useful in supporting hypotheses concerning the mechanism of bloodstaining and the location of the victim when the incident occurred. For example, if a bloodstain pattern whose size, shape, and distribution is consistent with an impact spatter associated with a gunshot and the pathology report indicates only an entrance gunshot wound, then one may deduce that the pattern was the result of backspatter rather than forward spatter.

A background in human anatomy can be of great assistance when extracting information from autopsy and hospital reports. If the bloodstain analyst is unclear about any of the information or terminology contained in any report, they need to contact the person who wrote the report for clarifications.

Examination of Crime Scene Sketches/Diagram(s)

Scene diagrams are useful in obtaining the size relationship within the scene as compared to the scene photographs. Scene diagrams also assist in establishing the location of physical evidence within the crime scene. It is helpful to use the scene diagrams when attempting to establish the location in which scene photographs were taken, especially if the bloodstain analyst was unable to attend the crime scene.

Examination of Crime Scene and Evidence Photographs

Prior to the examination of the scene photographs, they should have been numbered or lettered. Second, the photographs need to be organized in a manner that will allow for easy retrieval. Organize them first by overall type of photograph, and then organize each type by natural divisions within the scene.

An example of organizing photographs by general type follows:

1. Areas with bloodstaining
2. Areas without bloodstaining
3. Evidence photographs
4. Autopsy photographs
5. Photographs of defendant(s)
6. Additional photographs

An example of categorizing scene photographs within a particular type of residence:

I. Areas with Bloodstains
   A. Northwest bedroom
      1. North wall
         a. Window
      2. South wall
      3. East wall
      4. West wall
         a. Entry Door
      5. Ceiling
      6. Furniture
         a. Dresser (front, top, rear, left side, right side, bottom)

An example for categorizing photographs of bloodstains on a body:

I. Areas with Bloodstains
   A. Deceased (Jane Doe)
      1. Head
         a. Face
         b. Scalp
         c. Hair
      2. Arms
         a. Left anterior/left posterior
         b. Right anterior/right posterior
      3. Hands
         a. Left palm/left back
         b. Right palm/right back
      4. Torso
         a. Anterior
         b. Posterior
      5. Legs
         a. Left anterior/left posterior
         b. Right anterior/right posterior

The categorizing of the scene photographs for analysis is the same as dividing the actual scene into specific areas for analysis. These divisions of photographs often apply to how the final report on the case will be organized. Obviously, the analyst will need to modify the photographic categories to fit each individual crime scene. The greatest degree of modification will be necessary in cases involving vehicles, outside crime scenes, and cases involving multiple victims.

The bloodstain pattern analyst should extract the facts about the bloodstain patterns from each photograph as well as other relevant information. These facts may include the following information about the bloodstain: size, shape, distribution, location within the scene, directionality, overall stain pattern appearances, relationship of one stain pattern to another, and other physical evidence. Note any variables that may be evident within the scene photographs, such as the following:

- Excessive dirt and grime
- Movement of evidence between scene photographs
- Surface texture issues
- Photographic prospective concerns
- Overall quality of the photographs

When extracting information from scene photographs, note the number or letter previously assigned to the photograph, as well as the category the photograph was taken from. Information extracted from photographs may read as follows.

Photograph A: Northeast Bedroom, North Wall:

Spatters of 1 to 3 mm in diameter
With upward directionalities
Pattern located between 0 in. and 27 in. from the floor
Contact stain located between 4 and 9 in. from floor; center of pattern located 97 in. from East wall and 56 in. from the West wall.
Apparent hair within contact stain pattern
Equally distributed with no void areas

Note the only information being extracted from the photographs is visible facts about the bloodstain patterns. No theories or conclusions are drawn at this time. At this stage the analyst should only be collecting data for later synthesis.

Examination of Physical Evidence

It is advantageous to categorize the physical evidence by where it was collected prior to its examination.

- Evidence collected from scene
- Evidence collected from victim(s)
- Evidence collected from accused(s)

Examine all of the evidence collected from one location prior to moving to the articles of evidence in the next category. Note the size, shape, distribution, configuration, and location and establish the side in which the bloodstain originated, directionality, and overall appearance of the bloodstain patterns. Refer to the articles of evidence by their previously assigned item numbers wherever possible. Photograph all of the significant bloodstain patterns identified on each article of physical evidence. These photographs may be your only source of visual evidence if the bloodstain pattern(s) are consumed during serologic testing. These photographs of the bloodstain patterns on the physical evidence should be used when discussing the significance of the patterns within the analyst's final report. The bloodstain pattern analyst should make the report as visually orientated as possible. When presenting testimony in regard to the bloodstain patterns on garments, the analyst should use the actual garment when explaining the significance of the bloodstain patterns to the jury. It should be recognized that judges in some jurisdictions will not allow the garment to be displayed to the jury because they consider it inflammatory.

The bloodstain pattern analyst should use the scene photographs to verify the bloodstain patterns on the deceased's clothes. The removal and transport of the deceased from the crime scene to the morgue may result in the bloodstain patterns on their clothes being altered or destroyed, or new bloodstain patterns may be inadvertently created. It is highly recommended that if an apparently significant bloodstain pattern is identified on the clothing of the deceased at the crime scene, the pattern should be photographed excessively from various angles and with various lighting configurations, then the article of clothing containing the bloodstain pattern should be removed from the deceased at the scene to prevent it from being destroyed or altered after consulting with the medical examiner.

Sequencing Evidence Examinations

Items of evidence submitted for laboratory analysis will often require examinations by more than one analyst. The question of which analysis should be conducted first often arises. The answer is dependent on the nature of the evidence, the type of examinations requested, and to what degree the item of evidence will be affected by the analysis. The analysis of the evidence should proceed based on the destructiveness of the testing, least destructive analysis first. On complex cases, scientists from multiple laboratory sections will need to confer to establish the best sequence of analysis for a specific item of evidence. The analysis of bloodstain patterns is a nondestructive analysis. The trace evidence examination is typically the only examination that takes precedence over the bloodstain pattern analysis examinations. The following examples are designed to illustrate potential examination sequences:

> Example 1: Suspect's shirt has been submitted for trace analysis, DNA, and bloodstain pattern analysis.
>
> Step 1. Visual examination of garment for evidence
> Step 2. Collection of trace evidence
> Step 3. Bloodstain pattern analysis
> Step 4. Initial serology presumptive testing
> Step 5. Sampling for DNA
>
> Example 2: A revolver was seized from a homicide suspect's vehicle. The victim in the case was beaten and shot at close range with the same caliber projectile

as what could be discharged from the evidence revolver. The revolver was submitted for trace analysis, serologic testing, bloodstain pattern analysis, discharge distance testing, bullet comparison testing, and fingerprint analysis.

Step 1. Visual examination of revolver
Step 2. Collection of any trace evidence
Step 3. Examination of exterior for bloodstain patterns
Step 4. Collection of samples for DNA
Step 5. Processing for fingerprints
Step 6. Sampling the inside of the barrel and cylinder for blood/biological material
Step 7. Discharge distance and bullet comparison testing

Serology Reports

The bloodstain pattern analyst must work in conjunction with the forensic serologist and forensic biologist. The analyst needs to be aware of whether the pattern they have identified in fact is the blood of one of the parties involved in the incident. Bloodstain analysts should insist on serologic testing of any pattern they are attaching significance to within a particular case, prior to their issuing a final report. The pattern may only be referred to as an "apparent" bloodstain pattern until serologic testing confirms it in fact is blood.

To assist in evaluating serologic findings, the analyst should chart whose blood was found where and on what item. When dealing with multiple victims, multiple defendants, and multiple articles of physical evidence, a serology chart will be of assistance in synthesizing the data from the written matter with the bloodstain patterns you observed. A serology chart for the bloodstain pattern analysis would look similar to Table 17.1

By charting the serologic findings in a simple manner, as seen in Table 17.1, you will be better able to attach significance to a particular bloodstain pattern in regard to whose blood it is, as well as where it was located.

Crime Scene Investigator's Notes and Reports

Extract any and all information in regard to initial crime scene investigator's scene observations, the actions they took, any contact the victim and suspect had, climate conditions at the crime scene, and eyewitness statements. Establish whether there are any variables that may have a direct or indirect effect on the interpretation of the bloodstain patterns. Examples of such variables could be the following:

- The suspect was actively treating the victim
- Police or EMTs coming into contact with both the victim and the suspect
- The removal of the victim from their original location
- Overall poor handling of the bloodstain evidence

Table 17.1 Serology Chart

Item/Sample #	Evidence Location	Victim's Blood	Accused's Blood	Mixture of Victim's & Accused's	3rd Party Blood	Inconclusive
S-1	NE bedroom wall	XXX				
S-2	Accused blue jeans			XXX		

Responding Officer's Notes

The notes of the officer who first responded are often useful in devising a timeline. Their notes will often inform the reader of how people were moved and the original location of the victim and the suspect on their initial arrival.

Emergency Medical Technicians' Notes or Depositions

The EMTs can be a valuable source of information to the bloodstain pattern analyst. The EMTs' medical interventions and actions taken while at the scene often account for many of the variables the bloodstain pattern analyst must consider when conducting their case evaluation. The saving of a life obviously takes precedent over all other situations, but in doing so the bloodstain patterns may be jeopardized. It is critical that the bloodstain pattern analyst establish whether a particular bloodstain pattern is the result of the actual incident or EMT intervention.

The following is a list of questions the bloodstain analyst should have answered by the EMTs who intervened:

- Who was assisting the injured party on your arrival?
- Did you observe any bloodstaining on the person lending assistance to the injured party?
- With multiple injured parties, who treated who? Did the same EMT ever come in contact with both parties?
- How did you move or alter the position of the injured party (or parties)?
- What types of medical interventions were used?
- Was there any blood in air passages of the victim?
- Was the victim mobile on your arrival?
- Did you alter the location of any piece of physical evidence (e.g., move the firearm)?

These questions, as well as any others asked of an EMT, should be designed to determine whether the bloodstain patterns were from the actual incident or a result of their intervention.

In those cases where the activities of the EMT personnel and removal of the victim to the hospital precede the arrival of the investigator, it is important to consult with the medical personnel regarding the original position of the body and the undisturbed condition of the scene on their arrival. Many EMTs make observations and take notes regarding the environment of the victim as well as movement and treatment rendered at the scene. They should be able to produce records of interventions made for a victim that may have produced bloodstains or altered those already present.

Additional Forensic Reports

This category refers to the extracting of information from any other forensic reports, which may assist in the reconstruction of the bloodstain patterns in a particular case (e.g., the blood found within a firearm by a firearms examiner, the bloody shoe print reported on by a shoe print expert, etc.).

Synthesizing the Data

The data collected from the photographs, scene examination, physical evidence, and the reports and notes on the case must now be synthesized. Review your notes and establish correlation's between what was observed at the scene and in the scene photographs with the information extracted from the written matter. While assimilating information, the analyst must consider all of the variables, which surround the facts of the case.

An example of a data synthesis would be as follows:

Data Synthesis for Bloodstain Pattern (A)

EMTs' notes	The injured party was inhaling and exhaling blood when we arrived.
Pathology report	Frothy blood extruding from nose and mouth
Scene photographs	Spatters of blood 2 to 3 mm in diameter
	Located on the north wall of the master bedroom
	The north wall was adjacent to where the victim was discovered
	Some of the spatters had air bubbles within them

Formulation and Testing of Hypotheses

Based on the synthesis of the scene data, formulate hypotheses. Simply stated, the analyst is formulating an educated guess related to the synthesized data based on past experiences and the content of the previously extracted data. After the formulation of each hypotheses for each bloodstain pattern, all need to be tested. The testing may include one or all of the following: review of the scene photographs, review of the written matter, actual experimentation, review of the literature, review of past experiments, and personal experience.

If we continue with the last example:

Data Synthesis for Bloodstain Pattern (A)

EMTs' notes	The injured party was inhaling and exhaling blood when we arrived.
Pathology report	Frothy blood extruding from nose and mouth
Scene photographs	Spatters of blood 2 to 3 mm in diameter
	Located on the north wall of the master bedroom
	The north wall was adjacent to where the victim was discovered
	Some of the spatters had air bubbles within them

Hypothesis for Bloodstain Pattern (A)

The bloodstain pattern located on the wall adjacent to where the victim was located is consistent with being an expirated bloodstain pattern that originated from the mouth or nose of the victim, while the victim was in the vicinity of the master bedroom's north wall.

Testing of Hypothesis of Bloodstain Pattern (A)

1. A review of the literature indicates that blood in the air passages, air bubbles within the blood, and a spatter size of 1 to 3 mm are factors indicative of an expirated bloodstain pattern.
2. All possible variables have been accounted for and excluded as possibly creating the pattern.

3. Experimentation with models expirating blood indicates the observed pattern to be of similar size, shape, and configuration with an expirated bloodstain pattern observed at this crime scene.
4. The EMTs' observations and autopsy findings support the previously stated hypothesis.

Formulation of a Scenario

The development of a scenario or theory of the case should always be the last stage in any bloodstain pattern analysis. There is no guarantee the bloodstain pattern analyst will be able to derive a single scenario from the data. On numerous occasions, the analyst will have more than one scenario that may equally fit the same set of facts. Likewise, the analyst may not be able to exclude enough possible scenarios to render a meaningful opinion as to what actually occurred at a particular scene. Remember not every question derived from the bloodstain patterns will have a definitive answer.

Factors to Consider When Analyzing a Bloodstain Pattern Case

- Never form any type of opinion until all data have been collected and synthesized.
- If you are unable to maintain a completely objective mindset on the case, refer it to another bloodstain analyst.
- If you cannot substantiate your opinion, you do not have an opinion to render.
- How successful the analyst will be is dependent on how well they can collect, organize, and synthesize large amounts of data.
- Notes should be brief and concise statements of fact, not opinion.

Report Writing

18

Once your opinion is in writing you own it and everything about it, good, bad or indifferent.

Introduction

The purpose of the bloodstain pattern analysis report is to convey the opinions of the bloodstain analyst to the attorneys, to the court, and ultimately to the jury. The opinions expressed within the bloodstain report must be substantiated scientifically. Reports allow those in the judicial system opportunity to review the analyst's opinions as well as the basis of those opinions prior to actually presenting testimony. The report should be concise and worded in a manner that will allow it to be understood by those completely unfamiliar with the discipline. The bloodstain pattern report should be a completely objective and unbiased work product of the bloodstain pattern analyst. In reality, the only substantial difference between a report requested by the prosecution versus one requested by the defense should be the person making the request.

It is common practice within the American judicial system for the analysts, who have written the reports, to present expert testimony regarding their written opinions. Many analysts use their testimony to explain and solidify the opinions expressed within their written reports. This method of essentially *double reporting* our opinions to the court is done far less frequently in other legal systems outside of the United States. In some European countries, expert testimony is relatively uncommon. The courts rely solely on what is contained within written reports. Thus, the content of the bloodstain pattern report must be firmly grounded in scientific fact. For this reason, the analyst should write every report as a "stand-alone" document.

Bloodstain pattern analysis is unique in that it is a crime scene as well as a laboratory-based discipline, and it often requires the information from other forensic disciplines to render a complete and thorough report. Ideally, the bloodstain pattern analysis report is one of the last reports to be issued based on our need to have information from the other forensic disciplines (i.e., pathology, serology, etc). However, based on individual agency policies, it may not be possible for the analyst to wait until the end of the case to complete their report, necessitating the need for supplemental reports.

We have divided the reports into the three types we most commonly encounter:

1. Scene specific
2. Evidence specific — laboratory
3. Final report

Scene-Specific Reports

A crime scene–specific report regarding bloodstain pattern analysis will be referring to the stains as "bloodlike" or "red-brown stains visually consistent with blood" as a result of the lack of scene-related serologic testing. If any presumptive testing was conducted at the scene, the results of these presumptive tests should be included in the scene reports. Origin determinations will typically appear in scene-specific reports but will not specify whose blood created the spatter patterns. These reports normally concentrate on pattern analysis (size, shape, distribution, concentration, etc.) and documentation data. Only a limited amount of scene reconstruction conclusions will be reported in scene-specific reports as a result of the lack of laboratory reports such as pathologic and serologic findings.

Evidence-Specific Laboratory Reports

Laboratory-based reports usually deal with the evaluation of bloodstain patterns on specific items of evidence, but they may also deal with the evaluation of scene photographs. Within these reports, the analyst should deal with items of evidence based on where they originated (i.e., accused's clothing, victim's clothing, items collected from scene, etc.). Laboratory reports will also have the results of the serologic testing such as presumptive testing, species testing, as well as an indication as to which stains are being forwarded for DNA analysis. Once again, these reports normally concentrate on pattern analysis (size, shape, distribution, concentration, etc.) and documentation of stain patterns with only limited reconstruction conclusions.

Final Report

We will be dealing with the final report for the remainder of this chapter. We are taking into consideration the following facts prior to rendering our final report: scene analysis; laboratory analysis; information from all other forensic disciplines that may directly or indirectly affect our analysis; and all other variables such as actions taken by law enforcement and emergency medical technicians, weather conditions, evidence handling issues, etc. We are doing everything that was done within the scene and evidence-specific reports, but we are now reporting on the case in its entirety. This is the report where we may be able to render specific reconstruction opinions regarding the bloodletting events.

Report Content

A bloodstain pattern analysis report should deal specifically with the evaluation of the geometric appearance of the bloodstain patterns observed. Analysts should not include observations and opinions beyond the scope of the bloodstain pattern analysis discipline or the author of the report's areas of expertise. This should not be confused with including facts obtained from the reports of other forensic disciplines. You should refer to these findings with the "Reportedly, …" phrase that implies to the reader that you are relying on the findings of another expert rather than forming these opinions on your own. When including information from other disciplines, you should be certain that the information has relevancy to the bloodstain patterns and is accurate. Outside sources of information such as DNA results, autopsy reports, police reports, and the like are often needed for reconstruction opinions. These outside sources of information may allow us to state not only where the blood source was located when struck but also enable us to include which blood source was struck.

Report Phrases

The interpretation portion of the bloodstain pattern analysis discipline is where many people fail to realize that what is said is equal in importance to how it is said. In other words, the way the observations and opinions are phrased can dictate how well your testimony will proceed. The wording used within a report must be unbiased and accurately represent the information you intend to convey to the reader. The bloodstain pattern analyst must be selective about just how definitive or nondefinitive they are with their conclusions. The opposing counsel will and should zero in on any report that is excessively definitive or excessively inconclusive in nature.

Phrases commonly used within bloodstain reports are as follows:

Apparent bloodstains
Stain patterns
Consistent with
Indicative of
Would suggest
Appeared to be
Based on current information
In the condition received
At the time of examination

These phrases are generally used when rendering an opinion or theory that may need room for further interpretation with the discovery of additional facts. However, when discussing *facts* that were extracted from the case, such as the diameter of a stain, the analyst cannot write, "the diameter of the stain *appeared to be* 3 mm." When a definitive measurement is actually taken, it must be reported in a definitive manner: "the diameter of the bloodstain is 3 mm." The opposing counsel should attack the bloodstain analyst whenever they have "hedged" a definitive fact such as a measurement.

When using a phrase within a report, one must be aware that certain phrases or terms attach a degree of probability to an opinion, which the analyst may not have intended or even been aware of, such as the following:

More likely
Likely
More
Tends to be

Once again, the opposing counsel should rigorously cross-examine the bloodstain expert who is attaching a degree of probability to their opinions, especially if they have stated no foundation for such an opinion. These phrases definitely attach a degree of significance to your opinions. These phrases imply a percentage of likelihood, which may not be scientifically substantiated.

Introduction of the Report

The introduction of the report could also be referred to as the "getting acquainted" section of the report. The introduction should include the following types of information:

- Who contacted you and when
- The location of the crime scene
- The location of where the evidence examination occurred
- A list of items received for examination
- The chain of custody information in regard to the physical evidence received
- Any administrative issues such as dates and times

The introduction should allow the reader to determine from what perspective the report is being written, from laboratory, scene, or final report.

Sample Introduction

On 20 June 2002, my office was contacted in regards to examining the bloodstain patterns related to the above-cited matter. On 6 May 2003, I received a packet of documents including laboratory reports, police reports, scene video, and autopsy protocol. On 20 May 2003, I received a set of scene photographs. On 10 June 2003, I examined the following articles of evidence at the New Castle County Police Department, New Castle, DE:

Clothing of Accused:

1. Sunglasses
2. Adidas hat
3. Yellow T-shirt with logo
4. Blue jeans with brown belt
5. (1) Pair of brown shoes "Street Car"

Evidence Collected from Scene:

1. Seat pad
2. Desk calendar ("Staples")
3. High back chair

Clothing of Deceased:

1. Blue jeans with black belt
2. (1) Pair of "Stanley" work boots
3. Blue sweater-like vest
4. Blue/green baseball cap

The introduction is an appropriate location to include pathologic and serologic findings that have an effect on your findings. You should refer to these findings with the "Reportedly, …" phrase based on the fact that most analysts are not forensic pathologists. The typical information you will include is the types of bloodletting injuries (gunshot wounds, blunt-force trauma injuries, etc.), arterial injuries, damage to airways, blood identified within the airways, etc.

Example of reporting on pathological issues:

> *Reportedly*, Mrs. Martinez was shot in the back of the head. A single projectile entered the back of Mrs. Martinez's head (occipital region) and traveled through her head and exited the head through her left upper eyelid.

You may also want to refer to specific serologic findings during the body of your report when referring to stain patterns on specific items of evidence, such as the following:

> Reportedly, through DNA analysis of bloodstains removed from Item #4 Jeans it was determined the source of the blood was identified as Michael Ruger.

It is not feasible, nor is it expected, that *every* stain at a scene or from an article of evidence be tested. This is not to say that representative sampling should not occur from the scene as well as from articles of evidence. For this reason you should consider using the following phrase in the introduction of your report:

> Within this report, stain patterns within the scene photographs that are visually consistent with "blood" will be referred to as blood and/or bloodstains.

An optional addition to the report located near the end of the introduction is what may be referred to as a *Definitions Section*. Essentially, the bloodstain analyst defines the key bloodstain pattern analysis terms that will be used throughout their report. Some analysts will actually attach a glossary as an appendix to their report, whereas others choose to define their terms during testimony or use established terms such as those set forth by the International Association of Bloodstain Pattern Analysts.

Body of the Report

The body of the report should include the overall observations made by the analyst in regard to the bloodstain patterns. This is the foundation for all of the analyst's conclusions. The manner in which this section is approached is often dependent on the issues surrounding the case. If the report is dealing specifically with the clothes of the suspect, then categories should be made dealing with the observations made of each item of clothing. Often there will be several bloodstain patterns on a multitude of mediums and locations within the report including the accused, accused's clothing, victim, victim's clothing, articles of physical evidence, and multiple rooms within the scene. Each pertinent issue must be presented and discussed independently. The opportunity to draw them together will present itself within the conclusions portion of the report. Within the body of the report the following sequence, or a modification thereof, should be followed when addressing issues relating to a bloodstain pattern case:

A. Crime Scene
- Divided into a logical sequence (room–room)
- Stains on victim

B. Physical Evidence
- Collected at the crime scene
- Victim's clothes
- Accused's clothes

C. Special Issues
- Reconstruction with models
- Chemical enhancement techniques
- Experiments conducted

Crime Scene

The crime scene should be presented in a systematic manner. The specific locations of bloodstain patterns within the scene should be specified in great detail. The observations made by the bloodstain analyst should be depicted in a clear and concise manner. Scene photographs and diagrams should be used within the report to allow the reader to visualize what has been written. A great deal of clarity may be achieved by including photographs within your reports. A juror will be hard-pressed to believe what they were told if they cannot read the report and identify the particular stain pattern within the scene photographs. The analyst should refer to the scene photographs by whatever method the photographer had labeled them previously.

Physical Evidence

The physical evidence should be discussed in an obvious sequence (i.e., sequential evidence numbers or by the location in which it was collected). The bloodstain pattern analyst should indicate the type of examination conducted on each item of evidence such as microscopic analysis, visual analysis with high-intensity illumination, or use of a chemical-enhancing technique(s). It is not out of the ordinary for the analyst to remove small portions of unstained fabric from an evidence garment for the purpose of conducting experiments. Although if done, the analyst should be able to account for all of the samples removed because bloodstain experiments are generally nondestructive in nature. The analyst should report, in detail, on any experiments they conducted on the samples they removed from the evidence garment. Any sample taken from an article of evidence must be documented in regard to its size and the location from which it was removed from the garment, as well as the purpose for taking a sample.

In cases where you are reporting on items of clothing evidence from multiple people, you should pay particular attention within your report of delineating whose clothing is whose. This will alleviate confusion by those who are not as familiar with the case as you are. It is also recommended for the analyst to specify where each piece of evidence was recovered from within a scene. The use of evidence photographs within your report is beneficial in that it is a clear and concise manner of conveying your observations. Figure 18.1 is an example of using evidence photographs in a report. If you use photographs within the crime scene portion of your report, you often can refer the reader of your report back to an earlier photograph that depicts the location of the item within the scene.

Special Issues

Special issues are those areas that are not routinely addressed in every case or report. These issues would include the use of models, experiments conducted for specific case questions,

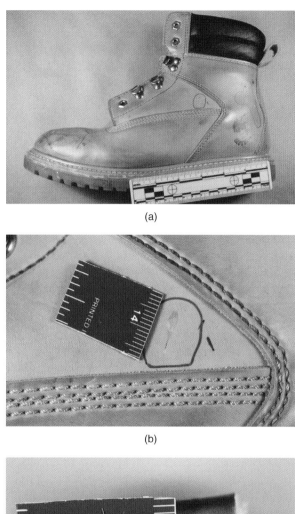

Figure 18.1 Photographic method of depicting the location and appearance of bloodstains identified on an accused left Gauchos boot. (a) Overall view of left Gauchos boot. (b) A closeup view of stain on left side of Gauchos boot. The stain exhibits a downward directionality. (c) Location, shape, and size of a bloodstain located on the back pull of the left Gauchos boot.

chemical-enhancement procedures, etc. Any procedures or techniques that are used to resolve specific questions surrounding the bloodstain patterns within a case should be reported in this section.

The examination of clothing evidence on a laboratory bench is one thing, but it does not allow for a three-dimensional perspective. Hence, it is often very useful to use a model of the same anatomic size as the original wearer of the clothes to model the evidence clothing. A model can be positioned in the exact fashion that best represents the position the original wearer of the clothes claims they were in when their garments were bloodstained. If a model is used in a reconstruction and discussed in a report, it is recommended that photographs of the model wearing the evidence garments be included in the report. This will tend to prevent any confusion about the analyst's opinion for the attorneys and for the jurors.

The bloodstain analyst will routinely have to develop and conduct experiments to answer specific questions about a case. If any experiments are conducted relative to the specific case for which the report is being prepared, then the experiments and their results should be included within the report. This is especially true in cases where case conclusions are based on case-related experimentation. Once again, it is helpful to provide photographs or a videotape of the experiments. Bloodstain experiments are routinely conducted to answer specific case questions, including the following:

- Drying time of blood
- Volume of blood
- Vertical height of satellite spatter
- Transfer pattern replications
- Side in which a bloodstain pattern originated on a garment
- Sequencing of bloodstain patterns

When blood searching or enhancing reagents such as luminol, Amido Black, and Leucocrystal Violet are used, the results should be discussed in their own separate section within the body of the report. The blood-enhanced patterns *should not* be discussed at the same time as the nonenhanced patterns are discussed. It can be very confusing to read a report where the analyst is discussing visible bloodstain patterns in one sentence and in the very next sentence they are discussing the enhanced bloodstain patterns. The use of a blood-enhancing technique should be treated as an actual experiment. The procedure, methodology, how variables were controlled, and the results should be discussed within the report. The results should be photographed and included in the report if any significance is to be attached to the blood enhancements. Analysts should always report when they have attempted any type of blood searching or enhancing techniques either at the scene or on any evidentiary items. The bloodstain pattern analyst is obligated to make the reader aware that blood-enhancing techniques are *only* presumptive indicators of blood, and confirmatory testing is necessary.

Conclusion(s) of the Report

The conclusions portion of the report is a compilation of the analyst's opinions based on the findings and should be listed in a sequential order, point by point. Conclusions should be clear and concise and above all, they should be substantiated by the information included in the body of the report. To formulate the conclusions, information may be drawn from multiple areas within the body of the report. The conclusions will often represent the hypotheses that the analyst was able to prove. Generally, because of the concise nature of the analyst's conclusions, this portion of the report will be relatively brief as compared to the introduction and the body of the report.

At the end of your report, it is advisable to add the following statement:

> This report is based solely on the information made available to me at this time and may be changed or modified with the discovery of new information.

In a discipline where you need to at least in part rely on the work of others (i.e., DNA/pathology), as well as data collected by others (police, evidence technician, etc.), you need to have a caveat that expresses to the reader that your opinions are based on the data you have had to work with at the time the report was written.

The Finished Report

When writing or reading a bloodstain report, the following should be remembered:

- The purpose of the report is not to convict or to exonerate but rather to inform.
- Bloodstain pattern analysis is only a forensic *tool* and should be used accordingly.
- The report should reflect the objective and unbiased opinion of the bloodstain pattern analyst.
- The report should be limited to only opinions that fall within the discipline of bloodstain pattern analysis.
- Written reports must be grammatically sound.
- The reports should be short and to the point. The excessively wordy or drawn out report will only facilitate a lengthy cross-examination.

The issues of who should be writing bloodstain pattern analysis reports and when it should be done are debated often. It is the view of the authors that opinions should be expressed in written form prior to any testimony. Those expressing written opinions should have sufficient qualifications prior to rendering and writing opinions. It is highly recommended that analysts have an internal review system in place in which another qualified expert reviews their opinions. Reports should not only be reviewed for accuracy of content but also for clarity and grammar.

In an effort to prevent future errors, we need to evaluate past errors. Common report errors occur when analysts overstate their opinions by "filling in the blanks" when reporting on a reconstruction with bloodstain pattern analysis. "Filling in the blanks" can occur while reporting on the reconstruction phase where the analyst either speculates or assumes information as fact or uses logic to link other pieces of factual information together. A scene reconstruction with bloodstain pattern analysis is similar to reading a book where various lines or paragraphs have been removed. The analyst must be careful not to fill in these blanks with information that is not scientifically supported. Discussing the shortcomings observed within the reports we have reviewed can be used as a learning tool.

The types of opinions that one should *avoid* would include:

1. Opinions discussing the victim's cognitive state at the time of their injury:
   The victim was sleeping when they were struck.
   The victim was reading at the time they were assaulted.
   The victim turned to the right to avoid further injury.
   The victim was frightened at the time of the assault.
2. Opinions that overstate or amplify the significance of the bloodstain patterns:
   "If the wrists were being held and the upper torso movement was restricted, as indicated by the bloodstain patterns, one might deduce that the assailant was twisting and rocking the blade of the knife in the victim."
   "The number and focus of the wounds, the apparent restriction of the victim by more than one person, the apparent tortuous, and tortuous nature of the wounds, and the confinement of the bloodletting to a small area suggests this act was deliberate and methodical. There are insufficient cast-off patterns, spatter patterns, and ambulation in the blood to infer this to be spontaneous and/or a frenzied act. The fact that there are no readily detectable trails of blood drops or transfers from the assailants as they departed the scene would indicate extraordinary care by the assailants in cleaning themselves before departure, or an inexplicable fluke."
3. Opinions that are not based on bloodstain pattern analysis at all:
   "The wound track and the victim's relative small physical size would greatly limit the victim's physical capability to self-inflict such a wound."
4. Opinions based on a speculation of what they believe should have occurred:
   "The gunshot wound would have created a substantial amount of backspatter."
   "The shooter would have had backspatter on their person."

The ultimate question an analyst must ask himself or herself on final review of any report is "Am I willing to spend the rest of my life incarcerated for the report I am about to submit?" Reports that reflect expert opinions have enormous weight and carry far-reaching implications to not only those directly involved in the violent events but to our entire criminal justice system.

Legal and Ethical Aspects of Bloodstain Pattern Evidence

19

CAROL HENDERSON AND BRITTAN MITCHELL

Introduction

The importance of expert testimony continues to increase, not only in the frequency of its use, but also in the new techniques that find their way into the courtroom. Data from the laboratory or crime scene have no real meaning in law until presented to the judge or jury. Lawyers must rely on expert testimony as a vehicle for communicating these data.

This chapter is designed to aid the expert and attorney in the following matters:

1. The evidentiary foundation required for the admissibility of the evidence
2. The rights and obligations regarding discovery of scientific evidence
3. An outline of attorneys' goals and methods of direct and cross-examination of the expert
4. Ethical consideration in the selection and use of experts

Legal Issues

Admissibility

The issue of admissibility of an expert's evidence is decided by a judge applying one of the two tests for admissibility. The oldest of these tests is the *Frye* test or "general acceptance" test of admissibility. This test is drawn from the often-quoted language of *Frye v. United States*,[1] which dealt with the admissibility of the precursor of the polygraph.

> The exact time a scientific principle or discovery crosses the line between the experimental and demonstrable stages is difficult to define. Somewhere in this twilight zone the evidential force of the principle must be recognized, and although courts will go a long way in admitting expert testimony deduced from a well-recognized scientific principle or discovery, the thing from which the deduction is made must be sufficiently established to have gained general acceptance in the particular field in which it belongs.[2]

---

[1] 293 F. 1012 (D.C. Cir. 1923).
[2] *Id.* at 1014.

The *Frye* Court held that the systolic blood pressure test had not yet gained recognition in the physiologic and psychological communities; therefore, the evidence was inadmissible.

The *Frye* test became the polestar to guide the admissibility of scientific evidence for many years;[3] however, it was not without its critics.[4] The test has been criticized as unworkable.[5] The problem of identifying the relevant field into which a particular scientific technique falls caused additional criticism.[6] Arguably, the field in which bloodstain pattern analysis belongs is the field of crime scene investigation and reconstruction;[7] however, it also uses principles from the fields of physics, chemistry, biology, and mathematics.[8]

The majority of courts, applying *Frye* to bloodstain pattern analysis (or "blood splatter" evidence,[9] as some have wrongly called it), have held that the evidence meets the general acceptance standard.[10] However, a minority of courts held that bloodstain pattern analysis did not satisfy the general acceptance test outlined in *Frye*. For example, in *People v. Owens*,[11] an Illinois appellate court held that a police officer's testimony regarding the

---

[3] Paul C. Giannelli, *The Admissibility of Novel Scientific Evidence: Frye v. United States, A Half Century Later*, 80 Colum. L. Rev., 1197, 1205 (1980).

[4] *Id*. at 1223–25; *See also* 3 BNA Criminal Practice Manual § 101.1901 [3] (Pike & Fischer 2000) (describing how some commentators and courts found dissatisfaction with *Frye* because it was too vague and too conservative).

[5] *Id*. at 1250.

[6] Andre A. Moenssens, James E. Starrs, Carol E. Henderson, and Fred E. Inbau, *Scientific Evidence in Civil and Criminal Cases*, 9 (4th ed., Foundation Press, N.Y., 1995). *See also* Andre A. Moenssens, *Admissibility of Scientific Evidence Under the Post-Daubert Rules*, 1 Forensic Sciences, 19-1, 19-7, (Cyril Wecht, Ed., Matthew Bender, N.Y. 1998) (discussing the dilemma that courts faced in trying to identify the proper scientific field).

[7] *State v. Moore*, 458 N.W.2d 90, 97 (Minn. 1990) (describing how a bloodstain pattern expert reconstructs the crime scene); *Farris v. State*, 670 P.2d 995, 997 (Okla. Crim. App. 1983) (defining bloodstain interpretation as a method used for the reconstruction of a crime scene); International Association of Bloodstain Pattern Analysts (IABPA) (last visited August 25, 2004) (explaining that bloodstain pattern analysis is useful in the reconstruction of a crime) *at* http://www.iabpa.org.

[8] *Holmes v. State*, 135 S.W.3d 178, 187–88 (Tex. Crim. App. 2004) quoting *Lewis v. State*, 737 S.W.2d 857 (Tex. Ct. App. 1987); See also *Franco v. State*, 25 S.W.3d 26, 34 (Tex. App. 2000) (McClure, J., concurring) (explaining the reliability of bloodstain pattern analysis because the methodology is based in "physics, chemistry, biology, and mathematics").

[9] *Gavin v. State*, No. 99–1127, 2003 Ala. Crim. App. LEXIS 262 *130 (Ala. Crim. App. Sept. 26, 2003) (defining the breadth of blood spatter analysis); *Holmes*, 135 S.W.3d at 181 n.1 (discussing the differences between blood spatter and blood splatter and opting to refer to the discipline as blood spatter analysis).

[10] *See* Danny R. Vielleux, Annotation, *Admissibility, In Criminal Prosecution, or Expert Opinion Evidence as to "Blood Spatter" Interpretation*, 9 A.L.R. 5th 369 at §6[a] (1993); *see, e.g., Robinson v. State*, 547 So.2d 910 (Ala. Crim. App. 1990); *People v. Knox*, 459 N.E.2d 1077 (Ill. App. Ct. 1984).

[11] 508 N.E.2d 1088 (Ill. App. Ct. 1987); *But see Holmes*, 135 S.W.3d at 192–95 (listing cases from all the jurisdictions that have found bloodstain analysis reliable and referencing only three cases that excluded the bloodstain evidence. The *Holmes* court in deciding whether to find bloodstain analysis reliable makes the following statement:

> Have any courts held blood spatter analysis to be invalid? The short answer is no. As noted earlier, the [*Franco*] Court skipped the reliability question and held the admission of the testimony was harmless... An Illinois court of appeals [in *Owens*] held blood spatter analysis to be inadmissible only because the State did not prove reliability.... And in Maine, the expert testified that he could determine the order of the gunshots by analyzing blood spatters. *State v. Philbrick*, 436 A.2d 844 (Me. 1981). Because the trial court held no hearing and had no evidence before it at all to demonstrate the field's reliability, the Supreme Court held that the trial court erred in admitting the specific testimony. Id. at 861. The Supreme Court did not conduct a comprehensive review or go out and look for any support of the reliability of blood spatter analysis from other jurisdictions.

> (None of these courts held that, after an extensive hearing, blood spatter analysis was unreliable. The courts either avoided the question or had nothing in the record before them to make a determination. *Id*. at 194–95.)

analysis of bloodstain evidence found at a murder scene was improperly admitted for failure to lay a foundation establishing that the technique was based on a well-recognized scientific principle or that it had gained general acceptance in the scientific community.

Some courts and commentators observed that the *Frye* test was superseded by Rule 702 when the Federal Rules of Evidence were enacted in 1975.[12] Federal Rule of Evidence 702 provides the following:

> If scientific, technical, or other specialized knowledge will assist the trier of fact to understand the evidence or to determine a fact in issue, a witness qualified as an expert by knowledge, skill, experience, training, or education, may testify thereto in the form of an opinion or otherwise, if (1) the testimony is based upon sufficient facts or data, (2) the testimony is the product of reliable principles and methods, and (3) the witness has applied the principles and methods reliably to the facts of the case.[13]

Therefore, scientific evidence, to be admissible pursuant to Federal Rule of Evidence 702, needs to be helpful, relevant, and reliable.

Bloodstain pattern interpretation evidence was admitted pursuant to this new test of admissibility. For example, in *United States v. Mustafa,* the prosecution introduced expert testimony that analyzed the trail of blood and patterns found near the victim at the scene of a rape–murder.[14] The appellate court, in affirming the lower court's decision, held that bloodstain pattern interpretation need not be generally accepted by the scientific community; rather, the testimony need only explain the evidence to the jury.[15] The court noted that the analysis was "grounded in established laws of physics and common sense." Thus, it was reliable and relevant, and the expert's testimony would assist the jury.[16]

After 1975 a split developed among the federal courts regarding whether the Federal Rules of Evidence "assist the trier of fact standard," seemingly more liberal than *Frye,* superseded the *Frye* "general acceptance" test. In 1993 the United States Supreme Court, in *Daubert v. Merrell Dow Pharmaceuticals, Inc.*, announced that Federal Rule of Evidence Rule 702 overruled the *Frye* test; however, the Court also explained that "general acceptance" is still a relevant inquiry.[17] In fact, "general acceptance" is one of the many factors listed in the *Daubert* opinion for judges to use in deciding scientific reliability.[18] Additionally in *Daubert*, the Supreme Court mandated that proof of scientific reliability of expert opinion evidence be produced prior to its admission into evidence.

The *Daubert* opinion states that the courts must assume a "gatekeeping" role regarding the admission of scientific evidence.[19] The Supreme Court construed Federal Rule of Evidence 702 as requiring the trial court to make a twofold inquiry: (1) whether the expert

---

[12] Moenssens, Starrs, Henderson, & Inbau, *supra* note 6 at 12. *See also* Moenssens, 1 Forensic Sciences, *supra* note 6 at 19–12 through 19–13 (2002, updated through 2004); *See e.g., U.S. v. Two Bulls,* 918 F.2d 56, 60–61 n. 7 (8th Cir. 1990) (mentioning that there is speculation with regard to Rule 702 superseding *Frye*); *Daubert v. Merrell Dow Pharmaceuticals, Inc.,* 113 S.Ct. 2786, 2792–93 n. 3–5 (1993) (exploring the controversy between jurisdictions and commentators regarding whether Rule 702 superseded *Frye*).
[13] Fed. R. Evid. 702 (2003).
[14] 22 M.J. 165 (C.M.A.), *cert. denied,* 479 U.S. 953 (1986) (Mil. R. Evid. 702 is similar to Fed. R. Evid. 702).
[15] *Id.* at 168.
[16] *Id.*
[17] 113 S.Ct. 2786 (1993), on remand 43 F.3d 1311 (9th Cir. 1995).
[18] *Id.* at 2796.
[19] *Id.* at 2798.

testimony will assist the trier of fact and (2) whether it amounts to "scientific knowledge." A proposition amounts to "scientific knowledge" if it is derived by the scientific method. Therefore, the judge's inquiry needs to focus on how the expert's conclusions or opinions were reached. The Court identified factors judges should consider in applying their definition of scientific knowledge. The Court stated that the list is not a definitive checklist.[20] Additionally, in listing these factors, the Court emphasized that this new test was flexible and that no particular factor should be determinative of the issue.[21] These factors include the following:[22]

1. Whether the proposition can be tested or has been tested
2. Whether the proposition has been subjected to peer review and publication
3. Whether the methodology or technique has a known error rate
4. Whether there are standards for using the methodology
5. Whether the methodology is generally accepted

The Court recognized that the general acceptance of a theory or technique is still relevant in two respects: "[F]irst, when the proposition is generally accepted ... it qualifies for judicial notice under the Federal Rule of Evidence 201.[23] Second, general acceptance of the methodology can be persuasive circumstantial evidence that the methodology is sound."[24]

When the *Daubert* case was remanded to the Ninth Circuit to apply the factors, the Ninth Circuit judges added an additional factor to their analysis—whether the technique or method had been developed solely for the litigation.[25] If the expert testimony was not based on independent research, the party proffering it must then come forward with other objective, verifiable evidence that the testimony is based on scientifically valid principles. One means of showing this is by demonstrating that research and analysis supporting the expert's conclusions have been subjected to normal scientific scrutiny through peer review and publication.

---

[20] *Id.* at 2796.
[21] For a further discussion on the issue of the flexibility of the factors *see infra* text accompanying notes 43–47.
[22] *Daubert*, 113 S.Ct. at 2796–97.
[23] Fed. R. Evid. 201 (2003) provides:
Judicial Notice of Adjudicative Facts
(a) **Scope of rule.** This rule governs only judicial notice of adjudicative facts.
(b) **Kinds of facts.** A judicially noticed fact must be one not subject to reasonable dispute in that it is either (1) generally known within the territorial jurisdiction of the trial court, or (2) capable of accurate and ready determination by resort to sources whose accuracy cannot reasonably be questioned.
(c) **When discretionary.** A court may take judicial notice, whether requested or not.
(d) **When mandatory.** A court shall take judicial notice if requested by a party and supplied with the necessary information.
(e) **Opportunity to be heard.** A party is entitled upon timely request to an opportunity to be heard as to the propriety of taking judicial notice and the tenor of the matter noticed. In the absence of prior notification, the request my be made after judicial notice has been taken.
(f) **Time of taking notice.** Judicial notice may be taken at any stage of the proceeding.
(g) **Instructing jury.** In a civil action or proceeding, the court shall instruct the jury to accept as conclusive any fact judicially noticed. In a criminal case, the court shall instruct the jury that it may, but is not required to, accept as conclusive any fact judicially noticed.

See, e.g., Holmes, 135 S.W.3d 178, 185–196 (Tex. Crim. App. 2004) (extensively discussing the use of judicial notice of the general acceptance of bloodstain analysis in the following jurisdictions: Texas, California, Oklahoma, North Carolina, Tennessee, Minnesota, Indiana, Illinois, Ohio, Alabama, Idaho, Virginia, Michigan, Washington, and New York).
[24] *Daubert*, 113 S.Ct. at 2797.
[25] *Daubert v. Merrell Dow Pharmaceuticals, Inc.*, 43 F.3d 1311 (9th Cir. 1995).

Although the Court based its analysis in *Daubert* on the statutory construction of the language of the Federal Rules of Evidence, it is important to note the decision had implications beyond federal practice. The *Daubert* ruling has had an impact in states that have evidence codes based on the Federal Rules of Evidence,[26] as well as those states that followed *Frye* in the absence of specific evidence rules or statutes similar to the federal rules. The minority of states continue to follow the *Frye* test.[27] However, many large jurisdictions are grouped within this minority, including California,[28] Florida,[29] Illinois,[30] New York,[31] Pennsylvania,[32] and Washington.[33,34]

[26] Moenssens, 1 Forensic Sciences, *supra* n. 6 at 19–35.

[27] Edward J. Imwinkelreid & William A. Tobin, *Forensic Symposium: The Use and Misuse of Forensic Evidence; Comparative Bullet Lead Analysis (CBLA) Evidence: Valid Inference or Ipse Dixit?*, 28 Okla. City U.L. Rev. 43, 55 (2003) (explaining that 18 states continue to use *Frye*); *But see Mississippi Transp. Com'n v. McLemore*, 863 S.2d 31, 35 (Miss. 2003) (converting Mississippi from a *Frye* jurisdiction to a modified *Daubert/Kumho* jurisdiction). *See also* Walter R. Lancaster, *Expert Witnesses in Civil Trials: Effective Preparation and Presentation*, 2nd ed., (West 2000 and Supp. 2003) (listing cases from 17 jurisdictions that use *Frye* in civil cases).

[28] Prior to *Daubert*, the adopted test of California was the Kelly/Frye test, *People v. Kelly*, 549 P.2d 1240 (1976). After *Daubert*, the California Supreme Court in *People v. Leahy*, 882 P.2d 321 (1994), held that the *Kelly* test of general acceptance remained controlling in California state courts. *See also People v. Farnam*, 47 P.3d 988 (2002) (holding that general acceptance within the appropriate scientific community is required for a new scientific technique). *See also* Alice B. Lustre, *Post-Daubert Standards for Admissibility of Scientific and Other Expert Evidence in State Courts*, 90 A.L.R. 5th 453 at §29 (2001) (explaining the controlling decisions in California on the issue of the admission of scientific evidence).

[29] *Ramirez v. State*, 810 So. 2d 836, 843 (Fla. 2001) (reiterating that Florida adheres to *Frye* because it is a more rigorous determinant of reliability than *Daubert*); *Butler v. State*, 842 So. 2d 817 (Fla. 2003) (explaining that the relevant scientific community needs to generally accept a scientific principle prior to allowing an expert to testify using the scientific principle or methodology in court); *Spann v. State*, 857 So.2d 845, 852 (Fla. 2003) (holding that because Florida follows *Frye* rather than *Daubert* and because handwriting identification is not new or novel scientific evidence, then the trial court did not need to reexamine the general reliability of the field under *Frye*).

[30] *See People v. Basler*, 740 N.E.2d 1, 4 (Ill. 2000) (holding that the courts in Illinois should follow *Frye* in determining the admissibility of novel scientific evidence); *See also* Lustre, *supra* note 28 at § 32 (discussing the cases in Illinois on the issue of *Frye* and *Daubert* from 1993 till 2004).

[31] *See Gallegos v. Elite Model Management Corp.*, 758 N.Y.S.2d 777, 778 (N.Y. Sup. Ct. 2003) (applying *Frye* to novel scientific evidence); *Zafran v. Zafran*, 740 N.Y.S. 2d 596 (N.Y. Sup. Ct. 2002) (explaining that *Frye* not only applies in criminal actions but also in civil actions). *See also* Lustre, *supra* note 28 at § 40 (discussing predominantly New York's post-2000 cases related to the court's use of *Frye* or *Daubert*).

[32] In Pennsylvania, courts still follow *Frye*. *Grady v. Frito-Lay, Inc.*, 839 A.2d 1038, 1044 (Pa. 2003). The *Grady* Court explained that the "general acceptance" test was more likely to "yield uniform, objective, and predictable results among the courts." The Court also provided an analysis on the burden the proponent of the evidence carried in proving the *Frye* standard. *Id.* at 1045. In particular, the court noted that the proponent was responsible for proving that the expert's methodology was generally accepted by the relevant scientific field. *Id.* However, the proponent was not required to demonstrate that the expert's conclusion was generally accepted by the scientific field. *Id.* The Court indicated that this last component was an important aspect of the *Frye* test because it would not stifle the creativity or innovative thought of the expert witness. *Id. See also* Lustre, *supra* note 28 §42 (discussing the Pennsylvania cases post-1994 on the issue of *Frye* and the admissibility of scientific evidence, also discussing the cases about the burden of the proponent in proving the *Frye* standard).

[33] Washington's Supreme Court has continually upheld the use of the *Frye* test in the admissibility of novel scientific evidence. *See, e.g., In re Detention of Thorell*, 72 P.3d 708, 724–25 (Wash. 2003). However, scientific evidence that is not new or novel is not subjected to a *Frye* examination. *State v. Roberts*, 14 P.3d 713, 740–41 (Wash. 2000) (holding that bloodstain pattern analysis evidence did not need a *Frye* examination because it was not a novel method of scientific evidence); *See also* Lustre, *supra* note 28 at § 43 (listing cases from Washington that have holdings related to *Frye* or *Daubert*).

[34] Imwinkelreid & Tobin, *supra* note 27 at 55. *See also* Lustre, *supra* note 28 (listing the predominant cases from each jurisdiction and explaining whether each state jurisdiction follows *Daubert* or *Frye*. Adding to the list of jurisdictions that continue to strictly adhere to *Frye* are Arizona, District of Columbia, Kansas, Maryland, Michigan, Minnesota, Missouri, and North Dakota).

Bloodstain pattern analysis testimony that is otherwise admissible may still be excluded if its probative value is outweighed by prejudice or if it confuses the jury.[35] Because there is an "aura of special reliability and trustworthiness" surrounding scientific expert testimony, parties should be aware of the increased danger of undue prejudice.[36] In fact, courts should be mindful of whether there are other, less prejudicial types of evidence.[37] For example, in *State v. Johnson*, the appellate court reversed the defendant's murder convictions and remanded the case for a new trial, concluding that the limited probative value of the prosecution's expert's testimony was outweighed by the risk of the danger of undue prejudice.[38] Although conceding that the testimony was not irrelevant, the court found that it was largely corroborative of other, essentially unchallenged testimony indicating the manner of death. The court noted that the prosecution's expert's lengthy presentation, which exposed the jury to numerous crime scene photos as well as 42 slides depicting bloodstain pattern exemplars, could not help but focus the jury's attention on the gruesome details of the conditions of the victims' bodies rather than on the defendant's guilt.[39]

Because *Daubert* makes general acceptance by the scientific community just one factor in the court's analysis, generally accepted techniques are now vulnerable to challenge. Lawyers proffering experts now have to elicit detailed testimony regarding the scientific methodology used, such as the tests conducted, the standards used, and the error rate. These new requirements even apply to well-accepted forensic techniques.

Although helpful, *Daubert* did not provide all the answers to the introduction of expert testimony. In the six years following *Daubert*, the federal circuits and commentators disagreed

---

[35] *Franco v. State*, 25 S.W.3d 26, 29 (Tex. App. 2000). In *Franco*, the defendant was convicted of murder. The state used a police officer to refute the defendant's claim of self-defense through the technique of bloodstain pattern analysis. *Id.* at 29. The state failed to qualify the police officer as a bloodstain pattern expert, and the state also failed to demonstrate that bloodstain pattern analysis was a field that was scientifically valid and reliable. The appellate court first questioned the validity and reliability of bloodstain pattern analysis:

> We can envision circumstances where the trier of fact may be quite properly aided by some evidence of blood splatter analysis, but we are dubious of the claim in this record that blood spatter evidence can "determine the aftermath of a violent incident of bloodshed and to try to determine the location of individuals before, during and after bloodshed and to try to determine, perhaps, a sequence of events that occurred based upon the bloodstain evidence available at the scene." Such evidence...would likely carry exceptional weight and an aura of reliability which could lead the jury to conclusions based more on speculation than scientific explanation.

Id. The court then explained that even if bloodstain evidence is reliable, the evidence still might be unhelpful "if its probative value is substantially outweighed by, e.g., the risk of unfair prejudice, confusion of the issues, misleading the jury, undue delay, or the presentation of cumulative evidence." *Id.* The court concluded, and the state conceded, that it was an error to allow the police officer to testify without qualifying his expertise in the area of bloodstain analysis and without offering evidence of the reliability of the scientific field of bloodstain pattern analysis. However, the state argued and the appellate court agreed that the police officer's testimony was harmless given the strength of the other evidence and testimony that also refuted the claim of self-defense. *Id.* at 29–30. Therefore, the appellate court affirmed the conviction of the defendant. *Id.* at 32. *See also* Fed. R. Evid. 403 (2003), which states:
Exclusion of Relevant Evidence on Grounds of Prejudice, Confusion, or Waste of Time
Although relevant, evidence may be excluded if its probative value is substantially outweighed by the danger of unfair prejudice, confusion of the issues, or misleading the jury or by considerations of undue delay, waste of time, or needless presentation of cumulative evidence.
[36] Frank E. Haddad, *Admissibility of Expert Testimony*, in 1 Forensic Sciences, 1–1, 1–20 (Cyril Wecht, Ed., Matthew Bender, N.Y. 1996).
[37] *Id.*
[38] 576 A.2d 834 (N.J. 1990).
[39] *Id.* at 853.

about the interpretation of certain *Daubert* issues. For example, some circuits interpreted *Daubert* to apply only to scientific testimony,[40] other circuits found *Daubert* applicable to all expert testimony,[41] and some circuits had not clearly defined the issue by 1999.[42] In 1999 the United States Supreme Court revisited certain *Daubert* issues in *Kumho Tire Company, Ltd. v. Carmichael*.[43] In *Kumho* one of the many issues before the Court was whether the *Daubert* factors applied to the testimony of experts who were not scientists.[44] The Court stated that the *Daubert* factors apply to all areas of expert testimony covered by Federal Rules of Evidence 702 and not simply to scientific testimony.[45] The Supreme Court also stressed that the *Daubert* factors are flexible in nature and that the list "neither necessarily nor exclusively applies to all experts or in every case."[46] The Court continued, "we can neither rule out, or rule in, for all cases and for all time the applicability of the factors mentioned in *Daubert*, nor can we now do so for subsets of cases categorized by category of expert or by kind of evidence. Too much depends upon the particular circumstances of the particular case at issue."[47]

The admissibility of scientific evidence is a question of law that the judge decides. The only time judges' decisions regarding admissibility will be overturned on appeal is if they abuse their discretion. A judge has broad discretion.[48] The United States Supreme Court in *General Electric Company v. Joiner* adopted the abuse of discretion standard for reviewing a trial court's admissibility decision under *Daubert v. Merrell Dow Pharmaceuticals, Inc.*[49] The Supreme Court reaffirmed the *Joiner* analysis in *Kumho*.[50] The Court emphasized that a trial judge has broad latitude in the determination of whether the *Daubert* factors are useful in assessing reliability and that the appropriate appellate review is the abuse-of-discretion standard.[51]

Weight of the Evidence

Once a judge decides to admit an expert's testimony, the jury must then decide what weight to give the expert's testimony. Studies have shown that jurors accord great weight to expert testimony. For example, a well-cited poll taken by the *National Law Journal* and Lexis/Nexis found that jurors not only find experts generally credible, but the experts often influenced the outcome of the case.[52] Overall, 71% of the jurors said the experts made a difference in

---

[40] Stewart Lee, *Evidence—Expert Witnesses—Daubert Applies to All Expert Testimony*, 69 Miss. L.J. 979, 986 n. 28 (Winter 1999) (listing cases from the Second, Fourth, Ninth, Tenth, and Eleventh Circuits).
[41] *Id.* (listing cases from the First, Third, Fifth, Sixth, and D.C. Circuits as examples of using *Daubert* for all expert witness testimony).
[42] *Id.* (listing cases from the Seventh and Eighth circuits on both sides of the issue).
[43] 119 S. Ct. 1167 (1999).
[44] *Id.* at 1171.
[45] *Id.*
[46] *Id.*
[47] *Id.* at 1175.
[48] Haddad, *supra* note 36 at 1–16 though 1–17. See also *Kumho Tire Company, Ltd. v. Carmichael*, 119 S.Ct. 1167, 1176 (1999).
[49] *General Electric Company v. Joiner*, 118 S.Ct. 512 (1997).
[50] 119 S.Ct. at 1176.
[51] *Id.*
[52] Joan Cheever & Joanne Naiman, *The View from the Jury Box*, Nat'l L.J., Feb. 22, 1993, at S4. The survey found that experts were thought believable by 89% of the criminal and civil jurors; See also Miles J. Vigilante, *Screening Expert Testimony after Kumho Tire Co. v. Carmichael*, 8 J.L & Pol'y 543, 543–44 n. 1 (2000) (citing statistics related to a juror's response to an expert witness); Regina A. Schuller, Deborah Terry, and Blake McKimmie Terry, *The Impact of an Expert's Gender on Juror's Decision*, 25 Law & Psychol. Rev. 59, 59–60 n. 2 (Spring 2001) (citing statistics related to a juror's response to an expert witness).

the verdict.[53] In criminal cases, 95% of the jurors thought the expert very believable or somewhat believable.[54] In spite of this inherent credibility, in most states, the judge will instruct the jurors that an expert's opinion is only reliable when given on a subject that the jurors believe him or her to be an expert and the jurors may choose to believe or disbelieve all or any part of an expert's testimony. For example, Florida's Standard Jury Instruction in Criminal Cases § 3.9(a) states the following:

> Expert witnesses are like other witnesses, with one exception—the law permits an expert witness to give [his] [her] opinion. However, an expert's opinion is reliable only when given on a subject about which you believe [him] [her] to be an expert. Like other witnesses, you may believe or disbelieve all or any part of an expert's testimony.[55]

Determining the weight to accord evidence is a question for the jury or for a judge in a bench trial. To determine the weight to accord scientific evidence, jurors will weigh factors such as acceptance of the scientific principles or theories, the basis of the expert's opinion, and the credibility of the expert. The weight accorded to an expert's testimony hinges in large part on the credibility of the expert. One way that the jurors judge an expert's credibility is by the expert's qualifications.[56]

Qualifications of Experts

Once the court decides whether expert testimony is admissible, it must then determine whether the proffered expert is qualified to render an opinion. At this point, the court will examine the expert's qualifications. According to the rules of evidence, a witness may qualify as an expert on the basis of knowledge, skill, training, experience, or education.[57]

In making this evaluation, the court may consider the expert's educational background, work experience, publications, awards, teaching or training positions, licenses or certifications, speaking or other professional engagements, prior expert witness testimony, and membership in professional associations and positions held in those associations.

In the field of bloodstain pattern interpretation, there is no universal agreement regarding adequate qualifications for one to be an expert. Herbert MacDonell, a well-known expert in the field of bloodstain interpretation, is of the view that such experts should

---

[53] Cheever & Naiman, Nat'l L.J. at S4.
[54] Id.
[55] Fla. Std. Jury Inst. in Crim. Cases § 3.9(a) (2003).
[56] Joel Cooper, Elizabeth A. Bennett, and Holly L. Sukel, *Complex Scientific Testimony: How Do Jurors Make Decisions?*, 20 (4) Law and Human Behavior 379, 386, 390 (1996) (providing a statistical analysis of scientific testimony). The authors of this article conducted a mock trial to test various hypotheses on jury members. One of the experiments was to test the jurors' perception of an expert's credibility based on credentials. In a trial where the testimony was complex, the jurors tended to accept the expert with the higher credentials as more credible and trustworthy. However, when the experts testified in "more comprehensible language," then both experts were considered equally trustworthy and credible. *See also* Neil Vidmar & Shari S. Diamond, *Juries and Expert Evidence*, 66 Brook. L. Rev. 1121 (2001) (reviewing many recent studies related to jurors' responses to expert witnesses).
[57] Fed. R. Evid. 702 (2003).

possess a Bachelor of Arts degree in either a science or in criminal justice with required science classes.[58] This view has been criticized by those who recognize that some individuals may lack formal education but are nonetheless qualified through skill, experience, and training.[59] A study of 1500 jurors revealed that jurors are most impressed by the amount and type of training an expert has received, and they expect such an expert to belong to a professional association.[60] Likewise, Professor MacDonell is of the view that experts in this field should be members of one or more of the professional associations dedicated to the forensic sciences, such as the American Academy of Forensic Sciences, the International Association for Identification (IAI), and the International Association of Bloodstain Pattern Analysts.[61]

Board certification exists for many of the forensic sciences.[62] The IAI offers certification for bloodstain pattern examiners. The IAI requires a basic 40-hour course that adheres to certain guidelines, 3 years of practice following the basic course, and a minimum of 240 hours of training in associated fields of study (e.g., photography, crime scene investigation, medical legal death investigation; and an examination).[63]

---

[58] Herbert L. MacDonell, *Bloodstain Pattern Interpretation*, in 3 Forensic Sciences, 37-1, 37-68 (Cyril Wecht, Ed., Matthew Bender, N.Y. 1991). *See also* Herbert L. MacDonell, *A Scientist's Comments on the Police Officer as a Bloodstain Pattern Analyst*, International Association of Blood Pattern Analysis News, 10(2), June 1994, 6–9 (responding to the commentary of Norman Reeves *infra* note 59, and reaffirming his belief that a bachelor degree with course work in physical sciences is necessary for an expert witness in bloodstain analysis). *See also* Amanda Onion, *Scientific Sleuthing*, ABC News (June 4, 2003) *at* http://web.archive.org/web/2004024113319/ http://www.abcnews.go.com/sections/scitech/DailyNews/forensics020603.html (providing commentary from Herbert MacDonell on bloodstain analysis: "I think bloodstain analysis is being used and abused more and more…[Bloodstain pattern analysis] sounds simple, but if you haven't got a lot of experience, you can make a lot of mistakes."

[59] Norman H. Reeves, *The Police Officer as a Bloodstain Pattern Analyst*, International Association of Blood Pattern Analysis News, 10(2), June 1994, 3–5 (specifically expressing and disagreeing with MacDonell's viewpoint that formal scientific education is necessary for the field).

[60] Sgt. Charles Illsley, U.S. Department of Justice, *Juries, Fingerprints, and the Expert Fingerprint Witness*, 26 (1987). Presented at the International Symposium on Latent Prints (FBI Academy, Quantico, Virginia, July 1987). A few highlights from the 1987 survey can be found at 9 (10) Expert Witness Journal 1, *FBI and Expert Witness*, (October 1997). Highlights include the following:

Sixty percent of jurors considered the amount of specialized knowledge as the most important factor in the expert's background.
Sixty-eight percent felt that the expert should have achieved at least a bachelor's degree.
Most jurors felt that the expert witness should have prior courtroom experience.
Sixty percent of the jurors felt that the expert should belong to professional associations related to his or her field of study.
Sixty percent of the jurors felt that the expert should use notes when testifying in court.

[61] *See* sources cited *supra* note 58. MacDonell expresses that educational background is important when considering the membership population for a professional association. MacDonell, *A Scientist's Comments on the Police Officer as a Bloodstain Pattern Analyst*, 10 (2) IABPA News. Specifically, he directed his comments towards the IABPA and mentioned that "[t]he absolute minimum educational requirement for membership in what professes to be a professional organization should be graduation from high school having passed courses in both science and mathematics." *Id*.

[62] Board certification presently exists for the following areas of forensic science, among others: forensic anthropology, crime scene reconstruction, criminalistics, document examination, forensic odontology, forensic pathology, forensic psychiatry, forensic psychology, forensic toxicology, footwear, forensic art, forensic photography, latent print, ten print fingerprint, forensic accounting, forensic nursing, trace evidence collection, forensic engineering, and medicolegal death investigators.

[63] International Association of Identification (IAI). Bloodstain Pattern Examiner Certification Requirements. (last visited August 25, 2004) *available at* http://www.theiai.org/certifications/bloodstain/requirements.html.

The Forensic Specialties Accreditation Board (FSAB) was created as a result of concerns regarding forensic credentialing.[64] "The goal of this program is to establish a mechanism whereby the forensic community can assess, recognize, and monitor organizations or professional boards, which certify individual forensic scientists or other forensic specialists."[65] FSAB has developed certain standards that explain the proper process that a board-certifying entity should use when assessing, documenting, and maintaining the knowledge, skills, and abilities of applicants.[66] FSAB will evaluate a credentialing organization to see if the organization meets the FSAB standards. If the organization does meet the appropriate standards, then FSAB will "confer [on it] the equivalent of a Good Housekeeping seal of approval."[67]

Just as experts in the field differ as to adequate expert qualifications,[68] so do the courts.[69] For example, in a case from Texas the court indicated that an expert witness must have special knowledge specific to the field of expertise.[70] However, the court defined this special knowledge as any knowledge that exceeds that of the average juror.[71] Although the proffered expert in *Holmes* had been in law enforcement positions for a number of years, he had only participated in 45 to 50 hours of instruction related to bloodstain pattern evidence.[72] The court found that the week-long training, combined with the detective's testimony that he had read books written by bloodstain experts and that he had practiced the principles of bloodstain analysis in the field, was sufficient to satisfy the court in qualifying the detective as an expert witness in the field of bloodstain pattern analysis.[73] Similarly, the Florida Supreme Court has overruled an assertion of error based on expert qualifications.[74] The Court affirmed that it was not an abuse of discretion to allow a bloodstain pattern analyst to testify as an expert because the analyst had attended one 40-hour course and had interpreted bloodstain evidence in the field, although the analyst had never previously testified in court.[75]

---

[64] "In a 1995 report, the Strategic Planning Committee of the American Academy of Forensic Sciences (AAFS) reported that the quality and standards applied by different forensic boards for granting certification varied widely." Forensic Specialties Accreditation Board, Inc., (last visited August 25, 2004) http://www.thefsab.org (providing FSAB's history); *See also* Mark Hansen, *Expertise to Go*, 86 ABA Journal 44, 51 (Feb. 2000) (providing information about the forensic credential crisis and the birth of FSAB) *reprinted at* 16(1) The Print 1 (January/February 2000) *available online at* http://www.scafo.org/The_Print/THE_PRINT _VOL_16_%20ISSUE_01.pdf; *See also*, Carol Henderson, *Expert Witness: Qualifications and Testimony*, in *Scientific Evidence Review: Admissibility and Use of Expert Evidence in the Courtroom*, 7–10 (Cynthia H. Cwik et al., Eds., ABA Publishing 2003) (exploring a few recent examples of credentialing boards providing certifications without doing a sufficient background check. For example, one man, who was serving a jail sentence for his wife's manslaughter, became board certified in forensic medicine, even after listing the state prison as his address on his application).
[65] Forensic Specialties Accreditation Board, *supra* n. 59.
[66] *Id.*
[67] Hansen, *supra* note 64 at 51 (relating the viewpoint of Barry Fisher, a past president of American Academy of Forensic Sciences).
[68] The current membership of the International Association of Bloodstain Pattern Analysts include the following backgrounds: "forensic serology and DNA testing, medical-legal death investigation, law enforcement, crime scene investigators and private consulting." *See generally* IABPA Membership website *at* http://www.iabpa.org/membership.htm (last visited August 23, 2004).
[69] See Veilleux, *supra* note 10 at §§ 11–16 *Admissibility Based on Qualifications of Alleged Expert* (exploring the reasoning of cases from different jurisdictions in the admissions of an expert witnesses' testimony in bloodstain analysis based on background).
[70] *Holmes v. State*, 135 S.W.3d 178, 182 (Tex. Crim. App. 2004).
[71] *Id.*
[72] *Id.* at 183.
[73] *Id.* at 183–184.
[74] *Anderson v. State*, 863 So. 2d 169, 179–180 (Fla. 2003).
[75] *Id.* at 180.

However, there are courts that have found that the trial court abused its discretion in allowing an alleged expert to testify about bloodstain analysis. For example, in *State v. Halake*,[76] the state used a police officer to testify that the blood spots found on the defendant's pants were consistent with gunshot blood spatter.[77] The state argued that the officer's expertise in the field was sufficient because he had seen a hundred crime scenes and various patterns of blood.[78] The appellate court explained that bloodstain analysis is unfamiliar to most lay jurors; therefore, expert testimony is necessary.[79] This is because of the complexity of bloodstain analysis.[80] The court weighed the fact that the officer had topically covered blood spatter in crime school; however, the officer never enrolled in a class specifically dedicated to bloodstain evidence.[81] The court found that the officer exceeded the scope of a lay witness when he entered the complex realm of bloodstain pattern analysis.[82] Therefore, the appellate court ordered a new trial because it was a prejudicial abuse of discretion to allow the officer to testify that the bloodstains on the victim's clothing resulted from "gunshot blood spatter."[83]

Courts have found people of varied backgrounds qualified to testify as experts in the area of bloodstain pattern interpretation. For example, chemists,[84] forensic scientists,[85] serologists,[86] and pathologists[87] have qualified as experts in bloodstain pattern analysis. Additionally, courts have accepted testimony from investigators,[88] forensic consultants,[89] criminalists,[90] and law enforcement agents.[91]

---

[76] 102 S.W.3d 661, 669 (Tenn. Crim. App. 2001).
[77] *Id.*
[78] *Id.*
[79] *Id.* at 670–671.
[80] *Id.* at 672.
[81] *Id.* at 671.
[82] *Id.*
[83] *Id.*
[84] *McCambridge v. Hall*, 303 F.3d 24 (1st Cir. 2002) (allowing the testimony of a forensic chemist as an expert in bloodstain pattern analysis).
[85] *State v. Hartman*, 754 N.E.2d 1150, 1166 (Ohio 2001) (allowing a forensic scientist to testify about certain attributes of the bloodstains); *State v. Roberts*, 14 P.3d 713, 720 (Wash. 2000) (allowing a forensic scientist to testify about the bleeding and movement of the victim based on the patterns of bloodstains).
[86] *Dye v. Hofbauer*, 45 Fed.Appx. 428, 430 (6th Cir. 2002) (introducing the testimony of a serologist to testify about the location of the shooter based on the pattern of bloodstains); *Railey v. Portuondo*, No. 97 Civ. 3878DABTHK, 2000 WL 1514787 *2 (S.D.N.Y July 26, 2000) (introducing the testimony of an expert serologist to testify about bloodstain analysis); *Woodal v. Commonwealth*, 63 S.W.3d 104, 129–130 (Ky. 2002) (affirming the appropriateness of a serologist to testify about crime scene reconstruction and bloodstain analysis).
[87] *State v. Dreiling*, 54 P.3d 475, 486–87 (Kan. 2002) (allowing a pathologist to testify about the relationship between the gunshots and the amount of blood spatter); *State v. Brown*, Nos. C-990348, B-9901823, 2000 WL 376403 *2 (Ohio App. 2000) (permitting the testimony of a board-certified coroner in general pathology to extrapolate from photographs of the blood spatter evidence); *Commonwealth v. Stallworth*, 781 A.2d 110, 121 (Pa. 2001) (holding that it was not an error to allow a pathologist to testify about crime scene reconstruction given that he was qualified in bloodstain analysis).
[88] *State v. Trotter*, 632 N.W.2d 325, 334 (Neb. 2001) (mentioning that a State Patrol criminal investigator testified about bloodstain evidence).
[89] *See Hartman*, 754 N.E.2d at 1167–1168 (holding that it was not an abuse of the trial court's discretion to allow someone who was not a medical doctor to testify about bloodstain analysis. The opinion details the specifics of the qualifications of this particular forensic consultant).
[90] *People v. Farnam*, 47 P.3d 988, 1028–29 (Cal. 2002) (holding a criminalist qualified to testify about crime scene reconstruction and bloodstain evidence).
[91] *State v. Abu-Fakherm*, 56 P.3d 166, 174 (Kan. 2002) (allowing the officer to testify to the existence of bloodstains and blood spatter at the crime scene); *State v. March*, No. A-01-755, 2002 WL 31300046 (Neb. Ct. App. Oct. 15, 2002) (Police officer testified regarding his opinion of the drying of the blood found at the crime scene).

A few words of caution regarding qualifications: An expert should be wary of joining professional associations that routinely provide board certification without rigorous testing or evaluation or that allow promotion to fellowship status without significant contribution to the scientific field and the professional association.[92] Membership in professional associations that are not perceived as credible will harm an expert's expertise and may result in a judge finding the expert not qualified to render an opinion. Furthermore, in some states, if an expert misrepresents association with or academic standing at a postsecondary educational institution or makes a false oral or written statement regarding an academic degree or title, the expert may be prosecuted for fraud.[93] Other sanctions for such violations include incarceration, fines, and suspension or revocation of the person's license or certification to practice his or her occupation or profession.[94] If the expert makes a false statement under oath with regard to a material matter, the expert may be prosecuted for perjury. Falsifying records is also a crime.

Burden of Proof

In criminal cases, the burden of proof (i.e., the burden of producing evidence of the elements of the crime and the burden of persuading the jury of the elements of the crime) is on the prosecution. The prosecutor must produce evidence and persuade the jury beyond a reasonable doubt that the elements of the crime are met. Thus, the bloodstain pattern expert may assist the prosecutor in producing and interpreting evidence and persuading the jury of the identity of the perpetrator or the victim.

The standard of proof of "beyond a reasonable doubt" does not mean beyond any doubt. A reasonable doubt is not a speculative, imaginary, or possible doubt. If the jury, after carefully considering, comparing, and weighing all the evidence, does not have an abiding conviction of guilt, or if, having a conviction, it is one that is not stable but wavers and vacillates, then there is a reasonable doubt.

The standard of proof in a civil case is by a preponderance of the evidence. This is a much less stringent standard of proof than beyond a reasonable doubt. You meet a preponderance of the evidence by the greater weight of the evidence or evidence that is more credible and convincing.

Chain of Custody

To lay a complete foundation for scientific evidence, lawyers need to accomplish four things:

1. Prove the chain of custody for any physical sample from the time of seizure until it is introduced in court.
2. Teach the trier of fact about the scientific area.
3. Describe when and how the analysis was done.
4. Interpret the evidence for the trier of fact.

---

[92] Hansen, *supra* note 64.
[93] *See* Fla. Stat. § 817.566 (2000 and Supp. 2004) "Misrepresentation of association with, or academic standing at, postsecondary educational institution." *See also* Henderson, *supra* note 64 at 8–11 (discussing the trend toward increased scrutiny of expert witnesses).
[94] *See* International Association of Identification (IAI). Code of Ethics for Bloodstain Pattern Examiners. at http://www.theiai.org/certifications/bloodstain/ethics.html "A bloodstain pattern examiner, whose professional or personal conduct becomes adverse to the best interests and purposes of the profession of Bloodstain Pattern Identification, shall be liable to censure, suspension or withdrawal of certification as a Bloodstain Pattern Examiner.... A Bloodstain Pattern Examiner who makes other than factual statements (false or misleading statements) is subject to suspension and revocation of his/her certification."

During trial, one who offers real evidence, such as bloodstains found on the victim's clothing at the crime scene, must account for the custody of the evidence.[95] Custody of the evidence must be documented from the crime scene to the moment at which it is offered in evidence. With the small samples common with blood evidence, chain of custody becomes an important issue. To lay the appropriate foundation for admissibility of real evidence, the attorney needs to demonstrate to the court, through witnesses' testimony, that the path of the evidence from the crime scene to the trial is well documented. Many experts mistakenly believe that once the evidence is safely at the laboratory, the chain of custody is no longer a problem. Not only must the experts establish the chain of custody of the evidence in the laboratory, they must also establish the chain of custody from the laboratory to the courthouse on the day of trial.

It is important to ensure that the laboratory has good documentation because chain of custody is a fertile area for lawyers to explore in discovery and during cross-examination. Opposing counsel will scrutinize the collection, preservation, and handling of blood evidence to probe the possibility of contamination or sample misidentification.

Discovery

Discovery is a pretrial process used in both criminal and civil cases to obtain facts and information about the case from the opposing party to assist in preparation for trial. The general policies underlying discovery are to prevent surprise at trial, to narrow the issues to be tried, and to speed the administration of justice by encouraging settlement (plea bargaining) of those cases where both sides know the strengths and weaknesses of the evidence.

The discovery of scientific information is, in one sense, merely a part of the total discovery of a party's case, and the same general purposes apply. However, special considerations arise with regard to expert discovery. The major reason given for expert discovery in both criminal and civil actions is the need for the opposing attorney to adequately prepare himself for cross-examination. The cross-examining lawyer often needs to have a very technical knowledge of the particular scientific field to put the expert's conclusions to a meaningful test. This is different from the cross-examination of a direct evidence witness (e.g., eyewitness), where the lawyer may draw from his own knowledge of the common fallibilities of sense perception and memory.[96]

Tools of discovery include depositions, interrogatories, and production of documents and reports. The process by which discovery is conducted is set forth in the civil and criminal rules of procedure. For example, Rule 16(a)(1)(F) of the Federal Rules of Criminal Procedure sets forth the prosecutor's discovery obligation to the defense, if requested, regarding scientific evidence:

> **(F) Reports of Examinations and Tests.** Upon a defendant's request, the government must permit a defendant to inspect and to copy or photograph the results or reports of any physical or mental examination and of any scientific test or experiment if:
>
> 1. the item is within the government's possession, custody, or control;
> 2. the attorney for the government knows — or through due diligence could know — that the item exists; and

---

[95] *See, e.g.*, Roger C. Park, *Trial Objections Handbook 2d* § 9.28 *Objections to Authentication of Real Evidence; Chain of Custody* (West. 2001).

[96] Moenssens, Starrs, Henderson, & Inbau, *supra* note 6 at 32.

3. the item is material to preparing the defense or the government intends to use the item in its case-in-chief at trial.[97]

The defense then has a reciprocal obligation to turn over its results or reports of tests or experiments.[98]

The prosecution must also disclose material exculpatory evidence (i.e., evidence material to a defendant's guilt or punishment that tends to justify, excuse, or clear the defendant from alleged fault or guilt).[99] In *Brady*, the United States Supreme Court held that a defendant is denied due process if the government suppresses exculpatory information, material to a defendant's guilt or punishment, that the government knew.[100] Therefore, if the prosecution expert's analysis shows that the accused did not commit the crime or is guilty of a lesser crime, the prosecution must disclose this information to the defense.[101] This requirement applies in criminal cases in federal court and in state court.

Another federal criminal discovery obligation is set forth in the Jencks Act.[102] The Jencks Act requires the government to produce, on the defendant's request, any statements by a government witness that relate to that witness' testimony. Statements include written statements signed, adopted, or approved by the witness; a contemporaneously recorded verbatim recital of a witness' oral statement; or a witness' statement to a grand jury. Therefore, if the prosecutors or investigators wrote down what the expert said and read it back to him or let him read it, they have created Jencks Act material.

In a civil case, discovery of experts and their opinions may be obtained by interrogatories. The interrogatories require the party to disclose each person that the party expects to call as an expert at trial. Furthermore, the party must state the subject matter and substance of the facts and opinions to which the expert is expected to testify as well as a summary of the grounds for each opinion.[103] Any person expected to be called as an expert witness at trial may be deposed. An expert who has been retained or specially employed by a party in anticipation of, or preparation for, trial who is not expected to be called as a trial witness may only be discoverable on a showing of exceptional circumstances.

A deposition is another method of pretrial discovery that consists of a statement of a witness under oath. It is taken in question-and-answer form with opportunity given to

---

[97] Fed. R. Crim. P. 16(a)(1)(F) (2002).
[98] Fed. R. Crim. P. 16(b)(B) (2002).
[99] *Brady v. Maryland*, 373 U.S. 83 (1963).
[100] *Id.*
[101] *See, e.g., Ex Parte Mowbray*, 943 S.W.2d 461 (Tex. Crim. App. 1996) in which the prosecution's knowing failure to disclose a blood pattern analyst's report supporting defense theory that deceased committed suicide was held to warrant a new trial. *But see Smith v. Massey*, 235 F.3d 1259 (10th Cir. 2000), *abrogated by Neill v. Gibson*, 278 F.3d 1044 (10th Cir. 2001) (abrogating a different point of law). The appellate court in *Smith* found that it was an error to allow the state's "expert witness" to testify in the field of bloodstain analysis because he lacked sufficient training. *Id.* at 1270–71. For example, the chemist testified that he had taken a week-long bloodstain analysis training course taught by Dr. MacDonell. *Id.* at 1271. However, postconviction, Mr. MacDonell indicated that this particular chemist had only taken a 3-day course and was not prepared to testify as an expert in court. *Id.* The defendant argued that the suppression of this information was a *Brady* violation. *Id.* at 1276. However, the defense could not demonstrate that the prosecution was aware that the chemist was falsifying his qualifications. *Id.* Additionally, the court was not persuaded that this information would satisfy *Brady's* materiality requirement because of the other "wealth of evidence pointing to [the defendant's] guilt. *Id.*
[102] 18 USCA § 3500 (2000).
[103] *E.g.*, Fla. R. Civ. P. 1.280 (2004) *General Provisions Governing Discovery*.

the adversary to be present and cross-examine. A deposition is reported and transcribed stenographically, or it may be videotaped in some instances.

The opposition's objectives in taking an expert's deposition include the following: (1) to gather additional information; (2) to attempt to impeach the expert; (3) to lock the expert into a position or story that will be difficult to maintain at trial; (4) to assess the expert's demeanor as a witness; and (5) to demonstrate to the expert's proponent the extent of knowledge and expertise the opposition possesses.

The majority of jurisdictions do not provide for discovery depositions in criminal trials.[104] However, Florida,[105] Iowa,[106] Indiana,[107] Nebraska,[108] Missouri,[109] North Dakota,[110] Vermont,[111] and Texas[112] allow for discovery depositions of experts in criminal cases. Depositions of experts in civil cases are universally allowed.

A federal civil procedure discovery rule requires a party to identify experts who are retained or specially employed and who may be used at trial.[113] The party proffering the expert must provide a written report containing the following types of information:

1. Complete statement of all opinions to be expressed and the basis and reasons therefore
2. The data or other information considered by the witness in forming the opinions
3. Any exhibits to be used as a summary of or support for the opinions
4. The qualifications of the witness, including a list of all publications authored by the witness within the preceding 10 years
5. The compensation to be paid for the study and testimony
6. A listing of any other cases in which the witness has testified as an expert at trial or by deposition within the preceding 4 years [114]

Federal Rule of Civil Procedure 26 permits each district court, by local rule or order, to "opt out" of the rule's requirements.[115] The majority of the 94 district courts have adopted Rule 26(a)(2); however, it is best to check with the local district court to determine the local discovery requirements.

The court's power in the discovery process is far-reaching. One court has held it was not an abuse of discretion for a trial court to order the plaintiff's expert to identify the accountant who prepared his tax return because it might lead to information regarding the amount of income he derived from testifying for the plaintiffs, which may be relevant

---

[104] Paul C. Giannelli, *Scientific Evidence in Civil and Criminal Cases*, 33 Ariz. St. L.J. 103, 111 (2001).
[105] Fla. R. Crim. P. 3.220(h) (1999).
[106] I.C.A. Rule 2.13 (2002) *formerly cited as* I.A. St. § 813.2 Rule 12; *See also* Moenssens, Starrs, Henderson, & Inbau, *supra* note 6 at 53.
[107] Indiana Code 35-37-4-3 (1998); *See also* Moenssens, Starrs, Henderson, & Inbau, *supra* note 6 at 53.
[108] Neb. Stat. 29–1917 (2003); *See also* Moenssens, Starrs, Henderson, & Inbau, *supra* note 6 at 53.
[109] Mo. Rule of Crim. Pro. 25.12(a) (2003); *See also* Moenssens, Starrs, Henderson, & Inbau, *supra* note 6 at 53.
[110] N.D.R. Crim. P., Rule 15 (2) (1990); *See also* John G. Douglass, *Balancing Hearsay and Criminal Discovery*, 68 Fordham L. Rev. 2097, 2188–89 n. 381 (2000) (explaining that North Dakota, as a matter of right, allows for criminal defense discovery depositions).
[111] Vt. R. Cr. P. 15 (1991). *See also* Douglass, *id.* (explaining that Vermont, as a matter of right, allows for criminal defense discovery depositions).
[112] Tex. Crim. Pro. Art. 39.02 (1979); *See also* Douglass, *id.* (explaining that in criminal cases Texas permits discovery depositions).
[113] Fed. R. Civ. P. 26(a) (2) (A-C) (2000).
[114] Fed. R. Civ. P. 26(a)(2)(B) (2000)
[115] *Id.*

to show bias.[116] The appellate court also upheld the lower court's decision to preclude the expert's testimony as a sanction for refusing to identify the accountant. Sanctions for discovery violations range from contempt to mistrial.

The process of discovery is a continuing obligation. Both parties must disclose discoverable information through the trial. Not all information is discoverable. Some information may be privileged. Privileges are governed by rules, statutes, or case law. Privileges were developed to encourage communication between parties in special relationships (e.g., doctor–patient, attorney–client). By protecting these communications from disclosure, communication between the parties is encouraged and the relationship is fostered.

A communication is protected by the attorney–client privilege if it is between privileged persons, in confidence, and for the purpose of seeking or obtaining legal assistance.[117] A communication is confidential if it is "not intended to be disclosed to third persons other

---

[116] *Pitt v. Griggs*, 585 So. 2d 1317 (Ala. 1991).
[117] The attorney–client privilege may be a common-law privilege (existing in case law only) in your jurisdiction or it may be codified. *See, e.g.*, Fla. Stat. § 90.502 (1999 & Supp. 2004), which provides:
(1) For purposes of this section:
(a) A "lawyer" is a person authorized, or reasonably believed by the client to be authorized, to practice law in any state or nation.
(b) A "client" is any person, public officer, corporation, association, or other organization or entity, either public or private, who consults a lawyer with the purpose of obtaining legal services or who is rendered legal services by a lawyer.
(c) A communication between lawyer and client is "confidential" if it is not intended to be disclosed to third persons other than the following:
1. Those to whom disclosure is in furtherance of the rendition of legal services to the client.
2. Those reasonably necessary for the transmission of the communication.
(2) A client has a privilege to refuse to disclose, and to prevent any other person from disclosing, the contents of confidential communications when such other person learned of the communications because they were made in the rendition of legal services to the client.
(3) The privilege may be claimed by the following:
(a) The client
(b) A guardian or conservator of the client
(c) The personal representative of a deceased client
(d) A successor, assignee, trustee in dissolution, or any similar representative of an organization, corporation, or association or other entity, either public or private, whether or not in existence
(e) The lawyer, but only on behalf of the client. The lawyer's authority to claim the privilege is presumed in the absence of contrary evidence.
(4) There is no lawyer–client privilege under this section when the following occurs:
(a) The services of the lawyer were sought or obtained to enable or aid anyone to commit or plan to commit what the client knew was a crime or fraud.
(b) A communication is relevant to an issue between parties who claim through the same deceased client.
(c) A communication is relevant to an issue of breach of duty by the lawyer to the client or by the client to the lawyer, arising from the lawyer–client relationship.
(d) A communication is relevant to an issue concerning the intention or competence of a client executing an attested document to which the lawyer is an attesting witness or concerning the execution or attestation of the document.
(e) A communication is relevant to a matter of common interest between two or more clients, or their successors in interest, if the communication was made by any of them to a lawyer retained or consulted in common when offered in a civil action between the clients or their successors in interest.
(5) Communications made by a person who seeks or receives services from the Department of Revenue under the child support enforcement program to the attorney representing the department shall be confidential and privileged as provided for in this section. Such communications shall not be disclosed to anyone other than the agency except as provided for in this section. Such disclosures shall be protected as if there were an attorney–client relationship between the attorney for the agency and the person who seeks services from the department.
(6) A discussion or activity that is not a meeting for purposes of s. 286.011 shall not be construed to waive the attorney–client privilege established in this section. This shall not be construed to constitute an exemption to either s. 119.07 or s. 286.011.

than (1) those to whom disclosure is in furtherance of the rendition of professional legal services to the client and (2) those reasonably necessary for the transmission of the communication."[118] Thus, expert witnesses may be included within the lawyer–client privilege.

Work product includes written statements, private memoranda, and personal recollections[119] prepared or formed by an attorney that reflect the attorney's efforts at investigating and preparing a case, assembling information, determining relevant facts, preparing legal theories, planning strategy, and recording mental impressions.[120] This privilege seeks to prevent opposing counsel from "free-riding" on the thoughts and research of the attorney.[121] Therefore, if any attorney discusses trial and strategy with the expert or writes letters or memos to the expert regarding trial strategy, it may be work product. Note that the inadvertent production of a work-product–protected document will constitute a waiver of the privilege.[122]

Direct and Cross-Examination of the Expert

Attorney's Goals and Methods of Direct Examination

The direct examination is the unfolding of the evidence told from each side's perspective.[123] The attorney should ask the expert witness short, direct, concise questions to elicit the following information:

1. Who are you? (e.g., your qualifications)
2. What did you do? (protocol)
3. Why did you do that? (did you consider alternative analyses, methodology, additional factors, underlying assumptions, and contrary theories)
4. What did your find? (basis of your opinion)
5. What does that mean to you? (your opinion)

The expert should educate the members of the jury and help them understand why and how the expert reached the opinion. The expert cannot do an adequate job of educating the jury unless there has been extensive preparation. To help in this preparation, an expert should insist on a conference with the attorney prior to trial.[124]

When testifying, experts must be wary of exceeding the scope of their expertise, or their testimony may be stricken from the record or the entire case may be overturned

---

[118] *E.g., Id.* at 1(c)(1).
[119] *See, e.g., Lopez v. State*, 696 So. 2d 725 (Fla. 1997) (holding that the attorney's handwritten notes that dealt specifically with the trial strategy and the techniques of cross-examination were exempt from disclosure because they were work product).
[120] *Hickman v. Taylor*, 329 U.S. 495 (1946). For an example of the privilege contained in state discovery rules, *see* Fla.R. Crim. P. 3220(g) (2000), which provides the following:
(1) *Work Product*. Disclosure shall not be required of legal research or of records, correspondence, reports, or memoranda to the extent that they contain the opinions, theories, or conclusions of the prosecuting or defense attorney or members of their legal staffs.
[121] *Cincinnati Ins. Co. v. Zurich Ins. Co.*, 198 F.R.D. 81 (W.D. N.C. 2000).
[122] *See, e.g., Amgen Inc. v. Hoechst Mario Roussel, Inc.*, 190 F.R.D. 287 (D. Mass. 2000) (finding that the defendant's attorney–client privilege and the work product privilege had been waived as to the information contained within the inadvertently disclosed 3821 pages of documents).
[123] For a helpful guide written for expert witnesses on the stages of litigation *See* American Association for the Advancement of Science, *Case Experts Handbook, Being an Expert* (January 2002) available online at http://case.aaas.org/handbooks/experts/beinganexpert.html (last visited August 15, 2004).
[124] For other helpful tips for experts preparing attorneys *See* Garcia *infra* note 159.

on appeal.[125] A case from South Carolina illustrates the hazard of testifying beyond one's expertise. The state's expert was properly qualified in crime scene processing and fingerprint identification.[126] This expertise allowed him to testify regarding the measurements from the crime scene, the identification of bloodstains, and finally to the collection of shell casings.[127] However, the court allowed the expert to also testify about his conclusion about the location of the defendant's body and the victim's body in relation to the crime scene.[128] On appeal, the defendant argued that the state failed to qualify the crime-scene-processing expert as a crime-scene-reconstruction expert.[129] The Supreme Court of South Carolina agreed with the defendant's argument, reversed the murder conviction, and remanded the case for a new trial.[130] The Court found that allowing the expert to testify beyond the scope of his expertise not only was a reversible error but also was prejudicial to the outcome of the case.[131]

Demonstrative Evidence

The modern world is increasingly oriented toward visual stimuli. Experts approximate that a high school student, on average, has completed 11,000 hours of classroom education in contrast to the estimated 15,000 hours of watched television.[132] Many surveys show that humans respond better to visual learning methods.[133] For example, one survey purports that whereas humans only retain about 10% of what they hear, they are able to retain approximately 87% of what they see.[134] Another researcher advocates that as "Generation Xers" fill the jury box, the use of high-quality visual aids and computer animations will be vital to retention of the attention of these members of the population that have always lived with television, action-packed video games, and mass media.[135] Therefore, to communicate effectively to judges and jurors, experts must give considerable thought to the types of exhibits they wish to use at trial. Real evidence is evidence furnished by things themselves, as distinguished from a description of them through a witness's testimony.[136]

---

[125] *State v. Gilliam*, 514 So.2d 1098 (Fla. 1987) (medical examiner was not an expert in shoe pattern analysis; therefore, it was erroneous to allow her to testify that the defendant's sneaker left marks on the decedent); *see also Behn v. State*, 621 So.2d 534 (Fla. Dist. Ct. App. 1993) (accident reconstructionist testified regarding extent of injuries that would have been suffered had truck's brakes been fully operative; it was beyond competence because he was not an expert in medicine, physiology, or biomechanics); *Kelvin v. State*, 610 So.2d 1359 (Fla. Dist. Ct. App. 1992) (held error to allow evidence technician to testify to trajectory of bullets depicted by dowels stuck into purported bullet holes in sofa at crime scene; the technician was not a reconstructionist and had no training in ballistics).

[126] *State v. Ellis*, 547 S.E.2d 490, 491 (S.C. 2001).

[127] *Id.*

[128] *Id.*

[129] *Id.*

[130] *Id.*

[131] *Id.*

[132] Daniel Wolfe, ATLA Annual Convention Reference Materials, *Strategic Uses of Jury Research and Use of Technology in Complex Cases*, 1 Ann.2001 ATLA-CLE 961 (July 2001).

[133] Betsy S. Fiedler, Note, *Are Your Eyes Deceiving You?: The Evidentiary Crisis Regarding the Admissibility of Computer Generated Evidence*, 48 N.Y.L. Sch. L. Rev. 295, 306 (2003).

[134] *Id.*

[135] Dean M. Harts, *Reel to Real: Should You Believe What You See? Keeping the Good and Eliminating the Bad of Computer-Generated Evidence Will Be Accomplished Through Methods of Self-Authentication and Vigilance*, 66 Def. Coun. J. 514, 520 (October 1999) (commenting that "jurors retained twice as much from a visual as from an oral presentation and more than six times retention when oral and visual presentation were combined, compared to an oral presentation").

[136] Black's Law Dictionary 599 (8th ed. 2004) ("Physical...that itself plays a direct part in the incident in question").

The dried bloodstain pattern on the victim's clothing is an example of real evidence. Demonstrative evidence illustrates, demonstrates, or helps explain oral testimony.[137] Examples of demonstrative evidence include models, diagrams, charts, and computer animations and simulations.[138]

Demonstrative evidence can be used to highlight salient points of an expert's testimony, increase the jurors' comprehension, illustrate information difficult to comprehend, permit the jury to digest large amounts of data, recreate or reconstruct critical events in an evidentiary chain, and add dramatic effect to oral testimony.[139]

Demonstrative evidence is typically admissible to assist jurors in comprehending a witness' testimony.[140] The prosecution, in *Doerr*, used a diagram of the defendant's apartment with color-coded dots to show the jury the location of certain pieces of evidence. The defendant was arguing before the appellate court that the diagram was misleading; however, the appellate court noted that the test for the admissibility of diagrams focuses on the diagram's helpfulness to the jurors rather than on the absolute correctness of the diagram.[141] A detective prepared the *Doerr* diagram and used it in his testimony. Each dot represented a bloodstain, and the color of the dot represented the blood type.[142] The appellate court found that the trial court had not abused its discretion in allowing the detective to use this diagram during testimony because there was a large area that encompassed many bloodstains.[143]

To be admitted, demonstrative evidence must illustrate or explain a relevant issue in the case. The admissibility of such evidence is within the court's discretion, and the judge's decision will only be overturned if the court abused its discretion.[144] In determining the admissibility of such evidence, the court must weigh the probative value against any prejudicial effect the evidence might cause. Studies have shown that jurors and judges are impressed by visual aids.[145] In fact, in a study of judges and their impressions of fingerprint experts, the judges noted that a lack of visual aids was one of the three major obstacles to jurors' understanding of expert testimony.[146] More than 75% of the judges surveyed felt an expert should make a charted enlargement even for nonjury trials.[147]

---

[137] *Id.* at 596 ("Physical evidence that one can see and inspect, and that…does not play a direct part in the incident in question").
[138] E. Scott Savage, Demonstrative Evidence: Seeing May Not Be Believing but it Beats Not Seeing At All, *Utah Bar Journal* 17 (November 8, 1995) (providing practical advice for the use of illustrative and real evidence in litigation); For an introduction to frequently used demonstrative evidence and the legal concerns in using such evidence, see Ch. 2, "Demonstrative Evidence," in Moenssens, Starrs, Henderson, & Inbau, supra note 6. There is an extensive bibliography of resources at the end of the chapter.
[139] Mark A. Dombroff, *Dombroff on Demonstrative Evidence*, 4 (Wiley & Sons, N.Y., 1983).
[140] State v. Doerr, 969 P.2d 1168, 1178 (Ariz. 1998).
[141] *Id.*
[142] *Id.*
[143] *Id.*
[144] Paul M. Wolff, *Demonstrative Evidence*, 1 Forensic Sciences 12-1, 12-4, Cyril Wecht, Ed. (1996), Matthew Bender, New York.
[145] Illsley, *supra* note 60 at 49.
[146] Telephone conversation with Charles Illsley; *See also* Derek Danois, *Computer Animation Helps Win Cases by Visualizing Complex Evidence, Juror Comprehension Rises*, 18 (29) Pennsylvania Law Weekly July 17, 1995 at S8.
[147] Telephone Conversation with Charles Illsley, *Id.*; *See also* Fred Chris Smith & Rebecca Gurley Bace, *A Guide to Forensic Testimony: The Art and Practice of Presenting Testimony as an Expert Technical Witness*, 84 (Addison-Wesley 2003) (informing experts that they should "construct simple and honest graphics to allow the expert and the attorney and then the jurors to put it all together as the elements are presented through the testimony of the witness").

Another study has found that jurors perceive an expert who demonstrates something, like a chart or diagram, in front of the jury to be more credible than an expert who does not leave the witness stand.[148]

Attorneys' Goals and Methods of Cross-Examination

The cross-examination can be the most frightening and intimidating part of the trial for the novice expert witness. The opposing lawyer's purpose on cross-examination is to undermine the expert's credibility, explanations, and opinions.[149] The opposing attorney will do this by emphasizing three points: What the expert is not, what the expert did or did not do, and what the expert does not know.

The scope of the cross-examination is limited to the subject matter of the direct examination and matters affecting the credibility of the witness.[150] However, the court does have discretion to permit inquiry into additional matters.[151] The attorney will ask leading questions to control the witness.[152]

Courts often allow attorneys wide latitude on the cross-examination of experts to test their opinions on matters that are not common knowledge. Cross-examination of an expert is allowed for the purpose of explaining, modifying, or discrediting the expert's testimony. The jury has a right to know the extent of the witness's expert knowledge and experience. The expert's credentials and past record are relevant areas of inquiry.

There are many methods of cross-examination. Nine methods will be discussed in this section. The first method is to attack the field of expertise. To diminish an expert's credibility, an attorney wants to demonstrate that the field of expertise is art, not science.

The second method of cross-examination is exposing bias — "Isn't it true you only work for the prosecution? Isn't it true you've never testified for the defense?" Through this line of cross-examination, the attorney is trying to make the jury believe the expert favors one side over the other. Another area where attorneys attempt to demonstrate bias is by inquiring into the amount of money an expert earns as an expert witness as well as the number of times certain firms retain the expert. Some courts have allowed the opposing counsel to inquire into an expert's income records for recent years prior to testifying as well as inquire into the amount of referrals the expert receives from the retaining law firm.[153]

The third method of cross-examination is attacking the chain of custody. The attorney will explore the crime laboratory's record keeping, evidence storage, and documentation.

---

[148] Richard Tanton, Jury Preconceptions and Their Effect on Expert Scientific Testimony, 24, *J. Forensic Sci.*, 691 (1979).

[149] American Association for the Advancement of Science, Case Experts Handbook, Being an Expert (January 2002) at http://case.aaas.org/handbooks/experts/beinganexpert.html (also providing 15 tips for experts in handling opposing counsel's questions on cross examination).

[150] Fed. R. Evid. 611(b) (2003)

[151] *Id.*

[152] Leading questions are those that usually limit the answer to a "yes" or "no" or those that suggest to the witness the answer desired. Black's Law Dictionary 906 (8th ed. 2004); See also Fed. R. Evid. 611 (c) (2003) for the appropriateness of leading questions.

[153] *See*, for example, *Flores v. Miami-Dade County*, 787 So.2d 955, 957–959 (Fla. Dist. Ct. App. 2001) (holding that exploring referral arrangements between the treating doctor and the plaintiff's law firm was an appropriate method of establishing bias. Additionally, instead of limiting the defense counsel to the referral arrangements relative to the specific suit, the court permitted the questioning to include compensation arrangements relative to other cases); *Gunn v. Robertson*, 801 So. 2d 555 (La. App. 2001) (allowing the defense to establish for the purpose of demonstrating bias, that 75% of the expert's medical practice stemmed from the referrals of the plaintiff's attorney); *See also Trower v. Jones*, 520 N.E.2d 297 (Ill. 1988).

Some courts have held that a case must be dismissed if crucial evidence, such as fingerprint evidence, has been lost or misplaced.[154] To avoid such disastrous results, some laboratories have developed evidence-tracking devices not unlike the bar-coding system used at grocery stores.[155]

The fourth method of cross-examination is inquiring into laboratory protocol. Each laboratory should have documented training and protocol for each area of forensic science. If no protocol exists, the lawyer may cross-examine by intimating sloppy laboratory procedure. If a protocol does exist and it was not followed, the attorney will establish this fact on cross-examination and will not ask the expert for an explanation. Rather, the attorney will argue to the jury in closing to disbelieve an expert who does not follow established procedures, thus casting doubt on the credibility of the analysis as well as the expert.

The fifth method of cross-examination is attacking the expert's factual basis for his or her opinion. Most experts would concede that their opinions are only as good as the factual and theoretic bases on which they are founded. When attorneys use this method of cross-examination; for example, an attorney might inquire into the fact that the expert was not at the crime scene or did not personally take the blood samples.

Attacking the instruments used by the expert to do the analysis is the sixth method of cross-examination. Lawyers will explore areas such as how data are entered, whether the equipment is calibrated, how it is maintained, and whether it is state of the art or a dinosaur-like relic.

The seventh method of cross-examination is an inquiry into proficiency testing. An expert's laboratory should participate in such testing on a periodic basis. During discovery, lawyers may inquire into the laboratory's and expert's results on such tests to unearth material to use in a cross-examination.

Laboratory accreditation and an expert's board certification is also a relevant area of inquiry on cross-examination. If a laboratory has been put on probation by an accrediting body, such an action may reflect adversely on the expert. Attorneys may cross-examine an expert if board certification exists in the expert's area of expertise and the expert is not certified, or the expert was certified and had the certification revoked or not renewed.

The eighth method of cross-examination is attacking an expert's qualifications. This may occur during a motion *in limine* when one side is asking the court to rule on the admissibility of the field and the expert's expertise, or during the *voir dire* of an expert.

The ninth method of cross-examination is impeachment with a learned treatise. When using this method of cross-examination, the attorney will ask the expert if he or she recognizes a text as a leading authority in the field.[156] Note that some courts will not accept a person as an expert unless the person is familiar with the leading texts and journals in their

---

[154] *See, e.g., Scoggins v. State,* 802 P.2d 631 (N.M. 1990); *But see* Dale Joseph Gilsinger, *Authentication of Blood Samples taken from Human Body for Purposes other than Determining Blood Alcohol Content,* 77 ALR 5th 201 §14 (2000) (providing numerous examples of proving chain of custody even with periods of time when the blood samples were misplaced).

[155] John Donnelly, *Death Goes High Tech: Body Toe Tags Get Bar Codes,* Miami Herald, Oct. 4, 1993, at A1; *See also* Dennis Tatz, *Dedham Prisoners Join State DNA Web; Mass. State Police Database expanded to help ID suspects; DNA database,* Patriot Ledger (Quincy Mass.), April 20, 2004, at 1 (discussing the use of bar codes to identify and verify blood samples used to gather DNA evidence); *Arlington to Update Evidence-Tracking System,* Chicago Daily Herald, March 17, 2004, at 1 (The Arlington Heights Police Department announced that it would spend approximately $12,000 to upgrade the evidence storage system by integrating bar codes on evidence for tracking purposes).

[156] William D. Farber, *Contradiction of Expert Witness Through Use of Authoritative Treatise,* 31 Am. Jur. Proof of Facts 443 (2d ed. August 2003).

field of expertise.[157] Once the expert admits that fact, the attorney is free to use that authority to contradict the expert's analysis or opinion, thus impeaching the expert's credibility.

When preparing for cross-examination of an expert, diligent attorneys will read all of the expert's publications and will locate deposition testimony, transcripts of trial testimony, and reported decisions of cases in which the expert testified. All of these items may reveal potential impeachment material, such as a change in theory, or a contrary opinion on a similar factual basis.

The key to reducing an expert's vulnerability on cross-examination is preparation by both expert and attorney. This means the expert must do his "homework" and help the attorney become familiar with the area of expertise, credentials, and potential areas of weakness. If the attorney knows these things in advance of trial, he or she can present the expert in a most favorable light and minimize or even eliminate the possibility of an uncomfortable cross-examination experience.[158]

Another way of reducing vulnerability on cross-examination is to maintain control. The adverse attorney is attempting to control the expert on cross-examination. That is why the attorney asks only leading questions. If the expert maintains a calm, thoughtful demeanor and does not lose his or her temper or composure in spite of the lawyer's tactics, the expert's credibility will be enhanced.

Advice to the Expert—Guidelines for Deposition and Trial Testimony

There are many steps than an expert witness can take to prepare an attorney for trial. The following list will aid the expert witness in the preparation of the attorney:

1. Insist on more than one witness conference.
2. Establish rapport with the attorney.
3. Provide the attorney with your qualifications. Think more broadly than just in terms of your education and publications. It is important to establish rapport with the jury. Realize, however, that not all jurors will have a natural empathy with you as a witness because of your special and unique skills. Some of them may resent you and may be looking forward to the opportunity to watch a lawyer impeach you. Therefore, you must attempt to establish a rapport with the laypersons on the jury and make your opinions meaningful to them.
4. Educate the attorney. Most attorneys do not have a significant educational background in science. Therefore, you must explain your theories in lay terms. Providing a glossary of terms is helpful to the attorney. [You] may even provide it to the jury if the court permits. It is extremely important that you explain your test procedures. For example, you may conduct certain tests before other tests because the former

---

[157] *See*, for example, Trial Transcript at 522–523, December 10, 1991, *United States v. Parks*, Case No. CR 91-358-JSL (C.D. Cal.).

[158] *See, e.g.*, Donald R. Wilson & Stanley E. Preiser, *Discovery and Inspection*, 1 Forensic Sciences, 11-1, 11-22, Cyril Wecht, Ed. (1997), Matthew Bender, New York *citing* the Federal Rules of Civil Procedure Advisory Committee of the United States Supreme Court:

> Effective cross-examination of an expert witness requires advance preparation. The lawyer even with the help of his own experts frequently cannot anticipate the particular approach his adversary's expert will take or the data on which he will base his judgment on the stand.... Similarly, effective rebuttal requires advance knowledge of the line of testimony of the other side. If the latter is foreclosed by a rule against discovery, then the narrowing of issues and elimination of surprise that discovery normally produces are frustrated.

tests are less destructive and thus you may be able to preserve more of the sample. This evidence can be used by the attorney to demonstrate your professionalism, thus enhancing your credibility.

5. Provide the attorney with your reports and notes. This will aid the attorney in complying with certain discovery obligations.
6. Inform the attorney if you have written any articles or treatises or testified in trials where you used a different theory or came to different results than you have in this particular case. This is important because it would be impeachment material.
7. Prepare the attorney for cross examination of the opposing expert. You can do this by informing the attorney of the different types of tests that could have been performed and by critiquing the other expert's work. You may even provide the attorney with additional information the attorney may use on *voir dire* and cross examination of the expert.
8. Inform the attorney of the limitations in your expertise so that the attorney will only use you to testify in those areas for which you are qualified.
9. Inform the attorney of the usefulness of certain pieces of evidence. Aid the attorney in creating demonstrative evidence graphs, charts, enlarged photographs, etc. The court will determine whether a piece of demonstrative evidence is relevant and useful to the jury's understanding of the case. The court will not admit such evidence if its use is prejudicial.
10. Emphasize your ethical obligations. You must make it clear to the attorneys what you ethically may or may not say on the witness stand. Remember it is your reputation that is at stake.[159]

To enhance the effectiveness of your testimony at deposition, trial, or hearing, we have included the following helpful guidelines, which are based on experience with many witnesses in different cases:[160]

1. Tell the truth.
2. Prepare yourself by review of the facts.
3. Remember that most questions can be answered:
   "Yes"
   "No"
   "I don't know"
   "I don't remember"
   "I don't understand" or
   By stating a single fact.
4. If "yes" or "no" will do, one of those should be your answer.
5. Limit your answer to the narrow question asked, then stop talking.
6. Never volunteer information or answers.
7. Do not assume you must have an answer for every question.
8. Be cautious of repeated questions about the same point.
9. Do not lose your temper.
10. Speak slowly, clearly, and naturally.

---

[159] Carol Henderson Garcia, 1 Criminal Practice Manual § 12: 6 Expert's checklist for attorneys (1987).
[160] Reprinted with permission. © *Tageh Press,* August 2004. *See also* Harold A. Feder, *Succeeding as an Expert Witness: Increasing Your Impact and Income,* Appendix 2.5 "More Testimonial Tips for the Experienced and the Novice Expert Witness" (Tageh Press, Colorado 2000) (listing over 100 useful tips for the expert witness).

11. Your posture should be forward, upright, and alert.
12. Do not nod or gesture in lieu of an answer.
13. Do not be afraid to ask for clarification of unclear questions.
14. Do not be afraid of the examining attorneys.
15. Be accurate about all fact conditions, damages, and injuries.
16. Restrict your answers to facts personally known to you.
17. State basic facts, not opinions or estimates, unless they are asked of you.
18. Be cautious of questions that include the word "absolutely" or "positively."
19. Remember, "absolute" means forever, without exception.
20. Be cautious about time, space, and distance estimates.
21. Do not guess if you do not know the answer.
22. Do not fence, argue, or second-guess the examining counsel.
23. Admit that you discussed your testimony previously if you did.
24. Do not memorize a story.
25. Avoid such phrases as "I think," "I guess," "I believe," or "I assume."
26. Maintain a relaxed but alert attitude at all times.
27. Do not answer too quickly—take a breath before answering each question.
28. Do not look to counsel for assistance.
29. Make sure you understand the question before answering.
30. Do not answer if you are told not to do so.
31. Never joke during a deposition or testimony.
32. Do not exaggerate, underestimate, or minimize.
33. Dress conservatively in clean clothes.
34. Be serious before, after, and during testimony.
35. If you make a mistake, correct it as soon as possible.
36. Remain silent if attorneys object during the examination.
37. Listen carefully to dialogue between attorneys.
38. Avoid mannerisms that signal nervousness.
39. Do not use technical language without translating it for your lay hearers.
40. Speak simply.
41. Do not discuss the case in the hallways or restrooms.
42. Do not converse with opposing parties, attorneys, or jurors.
43. Tell the truth.

Ethical Issues

Expert's Ethics

An expert witness has certain ethical obligations. They may be spelled out in an employee handbook, in the code of a professional association, or even by state statute.[161] For example, members of the American Academy of Forensic Sciences (AAFS) are prohibited from making

---

[161] *E.g.*, National Association of Medical Examiners Bylaws, Article Ten, Ethics, 1(a) (2001), *at* www.the-name.org/AboutNAME/Bylaws.htm (providing that members of N.A.M.E. shall conform to the published ethics of the American Medical Association); The National Society of Professional Engineers Code of Ethics (2003), *at* www.nspe.org/ethics/ch1-codepage.asp (setting forth fundamental canons, rules of practice, and professional obligations); The American Board of Criminalistics Bylaws, Article IV.5 (1992), *at* http://www.criminalistics.com/ethics.cfm (last visited March 7, 2005)(promulgating rules of professional conduct which must be complied with by applicants and diplomats).

misrepresentations of their education or of the data on which their professional opinions are based.[162] If an AAFS member is found to have violated the code, an ethics committee may impose sanctions, such as censure, suspension, or expulsion from the organization.[163]

Similarly, the International Association of Bloodstain Pattern Analysts (IABPA) has adopted a code of ethics:

1. Every member of the Association shall refrain from any material misrepresentation of their standing within the IABPA.
2. Every member of the Association shall refrain from any material misrepresentation of education, training, or experience in the area of bloodstain interpretation.
3. Every member of the Association shall refrain from material misrepresentation of data on which an expert opinion or conclusion is based.[164]

Violations of these rules or any of the other rules contained in IABPA's code of ethics could lead IABPA to censure, suspend, or expel a member of IABPA.[165]

Some courts have sanctioned experts for their unethical behavior. In *Schmidt v. Ford Motor Co.*,[166] the court banned the plaintiff's accident reconstruction expert from testifying in federal court in Colorado because he had conveyed intentionally misleading information in deposition and informal conversations with the defense expert. The expert also concealed his knowledge from the defendant that one of the plaintiffs had tampered with the evidence.

Attorneys' Ethics in Dealing with Experts

Attorneys' ethical obligations are contained in each state's Rules of Professional Conduct or Code of Professional Responsibility. While no specific rule deals directly with attorneys and expert witnesses, some of the American Bar Association Model Rules of Professional Conduct are applicable. The Model Rules have been adopted in a majority of states.[167] Some of the model rules that affect an attorney's use of expert witnesses are the following:[168]

> Rule 3.3 (Candor Toward the Tribunal) requires the attorney to investigate the background of expert witnesses to avoid putting on perjurious testimony regarding their credentials.

---

[162] The American Academy of Forensic Sciences Bylaw Code of Ethics and Conduct, Art. II, Section I (2004) www.aafs.org (last visited August 17, 2004).
[163] *Id.* at Art. II, Section 2.
[164] Constitution and Bylaws of the International Association of Bloodstain Pattern Analysts, Chapter 2 § 1 Code of Ethics, (Revised February 2001) Available online in IABPA Newsletter (June 2001). http://www.iabpa.org/June2001News.pdf (last visited August 25, 2004).
[165] *Id.* at Chapter 2 § 5 *Investigative Action*; *see also* IABPA *Code of Ethics*, supra note 94.
[166] 112 F.R.D. 216 (D. Colo. 1986); *See also* Paula F. Wolff, *Federal District Court's Power to Impose Sanctions on Non-Parties For Abusing Discovery Process*, 149 A.L.R. Fed. 589 (1998).
[167] By the end of 2002, 44 jurisdictions, including the District of Columbia and the Virgin Islands, had adopted some version of the Model Rules. Center for Professional Responsibility, The American Bar Association *at* www.abanet.org/cpr/mrpc/alpha_states.html (last visited August 30, 2004) (listing the jurisdictions and the date of adoption). The vast majority of U.S. jurisdictions used the pre-2002 version of the Model Rules to develop the state-specific rules. Irma S. Russell, *Keeping the Wheels on the Wagon: Observations on Issues of Legal Ethics for Lawyers Representing Business* Organizations, 3 Wyo. L. Rev. 513, 535 (2003). However, in 2002, the ABA House of Delegates adopted a new version of the Model Rules. *Id*. at 534–535. Few states have completed incorporating those amendments [2002–2003 amendments] into their rules. Thomas D. Morgan and Ronald D. Rotunda, *2004 Selected Standards on Professional Responsibility,* Preface (written in October 2003) (Foundation Press, N.Y. 2004). *See also Status of State Review of Professional Conduct Rules, at* www.abanet.org/cpr/jclr/ethics_2000_status_chart.pdf (last visited August 30, 2004) (listing each state's consideration of the newer version of the Model Rules).
[168] Model Rules of Professional Conduct (2003).

Rule 3.8 (Special Responsibilities of a Prosecutor) specifies that the prosecutor's role as a "minister of justice" requires him to make timely disclosure of evidence or information that will negate evidence of guilt or mitigate guilt. Therefore, if fraud is uncovered relating to the expert's acts or knowledge, it must be disclosed.

Rule 5.3 (Responsibilities Regarding Nonlawyer Assistants) applies to experts as well as paralegals and extends to situations where the lawyer is in essence ratifying the unethical conduct of the expert.

Rule 8.3 (Reporting Professional Misconduct) requires a lawyer to report unethical conduct of other attorneys. Therefore, if the opposing party's counsel knowingly uses an expert discovered to be a fraud, counsel is obligated to report the other lawyer to the grievance committee. If counsel does not report, counsel is in violation of this rule.

Rule 8.4(a–d) (Misconduct) states that it is professional misconduct to violate the Model Rules; commit a criminal act that reflects adversely on a lawyer's honesty, trustworthiness, or fitness; engage in conduct involving dishonesty, fraud, deceit, or misrepresentation; or engage in conduct that is prejudicial to the administration of justice.

Additionally, an attorney shall not fabricate evidence, counsel or assist a witness to testify falsely, offer an inducement to a witness that is prohibited by law, make frivolous discovery requests, or intentionally fail to comply with a legally proper discovery request.[169] A lawyer shall not make false statements of material fact or law to a third person, such as an expert witness.[170]

Model Rules 1.1 (Competence) and 1.3 (Diligence) require an attorney to seek out expert services, if needed by the client. Failure by a defense counsel in a criminal case to obtain the services of expert witnesses may later be deemed by courts to have resulted in the ineffective assistance of counsel.[171] Many other examples exist of ineffective assistance of counsel dealing with scientific evidence.[172] The 9th Circuit in *Baylor v. Estelle* stated the following:

> We have difficulty understanding how reasonably competent counsel would not recognize "the obvious exculpatory potential of semen evidence in a sexual assault case," particularly when the criminalist's report plainly indicates that the donor

---

[169] Model Rules of Professional Conduct, Rule 3.4 (2003).

[170] Model Rules of Professional Conduct, Rule 4.1 (2003).

[171] *Moore v. State*, 827 S.W.3d 213 (Mo. 1992) (holding counsel ineffective for failing to request serologic test); *United States v. Tarricone*, 996 F.2d 1414, 1418 (2d Cir. 1993) (finding that failure to consult with a handwriting expert could equate with ineffective assistance of counsel).

[172] Paul C. Giannelli, *Scientific Evidence in Civil and Criminal Cases*, 33 Ariz. St. L.J. 103, 108–11 (2001) (providing examples of cases and commentary regarding the plethora of ineffective assistance of counsel cases in the area of scientific testimony—including the following quote from *Driscoll v. Delo*, 71 F.3d 701, 709 [8th Cir. 1995]):

> In a capital murder case whether or not the alleged murder weapon...had blood matching the victim's constituted an issue of the utmost importance. Under these circumstances, a reasonable defense lawyer would take some measures to understand the laboratory tests performed and the inference that one could logically draw from the results. At the very least, any reasonable attorney under the circumstances would study the state's laboratory report with sufficient care so that if the prosecution advanced a theory at trial that was at odds with the serology evidence, the defense would be in a position to expose it on cross examination.

was an ABO nonsecretor whereas [the defendant] was an ABO type "O" secretor and that this "would thus eliminate" [the defendant] as the perpetrator unless a test…on a liquid semen sample showed that he mimicked a nonsecretor…. Whether or not [the criminalist's] report was itself conclusive, it was one test away from tilting the scale powerfully in [the defendant's] direction.[173]

Additionally, attorneys might be sanctioned by the state bar for failing to retain an expert witness.[174]

A lawyer may not promise an expert a fee contingent on the outcome of the case,[175] nor may an attorney share fees with an expert.[176] An attorney has been held to have an ethical obligation to pay an expert's fees unless he gives an express disclaimer of responsibility.[177]

Attorneys have been sanctioned by the bar for abusing an expert witness on cross-examination. A prosecutor was suspended from the practice of law for 30 days for improperly eliciting irrelevant testimony from the defense's expert witness, a psychiatrist.[178] The prosecutor insulted the witness; ignored the court's rulings on defense objections, which were sustained; and inserted his personal opinions on psychiatry and the insanity defense into his questioning. More recently, another prosecutor found himself before the Arizona court in a disciplinary matter regarding the abuse of an expert witness.[179] Additionally, experts might be able to recover civilly from attorneys who suffer damages based on abusive litigation strategies.[180]

---

[173] *Id.* at 108, *citing Baylor v. Estelle*, 94 F.3d 1321 (9th Cir. 1996).

[174] *In re Zimmerman*, 19 P.3d 160 (Kan. 2001) (failing to retain an expert witness was one of many factors used by the court in deciding that a 1-year suspension from the practice of law was an appropriate disciplinary action); *see also Matter of Ortiz*, 687 A.2d 267 (N.J. 1997) (failing to retain an expert in a medical malpractice case was one of many factors used by the court in deciding the appropriate reprimand).

[175] Model Rules of Professional Conduct Rule 3.4 cmt. 3 (2003) (indicating that the general common law rule amongst jurisdictions is that "it is improper to pay an expert witness a contingent fee"); *See*, for example, Michigan Complied Laws Annotated § 600.2169 (4) (West 2000) (making it a misdemeanor of experts to testify on a contingency basis in medical malpractice cases); *see, e.g.*, Florida Statutes Annotated Bar Rule 4-3.4 (b) (2003), *Fairness to the Opposing Party and Counsel*, ("a lawyer may pay…a reasonable, noncontingent fee for professional services of an expert witness); *See also* Frank E. Haddad, *Compensation of Experts*, 1 Forensic Sciences, 17-1, 17-6 (Cyril Wecht, Ed., Matthew Bender, N.Y., 1996) (explaining the history of why contingent fees are void according to public policy and why the judicial system will not enforce contingent fees).

[176] Sharing fees with nonlawyers violates ABA Model Rule 5.4, Model Rules of Professional Conduct (2003).

[177] Steve Morris and Christina C. Stipp, *2002 Civil Practice and Litigation Techniques in Federal and State Courts: Ethical Conflicts Facing Litigators*, American Law Institute — American Bar Association Continuing Legal Education (August 2002) (discussing the case law supporting a lawyer's duty to pay litigation expenses, including expert witnesses). *See, e.g., Copp v. Breskin*, 782 P.2d 1104 (Wash. Ct. App. 1989) (citing ABA Model Rules of Professional Conduct, 1.8(e) and 4.4 for authority to hold attorney responsible to pay expert fees).

[178] *The Florida Bar v. Schaub*, 618 So.2d 202 (Fla. 1993).

[179] *In re Zawada*, 92 P.3d 862 (Ariz. 2004). *See also Kentucky Bar Ass'n v. Mussler*, 19 S.W.3d 87 (Ky. 2000) (publicly reprimanding an attorney for mistreating an expert witness during a civil discovery deposition); *See also Matter of Genchi*, 861 P.2d 127 (Kan. 1993) (Holding that a 6-month suspension from the practice of law was appropriate for the professional misconduct of the attorney. The attorney's behavior included many issues worthy of discipline; however, the court noted that one of the misbehaviors was abusive behavior toward an expert witness).

[180] *See, e.g.*, Richard B. Schmitt, *Milberg Weiss Agrees to Pay $50 Million to Settle Lexecon Case, Wall Street Journal*, April 14, 1999, at B17 (detailing the results of an economic and legal consulting firm suing a successful law firm for an abusive litigation tactic of attempting to discredit the consulting firm and all of its testifying experts by targeting the consulting firm in a savings-and-loan case).

It should also be noted that Rule 11, Federal Rules of Civil Procedure, provides for sanctions to punish an attorney who knowingly files a false and misleading pleading. Compliance with Rule 11 requires an attorney to reasonably research both the law and the facts that support a pleading.[181] Courts have held that failure to disclose a contrary expert opinion alone is an insufficient basis for imposing Rule 11 sanctions.[182] In *Coffey v. Healthtrust, Inc.*, the court granted the defendant's motion for Rule 11 sanctions against a plaintiff's attorney claiming that, when the lawyer filed an economic study, expert's affidavit, and accompanying brief supporting plaintiff's position that the hospital did not have competitors in its geographic market, the study's authors had told him that the expert's use of the study to support plaintiff's position would be misguided. The Tenth Circuit reversed, noting that whereas the attorney lied at the Rule 11 hearing regarding whether the study's authors talked with him, it may be a disciplinary matter but not a subject for Rule 11 sanctions. The court reasoned that an attorney must be allowed to reasonably rely on an expert's opinion as the basis of the client's position without fear of punishment for the expert's errors in judgment.[183] Furthermore, the court found that the attorney's reliance on the expert's opinion had been reasonable because (1) the expert had sworn to his position in his affidavit, (2) the trial court had accepted the witness's expert status, and (3) the expert had held his position even when confronted with the contradictory conclusions of the study's authors. The court noted that opposing counsel had the opportunity and duty to expose weaknesses in evidence.

The American Bar Association Standards Relating to the Administration of Criminal Justice also set forth standards for prosecutors and defense counsel to follow when working with expert witnesses in criminal trials. The standards provide that the attorney should respect the expert's independence, not dictate the formation of the expert's opinion, and that paying excessive or contingent fees is unprofessional conduct.[184]

Expert Witness Malpractice

In recent years, the law has been developing a new cause of action designed to hold expert witnesses, such as doctors and lawyers, responsible for their negligent professional behavior.[185] The law of expert witness negligence has developed largely in response to a recent recognition

---

[181] *Del Sontro v. Cendant Corp.*, 223 F.Supp.2d 563, 582 (D.N.J. 2002).
[182] *Coffey v. Healthtrust, Inc.*, 1 F.3d 1101 (10th Cir. 1993).
[183] See also *Reliance on Expert May Be Reasonable Pre-Filing Inquiry*, 17(1) Fed. Litigator 10 (citing *Coffey* and *Dubois v. U.S. Dept. of Agriculture*, 270 F.3d 77 [1st Cir. 2001])as the leading cases about avoiding sanctions based on reliance on expert advice).
[184] American Bar Association Standards Relating to the Administration of Criminal Justice, Standards 3-3.3 Prosecutor's Relations with Expert Witnesses and 4-4.4 Defense Counsel's Relation with Expert Witnesses (3d ed. 1993) available online at http://www.abanet.org/crimjust/standards/home.html (last visited August 29, 2004) (providing the following guidelines for attorneys):
(a) A prosecutor [defense counsel] who engages an expert for an opinion should respect the independence of the expert and should not seek to dictate the formation of the expert's opinion on the subject. To the extent necessary, the prosecutor should explain to the expert his or her role in the trial as an impartial expert called to aid the fact finders and the manner in which the examination of witnesses is conducted.
(b) A prosecutor [defense counsel] should not pay an excessive fee for the purpose of influencing the expert's testimony or to fix the amount of the fee contingent upon the testimony the expert will give or the result in the case.
[185] Randall K. Hanson, *Witness Immunity Under Attack: Disarming "Hired Guns,"* 31 Wake Forest L. Rev. 497, 508–09 (1996) (reviewing the arguments in support of and in opposition to witness immunity for experts). See also Carol Henderson Garcia, *Expert Witness Malpractice: A Solution to the Problem of the Negligent Expert Witness*, 12 Miss. College L. Rev. 39 (1991).

that such negligence is not uncommon.[186] Erroneous conclusions have been reported even with well-accepted scientific techniques, such as fingerprint identification.[187]

Even if an expert is not negligent in rendering an opinion, the expert's opinion is only as good as the underlying basis for the opinion. That basis may depend on other expert's opinions. Whether engaged as a consulting or testifying expert, in a civil or criminal matter, an expert should be aware of the ethical responsibilities and potential liabilities in that role.

At present, the law does little to regulate the quality of expert testimony.[188] Solutions offered by the scientific, medical, and legal communities to curb expert abuses include capping expert witness fees,[189] prescreening experts, using only court-appointed experts,[190] adhering to a strict code of ethics, using peer review,[191] and having a science court.[192] Additionally, it

---

[186] Craig M. Cooley, *Reforming the Forensic Science Community to Avert the Ultimate Injustice*, 15 Stan. L. & Pol'y Rev. 381, 395–396 (2004) (listing 21 recent exonerations where the underlying conviction was based on inaccurate or completely false testimony by forensic examiners). See also Paul C. Giannelli, *Fabricated Reports*, 16 Crim. Just. 49, 49 (2002) (discussing some of the recent "experts" who consistently fabricated reports and testimonies to achieve a particular bias).

[187] David Heath, *Bungled prints expose problems at FBI; MAYFIELD CASE Fingerprints, long seen as infallible, may not be; standards, bureau's training criticized*, Seattle Times June 7, 2004 at A1 (discussing the controversy surrounding the Spanish police demonstrating that the FBI made a mistake in naming a suspect based on faulty fingerprint analysis though the FBI had proclaimed that it was 100 percent certain of its match); see also David Heath, *U.S. Loses Faith in Fingerprints*, 20(4) The Print 1 (July/August 2004) available online at http://www.scafo.org/The_Print/The_PRINT_VOL_20_ISSUE_04.pdf. See also David Weber and Kevin Rothstein, *Man Freed After 6 Years; Evidence Was Flawed*, Boston Herald, Jan. 24, 2004, at 4 (discussing the potential of a misidentified fingerprint leading to a wrongful conviction).

[188] Joseph Peterson & John Murdock, Forensic Science Ethics: Developing an Integrated System of Support and Enforcement, 34 J. Forensic Sci., 749 (1989); See also Austin v. American Association of Neurological Surgeons, 253 F.3d 967, 973 (7th Cir. (Ill.) 2001) (finding "it is well known that expert witnesses are often paid very handsome fees, and common sense suggests that a financial stake can influence an expert's testimony, especially when the testimony is technical and esoteric and hence difficult to refute in terms intelligible to judges and jurors. More policing of expert witnessing is required, not less").

[189] Garcia, supra note 185 at 59 (explaining different regulatory schemes aimed at curbing expert fraud; including that in the early 1990s Florida considered capping expert fees at $250 per hour). Haddad, supra note 175 at 17-11 (explaining that the statutory schemes surrounding expert witness fees generally deal with providing indigent people with professional services and whether the losing party has to pay expert witness fees in the assessment of costs).

[190] Justin P. Murphy, Note, Expert Witnesses at Trial: Where are the Ethics?, 14 Geo. J. Legal Ethics 217, 236–239 (2000); See also Catherine T. Struve, Doctors, the Adversary System, and Procedural Reform in Medical Liability Litigation, 72 Fordham L. Rev. 943 (March 2004) (discussing a 1998 survey that indicated that in litigation involving complicated technical or scientific testimony only 16% of federal judges used court-appointed experts); See also Oliver C. Schroeder & Theodore R. LeBlang, Court Appointed Experts, 1 Forensic Sciences, 18-1, 18-1 through 18-20 (Cyril Wecht, Ed., Matthew Bender, N.Y. 2000) (explaining the role of the court-appointed expert; the statutory authorization for the court-appointed expert; the procedure for the appointment of the expert; the methods to discover the court-appointed expert's opinion; the weight accorded to the court-appointed expert's testimony; the use of court-appointed experts in civil and criminal trials; and finally, the constitutional issues related to appointing an expert).

[191] Terry Carter, M.D. with a Mission: A Physician Battles Against Colleagues He Considers Rogue Expert Witnesses, 90 ABA Journal 41, 42 (August 2004) (advocating peer review in the courtroom for medical expert witnesses). See also Coalition and Center for Ethical Medical Testimony CCEMT at http://www.ccemt.org (visited March 7, 2005) (defining the mission as the elimination of the "ability of unethical experts to testify with impunity in medical-legal matters on the assumption or under any law or regulation that makes such testimony privileged or protected from scrutiny by peers").

[192] See Twenty-Five Year Retrospective on the Science Court: A Symposium, 4 RISK—Issues in Health & Safety 95–188 (1993), containing a series of articles by advocates and detractors of the science court, including one by its "inventor." See Arthur Kantrowitz, Elitism vs. Checks and Balances in Communicating Scientific Information to the Public, 4 RISK—Issues in Health & Safety 101 (Spring 1993); See Sven Timmerbeil, The Role of Expert Witnesses in German and U.S. Civil Litigation, 9 Ann. Surv. Int'l & Comp. Law 163, 170 (Spring 2003) (discussing several proposals, including the science court, that addressed the concern that hired experts hindered the truth-seeking process of the courts).

has been suggested that fraudulent experts be prosecuted[193] and that the principal safeguard against errant expert testimony is the opportunity for opposing counsel to cross-examine.[194] The reality is that most lawyers do a woefully inadequate job in cross-examining experts.[195] One reason for this is improper preparation.[196] Another reason may be that lawyers are often reluctant to incur the risks involved in challenging experts in their own fields. Many lawyers do not even avail themselves of the assistance of experts in preparing for cross-examination[197] and are therefore unable to challenge statements made by experts effectively. Finally, the vast majority of civil and criminal cases are settled or plea-bargained prior to trial so that the expert may never be subjected to rigorous questioning during the adversary process.

To date, none of the solutions offered to curb expert abuses has succeeded in accomplishing its goal. Arguably, attempts at monitoring expert testimony may serve to deter some expert negligence and also result in experts being held personally accountable. However, such steps do not necessarily provide for compensation to individuals harmed by an expert's negligence.

In the latter part of 1998, the Scientific Freedom, Responsibility and Law Program of the American Association for the Advancement of Science (AAAS) created a program, Court Appointed Scientific Experts (CASE), to assist federal trial court judges in using independent technical and scientific experts as court-appointed experts.[198] Federal judges formally began to use CASE early in 2001.[199] Some of the services promoted by CASE are the following:

1. Evaluating the qualifications and conflicts of interest of scientists proposed by parties to serve as court experts

---

[193] *In the Matter of an Investigation of the West Virginia State Police Crime Laboratory, Serology Division*, 438 S.E.2d 501, 509 (W. Va. 1993); *See also* Jane Harper "West Virginia Court Wants Forensics Expert Prosecuted," *Houston Post* (July 17, 1994) at A22. *See also* Henderson, *supra* note 64 at 8–10 (discussing a recent example where an expert witness pled guilty to perjury and spent 3 years in jail for falsifying credentials).

[194] Garcia, *supra* note 185 at 57–58 (specifically discussing the use of cross-examination to remedy expert witness fraud); *See, e.g., Wrobleski v. Lara*, 727 A.2d 930 (Md. 1999) (discussing the many jurisdictional approaches to the cross examination of an expert witness on issues of positional and financial bias); *See also* 3 No. 4 Criminal Practice Guide 10, Eliciting Proof of Bias (2002); *See also* Tim Talley, *Fingerprint Expert at Nichols' Trial Admits a Mistake*, 20(4) The Print 6 (July/August 2004) available online at http://www.scafo.org/The_Print/The_Print_VOL_20_ISSUE_04.pdf (providing an example where on cross-examination the defense attorney was able to get an FBI fingerprint examiner to retract a portion of his testimony on direct that tied his client to the crime scene).

[195] *See, e.g.*, Convicted by Juries, Exonerated by Science: Case Studies in the Use of DNA Evidence to Establish Innocence After Trial, *NIJ Research Report* (June 1996). Commentary by Walter Rowe arguing that even improved science will not remedy the problem of inadequate legal counsel. Twenty-eight cases are addressed in this study. Rowe points out that in many of these cases had the defense counsel sought the opinion of a competent scientific expert or had the defense counsel reviewed the case notes of the state's expert witnesses prior to trial, then the inconsistencies and inadequacies of these flawed testimonies could have been brought to light during the trial. Available at http://www.ncjrs.org/txtfiles/dnaevid.txt.

[196] Haddad, *supra* note 36 at 1–21and 1–23 (stressing that the keys to an effective cross examination are (1) preparation and (2) becoming knowledgeable in the particular field).

[197] Fritsch and Rohde, Legal Help Often Fails New York's Poor, *NY Times*, April 8, 2001 at A1; Fritsch and Rohde, For New York City's Poor, a Lawyer with 1,600 Clients, *N.Y. Times*, April 9, 2001 at A1.

[198] Court Appointed Scientific Experts: A Demonstration Projection of the AAAS, at http://case.aaas.org/ (last visited August 17, 2004).

[199] CASE Experience, http://www.aaas.org/spp/case/experience.htm (last visited August 17, 2004) (listing Michigan, Texas, Louisiana, Illinois, California, New York, Kansas, Maine, and Nebraska as states that have district court judges that have used CASE).

2. Advising on the management of litigation involving CASE-recommended experts
3. Assisting judges in refining the issues to be addressed by the expert and in determining what kinds of experts are most appropriate to address these issues[200]

Additionally, Duke University's Private Adjudication Center was working to develop a registry of experts, both scientific and technical, that can serve as court-appointed experts.[201]

The American Association of Neurological Surgeons (AANS) was one of the first of the medical societies to implement guidelines for its members when acting as expert witnesses.[202] The violation of these guidelines could result in sanctions, suspensions, or even expulsion.[203] However, one of the suspended neurosurgeons sued the society for the resulting damages of his six-month suspension for irresponsible expert witness testimony.[204] The suspended doctor appealed the district court's order granting the association's summary judgment.[205] He argued that it was against public policy for a voluntary association to sanction a member for "irresponsible" testimony.[206] His policy argument described the sanctioning of experts as a deterrent to experts from testifying.[207] The Seventh Circuit disagreed with the doctor's argument and found that the association was not responsible for the 35% drop of income the doctor argued resulted from the suspension.[208] Finally, the court indicated that professional self-regulation encourages rather than deters civil justice.[209]

In view of the inadequacy or unavailability of the solutions to expert negligence or intentional professional misconduct, including that professional sanctions against an expert do not make whole a person injured as a result of such misconduct, injured individuals are resulting to tort actions for damages against experts and their employers are being resorted to more and more frequently. In the West Virginia's serologist's case, a civil action for damages filed against the State resulted in a $1 million settlement.[210]

Of all the solutions offered to curb expert abuses, allowing for expert witness malpractice causes of action will better protect and compensate injured individuals, as well as deter future misconduct. It will ensure "quality control" of expert opinions by encouraging experts to be careful and accurate. The four elements of the expert witness malpractice

---

[200] *Id.*
[201] The Registry of Independent Scientific & Technical Advisors, A Program of the Private Adjudication Center: Duke University School of Law, at http://www.law.duke.edu/PAC/registry/index.html (last visited August 19, 2004). *See also* Stephen Breyer, *Science in the Courtroom*, Issues in Science and Technology Online (Summer 2000), at http://www.issues.org/16.4/breyer.htm (exploring both the AAAS and the Duke programs as useful tools in the understanding of the usefulness of court-appointed experts).
[202] Terry Carter, A Physician Battles Against Colleagues He Considers Rogue Expert Witnesses, 90 *ABA Journal* 41, 43 (August 2004).
[203] *Id.*
[204] *Austin v. American Association of Neurological Surgeons*, 253 F.3d 967 (7th Cir. 2001).
[205] *Id.* at 968, 974.
[206] *Id.* at 972.
[207] *Id.*
[208] *Id.* at 971–972, 974.
[209] *Id.* at 972–973.
[210] *In the Matter of an Investigation of the West Virginia State Police Crime Laboratory, Serology Division*, 438 S.E.2d 501, 509 (W. Va. 1993) (indicating that $1 million was the state's insurance policy limit, and that the case was settled for $1 million) available online at http://www.truthinjustice.org/zainreport.htm (last visited August 30, 2004).

action are (1) the existence of a duty owed to the plaintiff arising out of the relationship between the expert and the plaintiff; (2) a negligent act or omission by the expert in breach of that duty; (3) causation; and (4) damages.[211]

The premise of the expert witness malpractice cause of action is that, first of all, expert witnesses owe a duty to their clients. However, the duty does not end there. Expert witnesses also owe a duty to any foreseeable plaintiff who may be affected by the expert's conduct and who is likely to suffer damages as a result of a negligently rendered opinion. These duties, based on their professional knowledge and skills, are similar to those duties owed by a doctor to a patient and a lawyer to a client.

The standard of care for a forensic scientist is that of the reasonably prudent practitioner in the relevant scientific field.[212] Standards of professional practice and ethical codes, as promulgated by the discipline, may be used to help define the duty of care. Most disciplines within the forensic sciences have adopted such standards of conduct. For a plaintiff to prevail, it must be determined that the expert did not adhere to the standard of a reasonably prudent expert in rendering an opinion, in conducting an examination, or in giving testimony. Ordinarily, an independent evaluation by a disinterested expert skilled in the same field will be required to determine whether an expert deviated from the required standard of care.

A crucial element of the tort of malpractice is causation. Causation tests whether the defendant's actions were in fact connected by physical event to the plaintiff's injury and whether the connection was close enough to allow compensation to the injured party. As stated by Richard S. Frank, a past president of the AAFS, "[t]he impact of the forensic scientist's conclusions affords no room for error, because such an error may be the direct cause of an injustice."[213] In some cases, it will be readily apparent that an expert's testimony alone "caused" the wrong. This is especially true when the expert evidence is the only determinative evidence presented in the litigation. Many studies have demonstrated that, despite jury instructions to the contrary, jurors give expert testimony greater weight than other evidence.[214] Thus, it is clear that financial injury to a potential plaintiff or conviction and incarceration of a potentially innocent individual who is prosecuted on the basis of an expert's opinion evidence[215] are reasonably foreseeable consequences of negligence, incompetence, or intentional misconduct by an expert.

---

[211] W. Page Keeton et al., *Prosser and Keeton on the Law of Torts* § 30, at 164–65 (5th ed., 1984); Dan B. Dobbs, *The Law of Torts*, § 114, at 269 (West 2000); *See also* Kathleen L. Daerr-Bannon, *Cause of Action for Negligence or Malpractice of Expert Witness*, 17 Causes of Action 2d 263 (2003) (adding that in a malpractice case against a friendly expert, the plaintiff must also prove that witness immunity is not applicable to the facts to avoid dismissal).

[212] *See, e.g., LLMD of Michigan, Inc. v. Jackson-Cross Co.*, 740 A.2d 186, 191 (Pa. 1999) ("[t]he judicial process will be enhanced only by requiring that an expert witness render services to the degree of care, skill and proficiency commonly exercised by the ordinarily skillful, careful and prudent members of their profession"); *see also* Leslie R. Masterson, Witness Immunity or Malpractice Liability for Professionals Hired as Experts?, 17 *Rev. Litig.* 393, 393 (1998).

[213] Richard S. Frank, The Essential Commitment For a Forensic Scientist, 32 *J. Forensic Sci.* 5 (1987); *See also* Garcia, *supra* note 185 at 70.

[214] Kurt Ludwig & Gary Fontaine, Effects of Witnesses' Expertness and Manner of Delivery of Testimony on Verdicts of Simulated Jurors, 42 *Psychol. Rep.*, 955 (1978). There are many cases that express concern that the special aura of reliability and credibility that surrounds an expert witness will cause the jury to neglect their fact-finding role. *See, e.g., State v. Johnson*, 681 N.W.2d 901, 906 (Wis. 2001); *State v. Ward*, 138 S.W.3d 245, 270 (Tenn. Crim. App. 2003); *Franco v. State*, 25 S.W.3d 26, 29 (Tex. App. 2000) (discussing the aura of reliability surrounding a bloodstain expert).

[215] Courts have awarded plaintiffs damages of a certain amount per month for illegal confinement due to legal malpractice, rejecting the argument that estimating the value of a person's loss of liberty is speculative.

Where a claim of expert witness malpractice is proved to have occurred in civil litigation, the measure of direct damages could include (1) the difference between a full verdict of proved loss and the reduced verdict resulting from the expert's testimony; (2) the difference between a full settlement and the reduced settlement that resulted from the expert's misconduct; (3) the cost in the experts' investigations; and/or (4) the attorneys' fees for responding to the expert's testimony and in proving the misconduct.

Although traditionally experts were afforded absolute immunity in trial testimony and trial preparation, the expert witness malpractice causes of action are gaining momentum.[216] Courts in New Jersey,[217] Connecticut,[218] Texas,[219] California,[220] Pennsylvania,[221] Massachusetts,[222] Louisiana,[223] Vermont,[224] and Missouri[225] are among the growing number of jurisdictions that have allowed plaintiffs to sue experts for their malpractice. However, even with this trend some jurisdictions will likely continue to uphold the traditional policy of allowing absolute immunity for expert witnesses.[226] Such limitations are rare, however, and the general recognition that such actions will be recognized by courts has induced defendants in malpractice suits to agree to settlements in many cases.[227]

For example, Geddie v. St. Paul Fire and Marine Ins. Co., 354 So. 2d 718 (La. Ct. App. 1978). *See also Holliday v. Jones*, 264 Cal.Rptr. 448 (Ct. App. 1989) (awarding damages for emotional distress as a result of wrongful incarceration due to professional malpractice); *See also* Restatement (third) of the Law Governing Lawyers, *Liability for Professional Negligence and Breach of Fiduciary Duty*, § 53 cmt. (g) (American Law Institute 2000):

> *Damages for emotional distress.* General principles applicable to the recovery of damages for emotional distress apply to legal–malpractice actions. In general, such damages are inappropriate in types of cases in which emotional distress is unforeseeable. Thus, emotional-distress damages are ordinarily not recoverable when a lawyer's misconduct causes the client to lose profits from a commercial transaction but are ordinarily recoverable when misconduct causes a client's imprisonment. The law in some jurisdictions permits recovery for emotional-distress damages only when the defendant lawyer's conduct was clearly culpable.

[216] *Davis v. Wallace*, 565 S.E.2d 386, 389–90 (W. Va. 2002).
[217] *Levine v. Wiss & Co.*, 478 A.2d 397, 399 (N.J. 1984) (denying a court-appointed expert witness immunity). *Contra see* Laurie S. Weiss, Expert Witness Malpractice Actions: Emerging Trend or Aberration?, 15 (2) Practical Litigator 27, 37 (March 2004) (discussing cases from other jurisdictions that do apply witness immunity if the expert was appointed by the court).
[218] *Pollock v. Pahjabi*, 781 A.2d 518 (Conn. Super. Ct. 2000).
[219] *James v. Brown*, 637 S.W.2d 914 (Tex. 1982).
[220] *Mattco Forge, Inc. v. Arthur Young & Co.*, 6 Cal. Rptr. 2d 781 (Cal. Ct. App. 1992), on appeal after remand, 45 Cal. Rptr. 2d 581 (Cal. Ct. App. 1995), on subsequent appeal, 60 Cal. Rptr. 2d 780 (Cal. Ct. App. 1997).
[221] *LLMD of Michigan, Inc. v. Jackson-Cross Co.*, 740 A.2d 186 (Pa. 1999).
[222] *Boyes-Bogie v. Horvitz*, Case No. 991868F, 2001 WL 1771989 (Mass. Super. Oct. 31, 2001).
[223] *Marrogi v. Howard*, 805 So.2d 1118 (La. 2002) (holding that experts are not shielded by immunity for either pretrial preparations or trial testimony).
[224] *Politi v. Tyler*, 751 A.2d 788 (Vt. 2000).
[225] *Murphy v. A.A. Mathews*, 841 S.W.2d 671 (Mo. 1992) (holding that malpractice actions against experts are only appropriate for pretrial litigation support services and not for trial testimony).
[226] *Bruce v. Byrne-Stevens & Assoc. Engineers, Inc.*, 776 P.2d 666 (Wash. 1989); *see also* Weiss, *supra* note 217 at 30–31 (listing Washington as the only state where witness immunity still controls with "friendly" experts and providing a detailed description of the *Bruce* decision).
[227] *See*, Don DeBenedictis, Off-Target Opinions, *ABA Journal*, November 1994 at 76 (A hospital and a national drug laboratory misdiagnosed toxins in a baby's blood, leading a jury to send the baby's mother to prison for murder. The experts and the laboratory settled a civil suit on the eve of trial for undisclosed sums); *In the Matter of an Investigation of the West Virginia State Police Crime Laboratory, Serology Division*, 438 S.E.2d 501, 509 (W. Va. 1993) (a police serologist falsified evidence against an accused rapist, sending the man to a West Virginia prison for five years. When the extent of the supposed expert's misdeeds in two states became clear, the state exonerated the man and agreed to pay him $1 million).

Although no courts shield erring expert witnesses from perjury charges for willful deceptions[228] or from damage actions where the expert's conduct involved intentional or grossly negligent conduct, a few courts have shielded experts from civil liability for damages in ordinary negligence cases. The courts have arrived at this result in one of two ways: (1) by holding that negligent mistakes or inaccuracies do not constitute perjury or (2) by holding that testimony and reports provided to courts are privileged.[229] Some courts hold that the expert witness who gives opinion evidence is the court's witness and therefore enjoys immunity against all posttrial damage claims whether sued by a party or nonparty to the action. Two cases addressing such issues arose in California and Missouri. In *Mattco Forge, Inc. v. Arthur Young Co.*,[230] the court held that the litigation privilege in the California Civil Code does not protect a negligent expert witness from liability to the party who hired the witness. In so holding, the court stated "applying the privilege does not encourage witnesses to testify truthfully; indeed by shielding a negligent expert witness from liability, it has the opposite effect."[231] The case went to trial, and, in July 1994 the jury returned a verdict of $995,717 out-of-pocket expenses, $14,200,000 in compensatory damages, and $27,680,000 in punitive damages against the expert witnesses for their negligence.[232] However, the plaintiff appealed the verdict, and the appellate court reversed the verdict as to compensatory and punitive damages but affirmed the award for out-of-pocket expenses and interest.[233] Also, in *Murphy v. A.A. Mathews*,[234] the Missouri Supreme Court held that witness immunity does not bar an action against a professional who agrees to provide litigation-related services for compensation if the professional is negligent in providing the agreed services. The court stressed, however, that its holding would not subject an adverse expert to malpractice liability because the expert owes no professional duty to the adversary. The court also stated that an expert retained by the court, independent of the litigants, would not be subject to malpractice liability. The *Matthews* holding is limited to pretrial, litigation-support activities.[235]

Witness immunity is an exception to the general rules of liability. The rule is traditionally limited to defamation cases and is extremely narrow in scope. Immunity is not meant to bar a suit against a professional who negligently performs services.[236] The complaint is not with the testimony provided in court; it is with the out-of-court work product that was negligently produced. By testifying, the expert is merely publishing his or her negligence in court. Therefore, no absolute immunity should be afforded experts; they are

---

[228] Edward P. Richards and Charles Walter, *When are Expert Witnesses Liable for their Malpractice?*, Public Health Law Practice Project *at* http://biotech.law.lsu.edu/ieee/ieee33.htm.

[229] Michael J. Saks, Prevalence and Impact of Ethical Problems in Forensic Science, 34 *J. Forensic Sci.* 772 (1989) (containing a summary of some cases involving litigation against expert witnesses).

[230] *Mattco Forge, Inc. v. Arthur Young & Co.*, 6 Cal. Rptr. 2d 781 (Cal. Ct. App. 1992), on appeal after remand, 45 Cal. Rptr. 2d 581 (Cal. Ct. App. 1995), on subsequent appeal, 60 Cal. Rptr. 2d 780 (Cal. Ct. App. 1997).

[231] *Id.* at 788. The court stated, however, that the California litigation privilege would still shield experts that are court appointed and would also shield expert witnesses from suit by opposing parties. *Id.* at 789.

[232] *The Expert Witn. J.*, 1 (July 1994).

[233] *Mattco Forge, Inc. v. Arthur Young & Co.*, 60 Cal. Rptr. 2d 780, 798 (Cal. Ct. App. 1997).

[234] 841 S.W.2d 671 (Mo. 1992).

[235] Compare the *Matthews* holding with *Marrogi v. Howard*, 805 So.2d 1118 (La. 2002) (broadening the scope of expert witness malpractice to include not only pretrial litigation services but also the expert's actual testimony during trial).

[236] For a case supporting the suit against a party's own expert *see Marrogi v. Howard*, 805 So.2d 1118 (La. 2002) (holding that the policy behind witness immunity is not advanced by offering immunity for incompetent experts retained by a party to perform professional services including trial testimony); *see also Murphy v. A.A. Matthews*, 841 S.W.3d 671 (Mo. 1992). For a case supporting a suit against the opposing party's expert *see LLMD of Mich., Inc. v. Jackson-Cross-Co.*, 740 A.2d 186 (Pa. 1999).

neither judges nor their adjuncts, but merely third-party participants in litigation. The courts, the legal profession, and the forensic discipline recognize that the trend is toward permitting claims for damages resulting from negligent testimony by experts.

Experts need to be aware of this trend toward allowing malpractice suits. A court might allow the pursuit of the cause of action, even if your jurisdiction does not have any mandatory authority on the issue. For example, in *Davis v. Wallace*,[237] a convicted murder defendant was suing the state's doctors for negligence in preparing for and testifying in litigation. The trial court dismissed the lawsuit and ordered sanctions for the convicted defendant and her attorney. The trial court held that the lawsuit was frivolous and was filed with an oppressive purpose to intimidate the state-retained medical doctors.[238] However, the West Virginia Supreme Court held that the circuit court abused its discretion for sanctioning and dismissing a novel cause of action. The Court found that there was a good-faith basis for the lawsuit because of the trend of academics and other jurisdictions in allowing a suit against an expert witness for malpractice.[239] In light of this trend one attorney recommends that expert witness should consider the following pointers:

1. "Straightforwardly represent their qualifications and skills, and disclose any prior inconsistent testimony or writings relevant to the case in question;
2. Know and understand the legal rules of evidence and how those rules might effect [sic] their testimony or pretrial work. It is essential that firms providing litigation support, like accounting firms, provide adequate legal training to their employees;
3. Communicate and discuss with the hiring attorney any problems and weaknesses in their analyses;
4. Have enough time to prepare for the case, otherwise they might not receive all the facts until closer to trial, potentially leading to changed opinions;
5. Make sure that opinions are documented by relevant and accurate data. If the expert is relying upon other individuals' work product the expert should make sure it has been performed satisfactorily before testifying to it; and
6. Finally, given the trend of court decisions regarding expert witness immunity, expert witnesses should seriously consider taking out errors and omissions coverage."[240]

Following these pointers will help minimize the likelihood of success for the retaining party to challenge the negligence of the retained expert because the factors help to fully educate the attorney regarding the strengths and weaknesses of the expert's opinion. This education, in turn, minimizes the chances of surprise testimony during the course of direct or cross-examination. Another suggestion for experts is to carefully draft retainer agreements in light of the new trend toward professional malpractice. For example, an expert might want to establish that mere difference of opinion between experts does not result in professional malpractice.[241]

---

[237] 565 S.E. 2d 386 (W. Va. 2002); *see also* Sara H. Jurand, *Negligence Claim Against Experts is Not Frivolous*, 38 Trial 73 (Aug. 2002).
[238] 565 S.E.2d at 388.
[239] Id. at 389.
[240] Weiss, *supra* note 217 at 38; *See also* Mark Hansen, Experts Are Liable, Too, 86 *ABA Journal* 17, 18 (November 2000) (providing a similar list of helpful hints for the expert witness to avoid malpractice liability).
[241] Edward P. Richards, III, The Medical and Public Health Law Site, When are Expert Witnesses Liable for Malpractice? –LLMD of Michigan, Inc. v. Jackson-Cross Co., 559 Pa. 297, 740 A.2d 186 (Pa. 1999) at http://biotech.law.lsu.edu/cases/evidence/llmd_brief.htm (Last updated 8/15/04).

Conclusion

Some concern may be voiced over whether the growing recognition of a cause of action for expert malpractice will have a chilling effect on the willingness of persons to serve as forensic experts in litigation. The emergence of such a cause of action may, in fact, result in the disappearance of some experts who are habitually negligent or incompetent, but this is, of course, a salutary by-product of the legal trend. Even if the existence of a cause of action for expert malpractice has an effect on the availability of a number of experts, or results in an increase in the fees charged for their services, these results are not so compelling as to justify a public policy against recognizing a cause of action for expert witness malpractice. The very existence of the cause of action will ensure that experts are held accountable for their opinions. The full and accurate development of evidence in civil and criminal litigation is not served by protecting the negligent, incompetent, or dishonest expert witness. The justice system as a whole benefits when such causes of action are permitted. The forensic sciences themselves will enjoy greater respect and admiration when it is known their practitioners are accountable for misdeeds and that the professions favor eliminating the unworthy among them.

To be more effective, an expert should be aware of the legal and ethical responsibilities in that role. The expert should understand the tests for admissibility of expert evidence, the rules of discovery and procedure, and any statutory responsibilities. The expert should also be aware of the potential liability for malpractice as an expert witness.

20 Bloodstain Pattern Analysis: Postconviction and Appellate Application

MARIE ELENA SACCOCCIO

Introduction

This chapter is dedicated to postconviction attorneys dealing with issues concerning bloodstain pattern analysis. Although the emphasis is on postconviction cases, the law is the law. As such, the material should prove useful to trial attorneys dealing with issues of admissibility or expert qualifications, via a motion *in limine*, or during examination of the expert at trial. The coverage will overlap and complement Chapter 19 of this volume. Although not presumed to be the authoritative source, this chapter provides an overview of issues to be evaluated and addressed at the postconviction stage. Because there is a dearth of precedent on specific issues you may encounter in each jurisdiction, it will be beneficial to look to other jurisdictions for a case in point. Further, lists of appellate opinions dealing with admissibility of bloodstain pattern evidence and presumptive blood tests are presented in Appendices E and F to aid in your inquiry.

A Bit of History

> Any butcher is just as good an expert on that as this witness.
> **Commonwealth v. Sturtivant (1875)**[1]

At the outset, despite the verbiage in some contemporary opinions, there is nothing new about appellate courts addressing admissibility or expert qualifications with respect to bloodstain pattern interpretation.

In 1857 the Supreme Judicial Court of Maine held that bloodstain pattern analysis was admissible and was a proper subject for expert testimony. The case was *State v. Knight*.[2]

---

[1] *Commonwealth v. Sturtivant*, 117 Mass. 122, 126 (1875).
[2] *State v. Knight*, 43 Me. 11 (Me. 1857).

The expert proffered by the prosecution was Dr. Augustus Hayes, a scientist trained in the microscopic and chemical analysis of blood.[3]

The case was interesting, and the theme was familiar. The crucial issue before the jury was whether Mary Knight committed suicide or was murdered by her husband, George.[4] To be sure, George's defense was that Mary had killed herself. In support of this, his mother, Lydia, testified that on the fateful night, she was awoken by Mary. Mary wanted to join her in bed and Lydia complied. Then Lydia heard an "outcry" and Mary's arms fell on her. Lydia saw no other person in the room, nor did she see the final act.[5]

The prosecution's case mirrored that presented in a contemporary homicide. Two doctors testified that Mary's death resulted from one mortal wound—her throat had been slashed, and the slash was 5 inches long and 3 inches deep. She died instantly. She would not have been capable of any movement after such a wound so the blood found on the window stool and on the floor were not caused by her.[6] Further, such a wound would have necessarily caused blood on Lydia, who was sleeping next to her. The evidence at trial was that Lydia had no blood on her that evening.[7]

Especially significant was blood found on the defendant's undershirt, worn on the night of the murder.[8] Dr. Hayes testified that the bloodstain on the undershirt was not caused by the defendant bleeding because the stain was deeper in color on the outer side of the undershirt than the side touching his body. The defendant's position at trial was that he had cut his arm and had bled on the undershirt at some other time.[9] Further, Dr. Hayes testified that the stain likewise did not result from the spurting of a vein "from without," except through the "interstices of some other fabric which had strained it of some of its original coloring matter."[10] L.D. Rice, a witness for the state, testified that on the day after the killing, he saw George wearing a shirt with a bloodstain on the sleeve.[11] The shirt, worn over the undershirt, was never found.

On appeal, the defendant objected to the testimony of Dr. Hayes, arguing that "the opinion of experts should be limited and confined to such matters as are of a scientific nature."[12] "Since his testimony on the bloodstains was easily observable to the jury, expert testimony on the subject should have been excluded."[13] In response, the appellate court held that the opinion of the witness was based on chemical experiment and observations aided by the microscope and resulted from scientific knowledge and experience.[14]

In *Commonwealth v. Sturtivant*, an opinion rendered in 1875, the Massachusetts Supreme Judicial Court addressed the admissibility of blood spatter interpretation and the qualifications of the "expert."[15] The witness, a chemist accustomed to chemical and microscopic examination of blood and bloodstains, testified for the prosecution as to directionality: "If the force of a stream of liquid, whatever it may be, and especially blood,

---

[3] *Id.* at 19.
[4] *Id.* at 20, 61.
[5] *Id.* at 62.
[6] *Id.* at 41, 98–99, 130–131.
[7] *Id.* at 28–29 and 41.
[8] *Id.* at 24.
[9] *Id.* at 24.
[10] *Id.* at 102.
[11] *Id.* at 128.
[12] *Id.* at 85.
[13] *Id.*
[14] *Id.* at 133.
[15] *Sturtivant*, 117 Mass. at 130, 138.

be from below upward, the heaviest portion of the drop will stop at the further end of the stain; if from above downward, it will stop below."[16] Defense counsel objected to this testimony at trial, stating, "That is pure opinion as to a matter of mechanics, not chemistry. Any butcher is just as good an expert on that as this witness."[17]

On appeal, defense counsel argued that such testimony was inadmissible because the witness was allowed to opine as to directionality. This was incompetent evidence. The witness was not qualified as an expert.[18] Further, the "theory of the witness was incorrect in point of fact."[19]

In analogizing to the competency of lay witnesses to opine on identity of persons, things, animals, or handwriting; size, color, weight of objects; time and distance; sounds and their direction; and footprints, the court held, "It is within the range of common knowledge to observe and understand those appearances, in marks or stains caused by other fluids, which indicate the direction from which they came, if impelled by force...."[20] "There is no question of science or learning necessarily involved in the understanding.... Blood is a fluid which coagulates and stiffens rapidly when exposed to the air and might, therefore, more decidedly give indications of its direction."[21] If the theory of the witness were incorrect in point of fact, such a position should have been explored in cross-examination of the expert.

Massachusetts was not alone in its reasoning. Likewise in 1875, in *Lindsay v. People of the State of New York*, the Supreme Court of New York held admissible the "expert" testimony as to bloodstain evidence obtained 6 months after the body was discovered because such evidence tended to corroborate the testimony as to manner in which the body was taken from the murder scene to a loft.[22] The 6-month time lapse went to the weight of the evidence, not its admissibility.[23] Again, as in *Sturtivant*, the "expert," by training and experience, was a scientist who analyzed blood.[24]

Not every jurisdiction was as receptive to this type of evidence. In 1880 the Supreme Court of Mississippi held that the exclusion of expert testimony on bloodstain interpretation was correct.[25] The case was *Dillard v. State*, and the defendant had proffered testimony of several physicians about the relative positions of the combatants, as indicated by the bloodstains. Their proffered opinion was that, based on the bloodstain patterns, the defendant was probably prostrate on the ground and the deceased was on top of him when the bloodstains on the shirt were received.[26] The defense position at trial was that the deceased had first attacked the defendant by hitting him with a piece of railing; the men fought, and when the cutting was done by the defendant he was lying on the ground with the victim on top of him. After the fight, the defendant called some neighbors and explained that he had stabbed someone and the circumstances of the fight. He summoned them to the body to help.[27] Also noted at the time were scratches and bruises on the defendant's face, evidencing the fight.[28]

---

[16] *Id.* at 124, 126.
[17] *Id.*
[18] *Id.* at 130.
[19] *Id.* at 131.
[20] *Id.* at 134–135.
[21] *Id.* at 136–137.
[22] *Lindsay v. People of the State of New York*, 63 N.Y. 143, 155 (1875).
[23] *Id.*
[24] *Id.* at 147.
[25] *Dillard v. State*, 58 Miss. 368, 389 (1880).
[26] *Id.* at 380.
[27] *Id.* at 371.
[28] *Id.*

On appeal, the defendant cited to *Knight* as precedent for the admissibility of expert testimony on bloodstain pattern analysis.[29] The appellate court failed to address it as persuasive legal authority. Paradoxically, in affirming the exclusion of such highly relevant evidence, the appellate court cited to *Sturtivant*, which had unequivocally approved the admission of bloodstain pattern evidence via lay testimony.[30] Although charged with murder, *Dillard* was only found guilty of manslaughter.[31]

In *Wilson v. United States*, a most interesting opinion rendered in 1895, the United States Supreme Court held that "...the existence of bloodstains at or near a place where violence has been inflicted is always relevant and admissible."[32] In so holding, the Supreme Court cited to *Sturtivant*.[33] The issue as to the competency or qualifications of the witness was never raised. The bloodstain evidence admitted was bloody bedclothes; bloody soak-through on a bed; a pillow sewed over blood spots; a bloody axe; and charred pieces of cloth at a camping site, with some blood on the ground where the fire had been.[34] The victim's body was badly decomposed so that no cause of death could be determined, although a crushed skull was noted.[35] At trial Wilson called witnesses to show that the blood found on the bedclothes, and presumably the bed, had gotten there from the blood of a prairie chicken they had killed and also from the bleeding of sick horses.[36] What these men were doing in their bedclothes with a prairie chicken and a sick horse is beyond the scope of this chapter.

Although bloodstain interpretation has certainly developed as a recognized scientific expertise, capable of quantification and grounded in established laws of physics and "common sense,"[37] echoes of *Sturtivant* are alive and well in more recent decisions. In *State v. Hall*, a 1980 decision rendered by the Supreme Court of Iowa, the court ignored the "general acceptance" requirement and approved the admission of blood spatter interpretation, reasoning that "[s]uch observations are largely based on common sense, and in fact, lie close to the ken of an average layman....Such evidence need not wait an assessment of the scientific community, as it is inherently understandable and reliable."[38] Further, in *People v. Clark*, a 1993 opinion rendered by the Supreme Court of California, *en banc*, the court stated, "It is a matter of common knowledge, readily understood by the jury, that blood will be expelled from the human body if it is hit with sufficient force and that inferences can be drawn from the manner in which the expelled blood lands upon other objects."[39]

Despite these very liberal holdings on admissibility, it cannot be ignored that each court premised its reasoning on the inherent understandability, relevance, and reliability of such evidence. Further, most proponents therein would probably be qualified as experts today based on training and experience. Note, in *Sturtivant*, *Knight*, and *Lindsay* the witnesses were scientists, accustomed to chemical and microscopic examination of

---

[29] *Id.* at 383.
[30] *Id.* at 389.
[31] *Id.* at 370, 372.
[32] *Wilson v. United States*, 162 U.S. 613, 620 (1895).
[33] *Id.*
[34] *Id.* at 614.
[35] *Id.*
[36] *Id.* at 615.
[37] *United States v. Mustafa*, 22 M.J. 165, cert. denied, 479 U.S. 953 (CMA 1986).
[38] *State v. Hall*, 297 N.W.2d 80, 86 (Iowa 1980).
[39] *People v. Clark*, 857 P.2d 1099, 1142 (Cal. 1993)

bloodstains for the purpose of determining the origin of blood (i.e., whether it was human or animal).[40] To be sure, as scientists, their training also included mathematics and physics. In Hall, the "expert" turned out to be Herbert MacDonell, the "granddaddy" of bloodstain interpretation as we know it today.[41] In Clark, the court specifically found that the witness had (1) attended lectures and training seminars on blood dynamics in both California and Oregon; (2) read relevant literature; and (3) conducted relevant experiments.[42] Perhaps the only aberration is Wilson, but the qualifications of the witness were not raised in that appeal (i.e., the issue was simply not before the court).[43]

To the point, bloodstain analysis has a history of admissibility spanning more than a century. The objections addressed by the court in *Knight*, *Sturtivant*, and *Lindsay* are the same considerations presented in postconviction proceedings today. Starting from this vantage point will lend credibility and significance to postconviction issues on this topic. It is neither novel nor obscure.

The Postconviction Attorney's Role — The Daunting Task

In evaluating a case postconviction, one begins with the trial transcript, pleadings, and discovery provided. Of course, as a prosecutor, you just lie and wait; and if the issue is raised as to the competency of bloodstain evidence, or lack thereof, you respond. However, defense counsel's job is much more onerous. It is defense counsel's job, and indeed duty, to assess the bloodstain evidence as developed at trial. In its absence, the question then becomes should there have been such evidence? Both evaluations are equally important and determine the postconviction path taken.

The Appeal

In a case in which bloodstain pattern evidence was admitted, the following questions should be helpful:

1. Is there precedent in that jurisdiction for its admission?
2. Was the proponent a qualified expert by training or experience?
3. Was there the requisite foundation for the expert's testimony?
4. How was the defendant prejudiced by the admission of such evidence?

To breathe some life into this checklist and make it most useful for appellate counsel, the issues will be discussed in the following in the context of appellate opinions from various jurisdictions, whenever possible.

---

[40] *Knight*, 43 Me. At 19; *Sturtivant*, 117 Mass. At 24–26.
[41] *Hall*, 297 N.W.2d at 83; *Ex Parte Mowbray*, 943 S.W.2d 461 at 462, n.1 (Tex. Cr. App. 1996) (the prosecution sought agreement that MacDonnell is the "granddaddy of blood spatter").
[42] *Clark*, 857 P.2d at 1142.
[43] *Wilson*, 162 U.S. 619. (Further, it appears that Wilson, who was sentenced to death by hanging, was not represented by counsel before the United States Supreme Court as "no appearance for plaintiff in error" was noted).

Admissibility

To be sure, as expressed earlier, bloodstain analysis enjoys a history of admissibility spanning more than a century. This is true today not because it is perceived as "close to the ken of an average layman" as in *Sturtivant*, but rather because it is viewed as a scientific discipline, based on well-settled principles of chemistry and physics.[44] Not only has bloodstain evidence never been excluded from trial for lack of "general acceptance," its reliability is subject to judicial notice in some jurisdictions.[45] As recently stated in *Holmes v. Texas*, after a careful review of cases across the country, "None of the courts held that...blood spatter analysis was unreliable."[46] The shift from the *Frye* standard of admissibility to *Daubert* does not alter its admissibility.[47] In fact, *Daubert* has a much more liberal thrust aimed at admitting scientific evidence not yet deemed "generally accepted."[48] In a jurisdiction adhering to *Frye*, the "general acceptance" in the scientific community can certainly be satisfied by beginning with the list of cases, affirming the admissibility, provided in Appendix. However, *Daubert*, the first United States Supreme Court case addressing the admissibility of scientific evidence, presents a different obstacle. Now the trial judge must, *inter alia*, make a preliminary determination that the "reasoning or methodology underlying the testimony is scientifically valid."[49] However, the absence of *Frye*'s "general acceptance" in the relevant scientific community is no longer an impediment to admissibility, and "general acceptance" continues to be the significant and often the only issue.[50] As such, in jurisdictions holding bloodstain analysis admissible under *Frye*, it should now be admissible under *Daubert*.[51]

Expert Qualifications

> An expert, as the word imports, is one having had experience. No clearly defined rule is to be found in the books that constitutes an expert. Much depends upon the nature of the question in regard to which an opinion is asked.
> **The Ardesco Oil Co. v. Gilson (1870)**[52]

---

[44] *State v. Eason*, 687 A..2d 922, 925–926 (D.C. App. 1996); *State v. Moore*, 458 N.W.2d 90, 97 and n.6 (Minn. 1990).
[45] *Moore*, 458 N.W.2d 90, 97 (Minn. 1990); *Know*, 459 N.E.2d 1077, 1080 (Ill. App. 3 Dist. 1984).
[46] *Holmes v. State of Texas*, 135 S.W.3d 178 (2004).
[47] *Frye v. United States*, 293 F. 1013, 1014 (D.C. App. 1923) (employing "general acceptance" as prerequisite for admissibility of expert testimony); *Daubert v. Merrell Dow Pharmaceuticals, Inc.* 113 S.Ct. 2786, 2795 (1997) (rejecting Frye and adopting, *inter alia*, a "reliability" evaluation of: scientific technical and specialized knowledge "that might be helpful to the jury," consistent with F.R.E. 702).
[48] *Daubert*, 113 S.Ct. at 2794.
[49] *Id.* at 2796.
[50] *Commonwealth v. Lanigan*, 641 N.E.2d 1342, 1348–1349 (Mass. 1994); *but see, Commonwealth v. Sands*, 675 N.E.2d 370, 371 (Mass. 1997) (reaffirms *Lanigan* but states it has adopted *Daubert*, in part, with "general acceptance" the overruling factor); *State v. Porter*, 694 A.2d 1262, 1278 (Conn. 1997).
[51] ABA Section on Science and Technology, Scientific Evidence Review, Monograph No. 2, at 8, Professor Stephen Salsberg, *Daubert: Framework of the Issues*. ("Thus, while *Frye* no longer governs, the concept of general acceptance may retain importance. I suspect that a trial judge will have a difficult time excluding as unreliable evidence which is generally accepted . . . if it is met, reliability will be presumed, and if it is not, the burden of demonstrating reliability or unreliability must be born by the proponent or opponent of the evidence.").
[52] *The Ardesco Oil Co. v. Gilson*, 63 Pa. St. 146, 150 (1870).

The ground, upon which one is called as an expert, is, that the subject in controversy depends on science or peculiar skill, knowledge, or experience...and this must often depend upon circumstances of so peculiar a nature and character, that it is difficult to lay down any general rule to guide the decision of the judge in such a case.

Lincoln & wife v. Inhabitants of Barre (1850)[53]

Despite the passage of well over a century since the previously quoted opinions, there remains no clearly defined rule. One would presume, *ab initio*, that "experts" presenting the worst qualifications would probably be excluded pretrial, via a motion *in limine*, or during trial, on *voir dire* as to qualifications. However, from a survey of the case law on the issue, this presumption is questionable.[54]

To date, in cases in which appellate courts have addressed the qualifications of bloodstain pattern experts, there seems to be little uniformity, or for that matter comprehension, of what the expertise demands. As such, the courts have deemed qualified coroners, pathologists, medical examiners, crime scene investigators, serologists, firearms examiners, police officers, forensic science investigators, and a witness who stated no particular qualifications at all.[55] To exacerbate the problem, the majority of courts overwhelmingly review the trial court's decision as to qualifications for abuse of discretion, and, in the minority of jurisdictions, the appellate court reviews for clear error. To the point, this is a difficult road on appeal. For further explication of this topic, please refer to Chapter 19.

In sum, the various tests used for evaluating qualifications of an expert are vague and grant seemingly unbridled discretion to the trial judge. The 19th century quotations from *Oil Co. v. Gilson*[56] and *Lincoln & wife v. Inhabitants of Barre*[57] could easily be implanted into contemporary opinions. Although some jurisdictions use a "sliding-scale" approach to qualifications, correlated with the perceived complexity of the bloodstain evidence,[58] others qualify the expert with just a modicum of experience, beyond the average person.[59]

Despite this, there are some unifying principles. There seems to be "strict scrutiny" of the qualifications of police officers by the defense bar and the appellate courts. Presumably, this is because most officers lack academic credentials in the sciences to be testifying in matters dependent on physics, chemistry, and mathematics. Moreover, it is significant that empiric studies have shown that "jurors rated police appearing as experts to be the most honest, understandable, confident, likeable, and believable, and that these experts were discredited less often than other witnesses."[60] This being the case, strict scrutiny is appropriate.

Likewise, when the testimony is perceived by the appellate courts as "complex," as in a full-scale crime reconstruction or in-court reenactment/demonstration, "strict scrutiny"

---

[53] *Lincoln & wife v. Inhabitants of Barre*, 59 Mass. 590, 591 (1850).
[54] See Danny R. Vieulleux, Annotation, Admissibility, In Criminal Prosecution of Expert Opinion Evidence as to "Blood Spatter" Interpretation, 9 A.L.R. 5th 369, sections 6[a], and 13–16 and (Supp. 1997).
[55] *Id.*
[56] 63 Pa. St. at 150.
[57] 59 Mass. at 591.
[58] *Eason v. United States*, 687 A.2d 922, 926 (D.C. App. 1996); *People v. Knox*, 459 N.E.2d 1077, 1081 (Ill. App. 3 Dist. 1984).
[59] *State v. East*, 481 S.E.2d 652, 662–663 (N.C. 1997); *People v. Smith*, 633 N.E.2d 69 (Ill. App. 4 Dist. 1993).
[60] Daniel W. Shuman, Anthony Champagne and Elizabeth Whitaker, Juror Assessments of the Believability of Expert Witnesses, 36 Jurimetrics, *Journal of Law, Science and Technology*, No. 4, Summer, 1996, at 375.

is used. To the point, when raising the issue of the qualifications of a purported bloodstain expert on appeal, whether that expert is a police officer, technician, or anyone else, one needs to convince the appellate court that the bloodstain evidence was, indeed, complex. This is a challenging task in that bloodstain evidence is deceptively simple. Unlike testimony of blood typing, DNA analysis, pathology, etc., we have all witnessed "spatter," "smears," "transfers," and "wipes" in our daily lives. Yet, we have never sought to verify our categorizations for accuracy, nor engaged in sophisticated mathematic calculations, aimed at reenactment and dependent on angles, trajectories, velocity, size, shape, and distance. Nor are we qualified to do so. It is the verification and interpretation of the categorizations, based on specialized training, knowledge, and experience, that are essential to a qualified expert in this field.

Even in a case in which the appellate court clearly acknowledges the error, it may deem that error harmless. As such, whenever raising the issue of the qualifications of an expert, it is essential to demonstrate to the appellate court that the inadmissible expert testimony was not cumulative of other admissible expert testimony and that the testimony was somehow pivotal to the case. Further, be aware that the erroneous admission of bloodstain evidence can also prejudice the jury because it usually is the basis for the admission of gruesome and inflammatory photographs, perhaps otherwise inadmissible.

Noticeably absent from the cases discussed is any mention of the trial testimony of a blood spatter expert delineating the proper qualifications of an expert in that field. To the point, few cases really present a scenario in which one expert at trial deems the qualifications of another blood spatter expert insufficient. Usually this problem is resolved at the pretrial stage, and the incompetent expert, who truly lacks qualifications, simply does not testify. This may very well account for the lack of clarity and uniformity and the basic misunderstanding of the expertise in appellate opinions. Because on appeal the appellate court is confined to the record as follows, the only meaningful way of educating the appellate court is through a well-developed trial record. Where the battle has really taken place during discovery and outside the record on appeal, there is very little substantive information on the expertise available to the appellate court.

The Collateral Attack

Postconviction relief is especially onerous and circumscribed in scope. It is a collateral attack on the conviction, filed in the trial court. Although it is neither a new trial, nor a new appeal, it affords a unique opportunity to expand the trial record beyond that presented in a direct appeal. An evidentiary hearing will not always be afforded, but expert affidavits may be submitted in support of a Petition for Postconviction Relief. Thereafter, if a new trial is denied and the denial is appealed, the appellate record contains those affidavits. In the event an evidentiary hearing is granted, that transcript and the experts' reports, if any, are before the appellate court.

To be sure, a collateral attack on the conviction is the optimum vehicle when defense counsel at trial failed to retain or present the necessary bloodstain expert. Of course, the legal claim would be ineffective assistance of counsel, citing to the Sixth Amendment to the United States Constitution and its cognate provision in the state constitution. When this is the asserted claim posttrial, then defense counsel at trial is alleged to have been ineffective. When this is the asserted claim posttrial and postappeal, then defense counsel at trial and appellate counsel for the defense have been ineffective.

Although most collateral attacks dealing with bloodstain evidence are raised as ineffective assistance of counsel, bloodstain evidence can be the basis of a Brady violation, a claim of newly discovered evidence, expert perjury, or prosecutorial misconduct.

As in the previous sections of this chapter, these postconviction issues will be discussed in the context of actual cases. These cases were chosen because they are thorough and instructive on the legal avenues to explore, the procedural and substantive hurdles to be overcome, and the potential for bloodstain evidence to have a dramatic effect at this late stage.

Perjurious Expert Testimony — Newly Discovered Evidence

In *Clayton v. State*,[61] the defendant/petitioner applied to the trial court for postconviction relief; relief was denied. He had previously been convicted of first-degree murder and was sentenced to death. The judgment and sentence were affirmed on direct appeal.[62]

In applying for postconviction relief, the petitioner alleged that the prosecution's bloodstain expert had "overstated" his qualifications and that this evidence was newly discovered.[63] Also, because the expert's "lies" regarding his qualifications were revealed during the discovery phase of another case after Clayton's trial, but before his appeal, he also alleged ineffective assistance of appellate counsel.[64] The petitioner/defendant appended three affidavits attacking the prosecution expert's qualifications in the other case to his application for relief. Herbert MacDonell wrote one affidavit, asserting that the prosecution's expert was not qualified to testify to bloodstain pattern analysis. A police sergeant and police captain authored the other affidavits and asserted the same. In addition, the petitioner submitted three letters from the General Counsel for the Oklahoma State Bureau of Investigation disavowing the government witness as a blood spatter expert.[65]

The Oklahoma Court of Criminal Appeals held that the trial court abused its discretion in not determining that the admission of the expert testimony was erroneous. However, while acknowledging the "devastating impact of erroneous and incompetent expert testimony," the appellate court further held that the admission of the erroneous evidence was harmless in view of the overwhelming evidence, independent of the bloodstain testimony.

Prosecutorial Misconduct — Denial of Due Process

Ex Parte Mowbray, rendered in 1996 by the Court of Criminal Appeals of Texas, *en banc*, is an intriguing state *habeas corpus* case, replete with accusations of perjury and prosecutorial misconduct.[66] The petitioner/defendant was convicted of the murder of her husband, sentenced to life, and fined $10,000. The judgment and sentence were affirmed on appeal.[67] In her application for postconviction relief, the petitioner asserted that the State's expert

---

[61] *Clayton v. State*, 892 P.2d 646 (Okl. Cr. 1995).
[62] *Id.* at 651–652.
[63] *Id.*
[64] *Id.*
[65] *Id.* at 652.
[66] *Ex Parte Mowbray* 943 S.W.2d 461 (Tex. Cr. App. 1996).
[67] *Mowbray v. State*, 788 S.W.2d 658 (Tex. App. Corpus Christi 1990).

knowingly gave false and misleading testimony; that the state knowingly used false testimony; and that trial counsel for the defense was ineffective.[68]

The evidence at trial was that Susie Mowbray shot her husband while he was asleep in their bed.[69] They were alone in the bedroom and it was night.[70] The defense theory was that Mowbray and her husband were lying in bed with a pillow barrier between them. Mowbray saw her husband's elbow point upward. She reached to touch it and the gun went off.[71] Just as in Knight, was it suicide or was it homicide? Just as in Knight, the case turned on the blood spatter evidence.

At trial, Dusty Hesskew, a police sergeant, testified as bloodstain expert for the prosecution. In essence, he stated that he examined Susie Mowbray's nightgown and found "high velocity impact [blood] staining." Specifically, he counted 48 small stain areas around the stomach and chest of the nightgown, consistent with high-velocity spatter (i.e., a kind of mist). Thus, in his opinion Mowbray could not have been lying beside her husband at the time of his death. Luminol was used to identify and measure the bloodstains, otherwise invisible to the naked eye.[72]

Pursuant to court order, Herbert MacDonell testified at the *habeas* hearing. MacDonell had previously been retained by the prosecution to review the photographs and physical evidence 7 months prior to trial. His examination revealed no bloodstains either visible to the naked eye or under a microscope. MacDonell's opinion was that it was unlikely that the nightgown was in close proximity to the gunshot wound at the time of the shooting. As such, it was more probable than not that the deceased died from a suicide rather than a homicide. At the request of the prosecution, MacDonell memorialized his findings in a written report, which he delivered to the prosecution 2 weeks prior to trial. That report was then delivered to Mowbray's defense attorney. Very predictably, the prosecution did not call MacDonell to testify at trial.[73]

At the *habeas* hearing MacDonell was especially critical of Dusty Hesskew's exclusive use of luminol to identify the stains as blood and as a means to measure the pattern. MacDonell explained that luminol was merely a presumptive test in that it can react with various other substances and thus give a false-positive result. He opined that the luminescence from a luminol reaction cannot be accurately measured. He further stated that he had never heard of anyone trying to measure it, and that even if someone attempted to do so, the validity of the conclusion would be highly suspect. In MacDonell's opinion, the police officer did not understand the chemistry of luminol testing.[74]

In response, Hesskew admitted that his testimony at trial absolutely depended on his erroneous assumption that someone had performed confirmatory testing on the invisible stains to ensure that they were, indeed, blood. He conceded that his trial testimony was scientifically invalid and that his ultimate opinions, that the victim died as a result of a homicide and that her account of the incident was impossible, had no scientific basis.[75]

Tom Bevel, a captain with the Oklahoma City Police Department, also testified at the hearing. He had previously testified as a bloodstain expert for the defense at trial. His

---

[68] *Ex Parte Mowbray,* at 461.
[69] *Mowbray v. State,* 788 S.W.2d at 662.
[70] *Ex Parte Mowbray,* at 461.
[71] *Id.* at 462.
[72] *Id.* at 462–463.
[73] *Id.*
[74] *Id.* at 463.
[75] *Id.*

opinion then was that the deceased could have died by suicide and that it was impossible to measure high-velocity impact spatter in the manner used by the police officer. Like Hesskew, Bevel did not conduct any confirmatory testing to determine if the stains were blood. However, he stated that Hesskew informed him that the Department of Public Safety laboratory had confirmed human blood on the nightgown. He relied on this assertion.[76] A chemist in the Texas Department of Public Safety crime laboratory also testified at the hearing and stated that prior to trial he had conducted two confirmatory tests for the presence of blood on the nightgown and that both tests were negative.[77]

The prosecutor conceded that the state's case "depended upon" the blood spatter evidence and asserted that he did not call MacDonell to testify at trial because of the expense of securing his testimony. Defense counsel, a board-certified criminal law specialist, testified that he did not call MacDonell to testify because he did not know how MacDonell would respond, and he feared MacDonell might alter his favorable opinion.[78]

Following the hearing, the *habeas* judge entered Findings of Fact and Conclusions of Law, and determined that Susie Mowbray was denied due process. Thus, the judge recommended that the appellate court set aside the conviction and grant a new trial. As a basis, the judge entered the following findings.

The rationale for suicide was "equally persuasive." The deceased had attempted suicide twice before and had shot himself on one of those occasions. Also, he had vowed to kill himself. The "linchpin" of the state's case was the high-velocity impact spatter found on the nightgown. Hesskew "recanted" his trial testimony and admitted that his opinion was devoid of scientific validity. The State's conduct with respect to Herbert MacDonell was, "at best, questionable trial strategy, and, at worst, intentional deception" of Mowbray's trial counsel. The State was aware of the extent of MacDonell's opinion 7 months prior to trial and was under court order to disclose all Brady material then. Instead, on the eve of trial they provided the defense a scant report. To make matters worse, the state led defense counsel to believe that MacDonell would be a witness and available for cross-examination. The court found that defense counsel was not ineffective, but the prosecution was and this violated Mowbray's due process right to a fair trial.[79]

Citing to *Brady v. Maryland*, the appellate court noted that "the State has an affirmative duty to disclose favorable evidence under the Due Process Clause." Further, it held that the bloodstain evidence was favorable and material. Consistent with the *habeas* judge's recommendation, the appellate court set aside the verdict and remanded the case for a new trial.[80] On retrial, Susan Mowbray was acquitted.

Ineffective Assistance of Counsel

State of Florida v. Riechmann, a death penalty case, is still pending in the Florida state courts.[81] In 1988, Dieter Riechmann was convicted of murder and sentenced to death.[82] As a basis for the death penalty, the trial judge found that the "murder was committed

---

[76] *Id.*
[77] *Id.* at 463–464.
[78] *Id.*
[79] *Id.*
[80] *Id.* at 466.
[81] *State of Florida v. Dieter Riechmann*, Crim. No. 87-42355, Eleventh Circuit Court, Florida, Order on Motion to Vacate Judgment of Conviction and Sentence [hereinafter cited as Order on Motion to Vacate].
[82] *Riechmann v. State*, 581 So.2d 133 (Fla. 1991).

for pecuniary gain and was cold, calculated, and premeditated without any pretense of legal or moral justification." The defendant presented no mitigating evidence at the sentencing phase.[83]

The conviction and death sentence were affirmed on direct appeal by the Florida Supreme Court in 1991. Riechmann then filed a petition for a *writ of certiorari* to the United States Supreme Court. Certiorari was denied the following year.[84] Thereafter, in 1994, he filed a motion to vacate the original conviction and sentence in state trial court.[85] Riechmann's motion consisted of 14 separate claims, 11 of which asserted ineffective assistance of counsel, with 24 subparts alleging separate instances of deficient performance. The remaining claims alleged newly discovered evidence, a Brady violation, and a claim of judicial misconduct in that the trial judge did not write the findings supporting the death sentence; rather, they were the prosecution's creation, *per ex parte* order of the trial judge. Because the trial judge would be called as a witness on the issue of misconduct, the Florida Supreme Court appointed a successor judge to preside at the hearing and rule on the motion to vacate.[86] The facts presented by the Supreme Court of Florida to support the conviction and sentence on the direct appeal follow.

Dieter Riechmann and Kersten Kischnick, "life companions" of 13 years, were German citizens and residents. They were vacationing in Florida in October 1987. Kischnick was shot to death in Miami Beach on October 25 as she sat in the passenger seat of a rental car driven by Riechmann. The prosecution alleged that Riechmann was her pimp and he killed her to collect insurance proceeds because she was too sick to work and had wanted to quit the life. The insurance issued automatically when Riechmann used his Diner's Club card to rent the vehicle. Further, it provided double-indemnity coverage in the event of accidental death. Riechmann also had insurance policies on Kischnick in Germany. Those policies provided that murder was considered an accidental death. In all, the insurance proceeds totaled almost $1 million.

At trial Riechmann testified that he and Kischnick had been touring in the rental car in an effort to videotape the Miami sights. They got lost and stopped to ask a stranger for directions. Realizing that they were close to their destination, Riechmann unbuckled his seat belt, reached behind to grab the video camera, and placed it on Kischnick's lap. He handed her purse to her so that she could tip the stranger for his assistance. Riechmann saw the stranger reach behind and got scared, stretched out his right arm, with the palm facing outward, and hit the gas pedal. At that moment he heard an explosion and Kischnick slumped over. She was shot on the right side of her head. Riechmann drove around looking for help, driving 10 to 15 miles before he hailed a police officer for assistance.[87]

To be sure, the state's case was circumstantial. The State's gunpowder residue expert testified at trial that gunpowder residue was found on Riechmann's hands. Based on the number and nature of the particles, the State's expert concluded that there was a reasonable scientific probability that Riechmann had fired a gun. Moreover, the gunpowder residue was not consistent with Riechmann's account that he was merely sitting in the car when Kischnick was shot. An expert for the defense testified that the gunpowder residue merely indicated that Riechmann was in the vicinity of the gun when it was fired.[88] The State's

---

[83] Order on Motion to Vacate, at 2.
[84] *Id.*
[85] *Id.*
[86] *Id.* at 2–3.
[87] *Riechmann,* 581 So.2d at 135–136.
[88] *Id.* at 136.

bloodstain evidence was much more damaging. It was highlighted in the State's closing argument and relied on by the Florida Supreme Court in affirming the conviction and death sentence.[89] At trial the State presented expert testimony of a crime laboratory serologist on the blood found on Riechmann's clothes and in the rental car. He opined that because there was high-velocity blood spatter on the driver-side door, that blood could not have gotten there if Riechmann were in the driver's seat when Kischnick was shot. Further, the pattern of blood found on a blanket that had been folded in the driver's seat evidenced high-velocity spatter and aspirated blood.[90] Riechmann testified at trial that the blood had gotten on the blanket earlier that summer in Germany. His dog had had surgery and bled on the blanket on the trip home from the hospital. He brought the blanket along to Miami to use on the beach, after which he intended to discard it. The defense presented no blood spatter expert of its own.[91]

At the postconviction stage Riechmann's attorney, James Lohman, submitted two volumes, in excess of 400 pages, intricately setting forth the grounds for the motion to vacate. With respect to the bloodstain evidence, he asserted that "Defense counsel's failure to rebut in any way the devastating and patently erroneous bloodstain opinion testimony was unreasonable and contributed significantly to the adverse outcome of the trial."[92] In further support, he submitted a third volume containing expert reports and other documentary proof. Relevant to this chapter, and found to be the most troublesome new evidence for the hearing judge, was the report of Stuart James, blood spatter expert for the defense.

Excerpts of the highly exculpatory report follow:[93]

> The victim in this case, Kersten Kischnick, received a fatal gunshot wound of entrance to the right side of her head while sitting in the front passenger side seat of a rented, 1987, red Ford Thunderbird. The projectile entered 2H inches above and 1H inches behind the right external ear meatus. There was evidence of powder residue and stippling around the entrance wound, which according to the pathologist, indicated a range of discharge distance of from several inches to 18 to 24 inches. (Trial testimony of Dr. Villa, p. 2919). There was no wound of exit. The thrust of the analysis of the bloodstain evidence offered by the State at trial was to prove that the defendant, Dieter Riechmann, was not in the driver's seat at the time of the shooting of his girlfriend by an unidentified person as he stated to authorities and testified at trial. If he were not in the driver's seat, his account of the shooting was false and he was the murderer, allegedly for financial gain. The importance of proper analysis of bloodstain evidence in such a case cannot be overemphasized, especially when a defendant's version of events might be verified or contradicted by that evidence. The State here relied heavily upon the analysis of bloodstain evidence and experimentation of a serologist to show that Dieter Riechmann was not in the driver's seat of the Thunderbird at time of the shooting.
>
> Based upon my extensive review, examination and analysis of the bloodstain evidence in this case, I can state with a high degree of scientific certainty that

---

[89] Id.
[90] Riechmann, at 136.
[91] Id. at 136, n.6.
[92] Riechmann, Motion to Vacate, Vol. 1, at 156.
[93] Stuart James, Report of Evaluation of Bloodstain Evidence, Re: Dieter Riechmann, September 20, 1994.

the major conclusions reached and presented by the serologist in this case were grossly inaccurate. In general, the bloodstains were significantly misinterpreted. In addition, crucial bloodstains were completely ignored or overlooked. The serologist's primary conclusions are completely contrary to the overwhelming weight and significance of the bloodstain evidence, and the techniques he utilized in forming those conclusions are at odds with most basic professional norms and scientific principles upon which correct bloodstain analysis is founded.

James' conclusions follow:

In analyzing and discussing the photographs, physical evidence, reports and testimony in *State v. Riechmann*, I found a high degree of agreement and unanimity among my colleagues with regard to interpretations and conclusions that could be drawn from the bloodstain patterns in the automobile in which Kersten Kischnick was shot, on items within the vehicle, and on the clothing worn by her and the defendant, Dieter Riechmann. Based upon my review of the evidence, I make the following findings:

1. When the victim in this case, Kertsen Kischnick, received the gunshot wound to the right side of her head, her head went back to the headrest on its right side at which time blood and brain material accumulated on the headrest. Her head then either moved or was moved to its position on its left side as seen in the scene photographs. This is consistent with the Medical Examiner, Dr. Paul Villa's, deposition testimony on page 29, "that brain matter is more to the right of the body or the head."
2. At some point in time after bloodshed, the victim's seat was moved to the observed reclined position.
3. Transfer bloodstains on the face of the victim and the alteration of bloodstains on the chest of the victim indicate contact with her after bloodshed either by Mr. Riechmann, police personnel or paramedics, or some combination thereof.
4. Bloodstains on the hands of Dieter Riechmann are consistent with contact with a source of wet blood and are likely responsible for transfer bloodstains on the steering wheel of the vehicle. Blood from his hand may have been flicked onto the driver's door as well as possibly wiped on the blanket located on the driver's seat.
5. Bloodstains on the trousers of Dieter Riechmann are consistent with some blood transfer as well as some dripping of blood from above while he was in a sitting position. There are also some small spatters of blood on the right knee area that could have been produced by exhalation of blood droplets from the victim. These would appear to be the leftward extent of any possible exhaled blood within the vehicle.
6. Exhaled blood (not "aspirated blood") was deposited on the victim's left shoulder, left arm, thigh, pocketbook and money that is exposed on her left thigh. In reviewing photographs and examining Ms. Kischnick's white slacks, there is no question that at the time of bloodshed, victim had three (3) one-dollar bills on her left thigh, based upon the void or clean area within the exhaled

bloodstain pattern on left leg of the slacks. I find this to be one of the more significant aspects of the homicide scene, yet it was virtually ignored throughout the investigation and in those portions of trial testimony I have reviewed. In addition, Ms. Kischnick's left hand is directly below the currency in the few photographs depicting this part of the scene. This is consistent with the defendant's testimony that the assailant had turned away from the car momentarily, and that the victim was taking out some money to "tip" him at the time the fatal shot was fired. It would also appear that she may have been holding a cigarette in her left hand or mouth at the time she was shot, based upon the observation in the Medical Examiner's "Scene Investigation Report" of October 26, 1987. "A burned out cigarette is evident on the left side of the passenger's seat on the lateral side of her thigh." Other than Dr. Villa's reference, this fact was also ignored by any investigative efforts to reconstruct what was taking place at the time of the homicide.

7. It is impossible to determine the height of the passenger door window based on the location of the bloodstains because the stains most likely did not come from the entrance wound at the time of the gunshot. The window was completely closed in all the scene descriptions and photographs, and stains as likely as not were deposited on the window sometime after it was closed. Hence, the stains are obviously of no value whatsoever in determining the height of the window when Kischnick was shot.

8. The blood on the driver's door was not high-velocity blood spatter, i.e., it did not come directly from the gunshot wound of entrance on the right side of the victim's head. Even the serologist acknowledged several times that he "could not explain" how blood got from the entry wound on the right side of the head to the driver's door. (Trial testimony, pp. 3816, 3832, 3932). It is also extremely unlikely that this was exhaled blood from the victim's left nostril due to the unlikely position of the nose, in addition to the force that would be required to direct the exhaled blood to the driver's door. The six stains are too few in number and were not documented photographically. It appears that the stains were removed for presumptive blood testing and then the circled areas where they had been were photographed. Small circular stains as described by the serologist can be produced by other mechanisms such as flicking of blood from the hands or some object or other activity involving the driver, the paramedics or police personnel. It is also possible that these stains represent activity due to flies occurring at a later time but prior to the serologist's bloodstain examination done as late as one week after incident.

9. Based upon my analysis of all available crime scene evidence, descriptions, photographs and experiments, it is my opinion to a reasonable degree of scientific certainty, that there is no basis on which to conclude, as the serologist did, that the driver's seat of the automobile was not occupied at the time the victim was shot in the right rear of her head through the passenger side window. First, six spots on the driver's door are not high-velocity impact blood spatter resulting from the gunshot wound. Second, it is unlikely that exhaled blood from the victim's nose traveled farther to her left than the right edge of the driver's seat or right trouser of the driver. The small blood spatters on the right trouser leg of Mr. Riechmann's trousers could be the result of exhaled

blood. Third, the method utilized to identify high-velocity blood spatter or exhaled bloodstains on the blanket as high-velocity spatter or exhaled blood by the serologist, after a negative stereo microscopic examination, was not scientifically valid and was a gross misinterpretation. My analysis of bloodstain evidence would indicate that there is no more likelihood that the driver's seat was unoccupied than occupied at the time Ms. Kischnick was shot. It is impossible to conclude with any degree of scientific certainty that the seat was unoccupied....

I would have been available to provide expert assistance and opinion in 1987–1988. In discussing the evidence herein with several highly qualified professional colleagues, I can assert with confidence that any number of experts in my field would have been available at that time to present similar conclusions, in particular with regard to the poorly substantiated and inaccurate conclusions that were presented at trial.

In responding to the 14 claims set forth in Riechmann's motion to vacate, the trial court "took the liberty of restating the order of claims for purposes of addressing the most substantive ones first."[94] The bloodstain interpretation and analysis by Stuart James was first to be addressed. The postconviction judge noted that bloodstain evidence was crucial in that it was the focal point of the state's closing argument at trial and formed the basis for affirmance of the conviction on direct appeal.[95] The postconviction judge also noted that, consistent with community standards as testified to by Steven Potolski, Esquire, an expert in capital defense, a reasonable attorney would have taken steps to present an "available" expert to rebut the serologist's conclusions. Notwithstanding this, the postconviction judge stated the following:

[T]he Defendant failed to sufficiently meet his burden by demonstrating that, based on reasonable probability, Mr. James, or a similar expert, would have been found by an ordinary competent attorney using diligent efforts, and that such expert would have been prepared to rebut the State's serologist at trial. The fact that Mr. James was found years later, and has worked on the case since 1994, is irrelevant. Rather, the "reasonable probability" standard must be measured from trial counsel's perspective at the time, without resort to distorting hindsight. No testimony was offered that, given time limitations immediately before trial, Mr. James could have rendered the same opinions as offered at the postconviction hearing. At best, Mr. James merely "presumed" that he would have been available if contacted "depending on scheduling." (T. May 14, 1996, Stuart James, page 74.)[96]

The court also concludes that trial counsel's comprehensive and extensive cross-examination of the serologist...was effective in showing the weaknesses of the witnesses' testimony, and that such weaknesses were argued to the jury at closing.[97]

---

[94] Order on Motion to Vacate, at 10, n.3.
[95] *Id.* at 10–11.
[96] *Id.* at 15.
[97] *Id.* at 15–16.

The court's reasoning was and is illogical. James never testified that he would not be available for consultation and testimony. His response was merely an honest and accurate attempt at extemporaneously recollecting what he had scheduled on certain days of trial, years prior. Moreover, in his report, excerpted *supra*, James explicitly stated that he, as well as other experts, would have been available to testify. Further, James has been called to testify during trial on numerous occasions.[98] In a notorious multidefendant murder case in Massachusetts, with only 1 day's notice, he flew from Florida to Boston, reviewed the testimony of the state's bloodstain expert at trial and the available evidence, and testified. Consequently, that defendant was acquitted of first-degree murder.[99]

The judge found that trial counsel's performance was not deficient because a reasonably competent attorney would not have been able to *find* a bloodstain expert.[100] Implicit in this finding is the judge's misperception of bloodstain pattern interpretation. It is not novel; it is not obscure. To be sure, he should have read *Knight*,[101] *Sturtivant*,[102] and *Lindsay*,[103] all decided much more than a century ago. A superficial search would have revealed hundreds of cases, presenting hundreds of different bloodstain experts. Perhaps he would have also wanted them to extemporaneously testify, under penalties of perjury, whether they were available to testify on certain days years prior.

Fortunately for Dieter Riechmann, the postconviction judge found the trial judge's sentencing order invalid. Contrary to Florida law, and *per ex parte* order of the trial judge, the prosecutor, not the trial judge, drafted the death penalty sentencing order. Neither the *ex parte* communication nor the sentencing order were disclosed to defense counsel during the penalty phase.[104] Further, at the postconviction hearing Riechmann's attorney, James Lohman, presented 15 witnesses from Germany who would have been willing and able to testify at Riechmann's sentencing hearing had they been contacted and requested to do so. The court heard from landladies, friends, neighbors, and former lovers. All traveled from Germany at their own expense to speak on Riechmann's behalf. The court also received written statements from many others, including Riechmann's mother and brother, who would have been able to testify on his behalf at trial.[105]

Based on judicial misconduct, coupled with trial counsel's failure to present any testimony on mitigation, the postconviction judge granted Riechmann's motion to vacate the death sentence. In 2000 the Florida Supreme Court affirmed the denial of the Motion to Vacate the Conviction and remanded the case for resentencing before a different judge.[106] Subsequently,

---

[98] Telephone conversations with Stuart James.

[99] The author was present during trial. James was called to testify as a bloodstain expert for defendant, Ricardo Parks, and this request came during trial. The prosecution "sandbagged" the defense by only providing notice that a serologist would be testifying to blood typing. While on the stand that serologist engaged in a full-scale bloodstain interpretation, quoting from a publication authored by James. The defense moved for a mistrial based on unfair surprise. The trial judge denied the request and granted a short continuance. For a full discussion of the issue, see *Commonwealth v. Ventry Gordon*, 666 N.E.2d 122 (Mass. 1996) (affirming convictions of the other codefendants and discussing remedial measured afforded when there is unfair surprise).

[100] Order on Motion to Vacate, at 16.

[101] 43 Me. 11 (Me. 1857).

[102] 117 Mass. 122 (1875).

[103] 63 N.Y. 143 (1875).

[104] Order on Motion to Vacate, at 48–51.

[105] *Id.* at 53–55.

[106] *State of Florida v. Dieter Riechmann*, 777 So.2d 342 (Fla. 2000).

Dieter Riechmann, via new counsel, Terri Backhus, filed two separate postconviction attacks; both were denied and are consolidated for appeal before the Florida Supreme Court.[107]

Conclusion

This journey has spanned more than a century, from jurisdiction to jurisdiction, from trial court to appellate court, and often extending as high as the United States Supreme Court. It has delved into the quagmires of *Frye* and *Daubert*. Bloodstain pattern evidence has been the basis of issues of admissibility and expert qualifications on direct appeal. Bloodstain pattern evidence has formed the basis of ineffective assistance of counsel claims, Brady violations, newly discovered evidence, and expert perjury at the postconviction/collateral attack stage.

To be sure, an attorney with the burdensome task of attacking a murder conviction, posttrial, should sift through the trial transcript and discovery. In a case in which no bloodstain evidence was presented, should it have been? What do the crime scene photographs show? Can the defendant shed any light on the subject? Did anybody examine the defendant's clothes? A bloodstain expert would be most helpful in these inquiries. In the case in which bloodstain evidence was admitted, was it accurate and effectively presented, or did counsel merely rely on the opportunity for cross-examination? If cross-examination was defense counsel's strategic vehicle, was there something more that could have been elucidated by retaining an expert for the defense.

Whether one is proceeding on direct appeal or via collateral attack in trial court, having the court agree that there was error is only the first hurdle. It is necessary to prove to the reviewing court that the defendant was harmed. As such, one should pay special attention to the trial record on appeal in an effort to convince the appellate court that, although there may or may not be "overwhelming evidence of guilt," the bloodstain evidence was crucial. Likewise, when dealing with the conviction on collateral attack, an essential part of the presentation of proof should be how the new expert evidence is different from that educed at trial. Have the expert explicitly state how his or her conclusions differ from all the bloodstain evidence elicited at trial, fully anticipating perceived "concessions" achieved on cross-examination, which could later render your expert testimony cumulative.

Aside from this diatribe, always be mindful that sympathetic fact patterns consistently help. Having a client accused of the murder of a police officer in the line of duty does not help. Having a client who is an alleged pimp who killed for pecuniary gain likewise does not help. This is not to posit that such cases are hopeless; rather they present the special challenge.

---

[107] Florida Supreme Court Case Docket, Case Number: SC03-760. Results of DNA testing may be one of the issues on appeal. http://ccadp.org/dieterriechmann-motionDNA.htm

References

Bloodstain Pattern Analysis

Balthazard, V., Piedelievre, R., DeSoille, H., and DeRobert, L.. *Etude des Gouttes de Sang Projecte*. Presented at the 22nd Congress of Forensic Medicine, Paris, France, 1939.

Beneke, M., and Barksdale, L., Distinction of Bloodstain Patterns from Fly Artifacts, *Forensic Science International*, 2003.

Betz, P., Peschel, O., Stiefel, D., and Eisenmenger, W., Frequency of Blood Spatters on the Shooting Hand and of Conjunctival Petechiae Following Suicidal Gunshot Wounds to the Head, *Forensic Science International*, 76, 1995, pp. 47–53.

Bevel, T., and Gardner, R.M., *Bloodstain Pattern Analysis: With an Introduction to Crime Scene Reconstruction*, 2nd Edition, CRC Press, LLC, Boca Raton, FL, 2002.

Burnett, B.R., Detection of Bone and Bone-Plus-Bullet Particles in Backspatter from Close-Range Shots to the Head, *Journal of Forensic Sciences*, Vol. 36, No. 6, November, 1991, pp. 1745–1752.

Byrd, J.H., and Castner, J.L., *Forensic Entomology: The Utility of Arthropods in Forensic Investigations*, CRC Press, Boca Raton, FL, 2001.

DiMaio, V.J.M., *Gunshot Wounds: Practical Aspects of Firearms, Ballistics and Forensic Techniques*, 2nd Ed, CRC Press, Boca Raton, FL, 1999.

DeForest, P.R., Gaensslen, R.E., and Lee, H.C., *Forensic Science: An Introduction to Criminalistics*, McGraw-Hill, New York, 1983, pp. 295–308.

Evans, G.A., Evans, G.M., Seal, R.M., and Craven, J.L., Spontaneous Fatal Haemorrhage Caused by Varicose Veins, *Lancet*, 2:1359–1361, 1973.

Gardner, R.M., *Practical Crime Scene Processing and Investigation*, CRC Press, LLC, Boca Raton, FL, 2004.

Gray, Henry, *Gray's Anatomy*, Crown Publishers, Bounty Brooks, New York, 1977.

Haglund, W.D., and Sorg, M.H., *Forensic Taphonomy: The Postmortem Fate of Human Remains*, CRC Press, LLC, Boca Raton, FL, 1997, pp 436–437.

Hueske, E.E., Some Observations Concerning the Deposition of Blood on Handguns as a Result of Back Spatter from Gunshot Wounds, *Southwestern Association of Forensic Science Journal*, 19-1, 1997, pp. 19–20.

James, S.H., Ed., *Scientific and Legal Applications of Bloodstain Pattern Interpretation*, CRC Press, LLC, Boca Raton Florida, 1999.

James, S.H., Eckert, W.G., *Interpretation of Bloodstain Evidence at Crime Scenes*, 2nd Edition, CRC Press, LLC, Boca Raton, FL, 1999.

James, S.H., Edel, C.F., *Bloodstain Pattern Interpretation, Introduction to Forensic Sciences*, W.E. Eckert, Ed., CRC Press, LLC, Boca Raton, FL, 1997.

James, S.H., and Nordby, J.J., Eds., *Forensic Science: An Introduction to Scientific and Investigative Techniques*, CRC Press, LLC, Boca Raton, FL, 2002.

Karger, B., Nusse, R., Schroeder, G., Wustenbecker, S., and Brinkmann B., Backspatter from Experimental Close-Range Shots to the Head: I. Microbackspatter, *International Journal of Legal Medicine*, 109, 1996, p. 66–74.

Karger, B., Nusse, R., Troger, H.D., and Brinkmann, B., Backspatter from Experimental Close-Range Shots to the Head: II. Microbackspatter and the Morphology of Bloodstains, *International Journal of Legal Medicine*, 110, 1997, pp. 27–30.

Karger, B., Nusse, R., and Banganowski, T. Backspatter on the Firearm and Hand in Experimental Close-Range Gunshots to the Head, *American Journal of Forensic Medicine and Pathology*, 23, 2002, pp. 211–213.

Kirk, P.L., *Crime Investigation*, 2nd ed., John Wiley and Sons, New York, 1974, pp. 167–181.

Kish, P.E., and MacDonell, H.L., Absence of Evidence Is not Evidence of Absence, *Journal of Forensic Identification*, 46, No. 2, March/April, 1996, pp. 160–164.

Laber, T.L., Diameter of a Bloodstain as a Function of Origin, Distance Fallen and Volume of Drop, *IABPA News*, Vol. 2, No.1, 1985, pp. 12–16.

Laber, T.L., and Epstein, B.P., *Bloodstain Pattern Analysis*, Callen Publishing Company, Minneapolis, 1983.

Lee, H.C., Palmbach, T.M., and Miller, M.T., *Henry Lee's Crime Scene Handbook*, Academic Press, San Diego, 2001.

Lehrman, R.L., *Physics: The Easy Way*, 3rd ed., Barron's Educational Series, Inc., Hauppaugh, New York, 1998.

MacDonell, H.L., Interpretation of Bloodstains: Physical Considerations, *Legal Medicine Annual*, Wecht, C., Ed., Appleton, Century Crofts, New York, 1971, pp. 91–136.

MacDonell, H.L., and Bialousz, L., *Flight Characteristics and Stain Patterns of Human Blood*, United States Department of Justice, Law Enforcement Assistance Administration, Washington, DC, 1971.

MacDonell, H.L., and Bialousz, L. *Laboratory Manual on the Geometric Interpretation of Human Bloodstain Evidence*, Laboratory of Forensic Science, Corning, New York, 1973.

MacDonell, H.L., and Brooks, B., Detection and Significance of Blood in Firearms, *Legal Medicine Annual*, Wecht, C., Ed., Appleton-Century-Crofts, New York, 1977, pp. 185–199.

MacDonell, H.L., and Panchou, C., Bloodstain Patterns on Human Skin, *Journal of the Canadian Society of Forensic Science*, Vol. 12, No. 3, Sept. 1979, pp. 134–141.

MacDonell, H.L., Criminalistics, Bloodstain Examination, *Forensic Sciences*, Vol. 3, Wecht, C., Ed. Matthew Bender, New York, 1981, 37.1–37.26.

MacDonell, H.L., *Bloodstain Patterns — Revised*, Laboratory of Forensic Science, Corning, NY, 1997.

MacDonell, H.L., Another Confusing Bloodstain Pattern, *IABPA News*, Vol. 20, No.3, September 2004, pp. 11–14.

Pex, J.O., and Vaughn, C.H., Observations of High Velocity Blood Spatter on Adjacent Objects, *Journal of Forensic Sciences*, Vol. 32, No. 6, November 1987, pp. 1587–1594.

Piotrowski, E., *Uber Entstehung, Form, Richtung und Ausbreitung der Blutspuren nach Heibwundendes Kopfes*, K.K. Universitat, Wein, 1895.

Pizzola, P.A., Roth, S., and DeForest, P.R., Blood Droplet Dynamics — I, *Journal of Forensic Sciences*, Vol. 31, No. 1, pp. 36–49, Jan. 1986.

Pizzola, P.A., Roth, S., and DeForest, P.R., Blood Droplet Dynamics — II, *Journal of Forensic Sciences,* Vol. 31, No. 1, January 1986, pp. 50–64.

Raymond, M.A., Smith, E.R., and Liesegang, J., The Physical Properties of Blood — Forensic Considerations, *Science and Justice,* Vol. 36, 1996, pp. 153–160.

Reynolds, M., The ABA Card® Hematrace® — A Confirmatory Identification of Human Blood Located at Crime Scenes, *IABPA News,* Vol. 20, No. 2, June 2004, pp. 4–10.

Rolling and Rolling, *Facts and Formulas,* McNaughten and Gunn, Nashville, TN, 1973.

Sallah, S., and Bell, A., *The Morphology of Human Blood Cells,* 6th ed., Abbott Diagnostics, 2003.

Sparks, R. Chronic Venous Insufficiency Syndrome, *IABPA News,* Vol. 20, No. 3, September 2004.

Stephens, B.G., and Allen, T.B., Back Spatter of Blood from Gunshot Wounds — Observations and Experimental Simulation, *Journal of Forensic Sciences,* Vol. 28, No. 2, April 1983, pp. 437–439.

Sutton, T.P., *Bloodstain Pattern Analysis in Violent Crimes,* Division of Forensic Pathology, University of Tennessee, Memphis, TN, 1993.

White, R.B., Bloodstain Patterns of Fabrics — The Effect of Drop Volume, Dropping Height and Impact Angle, *Journal of the Canadian Society of Forensic Science,* Vol. 19, No. 1, 1986, pp. 3–36.

Wonder, A.Y., *Blood Dynamics,* Academic Press, San Diego, 2001.

Yen, K., Thali, M.J., Kneubuehl, B.P., Peschel, O., Zollinger, U., Dirnhofer, R. Blood Spatter Patterns — Hands Hold Clues for the Forensic Reconstruction of the Sequence of Events, *American Journal of Forensic Medicine and Pathology,* Vol. 24, No.2, June 2003, pp. 132–140.

Medical and Medical–Legal Aspects of Bloodshed

DiMaio, V.J., and DiMaio, D., *Forensic Pathology,* 2nd ed., CRC Press, Boca Raton, FL, 2001.

DiMaio, V.J., *Gunshot Wounds: Practical Aspects of Firearms, Ballistics and Forensic Techniques,* 2nd ed., CRC Press, Boca Raton, FL, 1999.

Moritz, A.R. *The Pathology of Trauma,* 2nd ed., Lee and Febiger, Philadelphia, 1954.

Knight, B., *Forensic Pathology,* 2nd ed., Oxford University Press, New York, 1996.

Spitz, W.U., Ed., *Medicolegal Investigation of Death,* 3rd ed., Charles C. Thomas, Springfield, IL, 1993.

Presumptive Testing and Species Determination of Blood and Bloodstains

General References

Deforrest, P., Gaensslen, R.E., and Lee, H.C., *Forensic Science: An Introduction to Criminalistics,* McGraw-Hill Book Company, New York, Chapter 9, 1983.

Gaensslen, R.E. *Sourcebook in Forensic Serology, Immunology and Biochemistry,* U.S. Government Printing Office, Washington, DC, Units 2, 4, 5, 6, 7, 9, 1983.

Saferstein, R. *Criminalistics: An Introduction to Forensic Science,* 6th ed., Prentice-Hall Inc., Englewood Cliffs, NJ, 1997.

Spalding, R.P., *FBI Laboratory Serology Unit Protocol Manual,* U.S. Department of Justice, Federal Bureau of Investigation, Sections 2, 3, 4, 5 and 11, May, 1989.

Sutton, T.P., Presumptive Blood Testing, In James, S. H., Ed., *Scientific and Legal Applications of Bloodstain Pattern Interpretation,* CRC Press, Boca Raton, Chapter 4, 1999.

Specific References

Adler, O., and Adler, R., The Reaction of Certain Organic Compounds with Blood, With Particular Reference to Blood Identification, Hoppe-Seyler's *Z. Physiol. Chem.*, 41, 59, 1904. Translation in Gaenssslen, R.E. *Sourcebook in Forensic Serology, Immunology and Biochemistry,* U.S. Government Printing Office, Washington, DC, Unit 9, 1983.

ABACard, HemaTrace, Abacus Diagnostics, product literature, 1998.

Bell, G.H., Breslin, P., and Lemen, R., Eds., Health Hazard Alert: Benzidine-, O-tolidine, and O-dianisidine-based Dyes, DHHS (NIOSH) Publication No. 81-106, U.S. Department of Labor and U.S. Department of Health and Human Services, 1980.

Budowle, B., Leggitt, J.L., Defanbaugh, D.A., Keys, K.M., Malkiewicz, S.F., The Presumptive Reagent Fluorescein for the Detection of Dilute Bloodstains and Subsequent STR Typing of Recovered DNA, *J. Forensic Sci.*, 45, 1090, 2000.

Cheeseman, R., and DiMeo, L.A., Fluorescein as a Field-Worthy Latent Bloodstain Detection System, *J. Forensic Ident.*, 45, 631–646, 1995.

Cheeseman, R., Direct Sensitivity Comparison of the Fluorescein and Luminol Bloodstain Enhancement Techniques, *J. Forensic. Ident.*, 49, 261–268, 1999.

Cheeseman, R., and Tomboc, R., Fluorescein Techinque Performance Study on Bloody Foot Trails, *J. Forensic Ident.*, 51, 16–27, 2001.

Cox, M., A Study of the Sensitivity and Specificity of Four Presumptive Tests for Blood, *J For. Sci.*, 36, 1503, 1991.

Curtius, T., and Semper, A., Verhalten des 1-äthylesters der 3-notro-benzol-1,2-dicarbonsäure gegen hydrazin, *Ber. Dtsch. Chem. Ges.*, 46, 1162, 1913.

Dilling, W.J., Haemochromogen crystal test for blood, *Brit. Med. J.*, 1, 219, 1926.

Fleig, C., Nouvelle reaction, a la fluorescine, pour la recherche du sang, en particulier dans l=urine, *C.R. Soc. Biol.* 69, 192–194, 1910.

Garner, D.D., Cano, K.M., Peimer, R.S., and Yeshion, T.E., An Evaluation of Tetramethylbenzidine as a Presumptive Test for Blood, *J. Forensic Sci.*, 21, 816–821, 1979.

Gimeno, F.E., and Rini, G.A., Fill Flash Luminescence to Photograph Luminol Blood Stain Patterns, *J. Forensic Ident.*, 39, 149–156, 1989a.

Gimeno, F.E., Fill Flash Color Photography to Photograph Luminol Blood Stain Patterns, *J. Forensic Ident.*, 39, 305–306, 1989b.

Grispino, R.R.J., The Effect of Luminol on the Serological Analysis of Dried Human Bloodstains, *Crime Laboratory Digest*, 17, 13–23, 1990.

Gross, A.M., Harris, K.A., and Kaldun,G.L., The Effect of Luminol on Presumptive Tests and DNA Analysis Using the Polymerase Chain Reaction, *J. For. Sci.*, 44, 8837–840, 1999.

Harrow, B., Borek, E., Mazur, A., Stone, G.C.H., Wagreich, H., *Laboratory Manual of Biochemistry,* 5th ed., W.B. Saunders Co., Philadelphia, 75, 1960.

Hatch, A.L., A Modified Reagent for the Confirmation of Blood, *J. For. Sci.*, 38, 1502–1506, 1993.

Higaki, R.S., and Philp, W.M.S., A Study of the Sensitivity, Stability and Specificity of Phenolphthalein as an Indicator Test for Blood, *Can. Soc. Forens. Sci. J.*, 9, 97, 1976.

Hochmeister, M.N., Budowle, B., and Baechtel, F.S., Effects of Presumptive test reagents on the Ability to Obtain Restriction Fragment Length Polymorphism (RFLP) Patterns from Human Blood and Semen Stains, *J. Forensic Sci.*, 35, 656, 1991.

Hochmeister, M.N., Budowle, B., Sparkes, R., Rudin, O., Gehrig, C., Thali, M., Schmidt, L., Cordier, A., and Dirnhofer, R., Validation Studies of an Immunochromatographic 1-Step Test for the Forensic Identification of Human Blood, *J. Forensic Sci.*, 44, 597–602, 1999.

Holland, V.R., Saunders, B.C., Rose, F. and Walpole, A., A Safer Substitute for Benzidine in the Detection of Blood, *Tetrahedron*, 30, 3299, 1974.

Hunt, A.C., Corby, C., Dodd, B.E., and Camps, F.E., The Identification of Human Blood Stains: A Critical Survey, *J. For. Med.*, 7, 112, 1960.

Huntress, E.H., Stanley, L.N., and Parker, A.S., The preparation of 3-aminophthalhydrazide for use in the demonstration of chemiluminescence, *J. Am. Chem. Soc.*, 56, 241, 1934.

Johnston, S., Newman, J., and Frappier, R., Validation Study of the Abacus Diagnostics ABAcard® HemaTrace® Membrane Test for the Forensic Identification of Human Blood, *Can. Soc. For. Sci. J.*, 36, 2003.

Kastle, J.H., and Shedd, O.M., Phenolphthalein as a Reagent for the Oxidizing Ferments, *Am. Chem. J.*, 26, 526, 1901.

Kastle, J.H., and Amoss, H.L., *Variations in the Peroxidase Activity of the Blood in Health and Disease*, Bulletin No. 31, U.S. Hygienic Laboratory, U.S. Government Printing Office, Washington, DC, 1906.

Kerr, D.J.A., and Mason, V.H., The haemochromogen crystal test for blood, *Brit. Med. J.*, 1, 134, 1926.

Kirk, P.L., *Crime Investigation*, Interscience Publishers, Inc., New York, 105, 1963.

Laux, D.L., Effects of Luminol on Subsequent Analysis of Bloodstains, *J. Forensic Sci.*, 36, 1512–1520, 1991.

Lee, H.C., and DeForest, P.R., The Use of Anti-human Hb Serum for Bloodstain Identification, Presented at 29th Annual Meeting, American Academy of Forensic Sciences, San Diego, 1977.

Lee, H.C., Identification and Grouping of Bloodstains, In Saferstein, R., Ed., *Forensic Science Handbook*, Prentice Hall, Englewood Cliffs, NJ, Chapter 7, 1982.

Loy, T.H., and Wood, A.R., Blood Analysis at Çayönü Tepesi, Turkey, *J. Field Archeol.*, 16, 451–460, 1989.

Lytle, L.T., and Hedgecock, D.G., Chemiluminescence in the Visualization of Forensic Bloodstains, *J. Forensic Sci.*, 23, 550, 1978.

Macey, H.L., The Identification of Human Blood in a 166 Year Old Stain, *Can Soc. For. Sci. J.*, 12, 191–193, 1979.

Material Safety Data Sheet, Fluorescein, Mallinckrodt Baker, Inc., Phillipsburg, NJ, 1999.

Material Safety Data Sheet, 3 aminophthalhydrazide (Luminol), Fluka Chemical Corp., Milwaukee, WI, 2001.

Miles Laboratories, Product Literature, Hemastix, 1981 and 1992.

Ouchterlony, O., Antigen-antibody Reactions in Gels, *Acta Pathol. Microbiol. Scand.*, 26, 507, 1949.

Ouchterlony, O., *Handbook of Immunodiffusion and Immunoelectrophoresis*, An Arbor Science Publishers, Inc., Ann Arbor, Chapter 5, 1968.

Ponce, A.C., and Verdu Pascual, F.A., Critical Revision of Presumptive Tests for Bloodstains, *Forensic Science Communications*, 1, #2, 1999.

Proescher, F., and Moody, A.M., Detection of blood by means of chemiluminescence, *J. Lab. Clin. Med.*, 24, 1183, 1939.

RCMP, Fingerprint Development Techniques, www.rcmp.ca/firs/recipes/leucomalachite_e.htm, 2001.

Reynolds, M., The ABAcard® Hematrace®: Confirmatory Identification of Human Blood Located at Crime Scenes, *International Association of Bloodstain Pattern Analysts News*, 20(2), June 2004.

Silenieks, E., Atkinson, C., and Pearman, C., Use of the ABAcard® Hematrace® for the Detection of Higher Primate Blood in Bloodstains, 17th International Symposium on the Forensic Sciences, Wellington, New Zealand, 2004.

Spear, T.F., and Brinkley, S.A., The HemeSelect™ Test: A Simple and Sensitive Forensic Species Test, *J. Forensic Sci. Soc.*, 34, 41–46, 1994.

Sutton, T.P., Presumptive Blood Testing, In James, S.H., Ed., *Scientific and Legal Applications of Bloodstain Pattern Interpretation*, CRC Press, Boca Raton, FL, Chapter 4, 1999.

Takayama, M., A Method for Identifying Blood by Hemochromogen Crystallization, *Kokka Igakkai Zasshi*, No. 306, 463–481, 1912, Translation in Gaensslen, R.E., *Sourcebook in Forensic Serology, Immunology and Biochemistry*, U.S. Government Printing Office, Washington, DC, Unit 9, 1983.

Teichmann, L., Concerning the Crystallization of Organic Components of Blood, *Z. Ration. Med.*, 3, 375–388, 1853. Translation printed in presented in Gaensslen, R.E., *Sourcebook in Forensic Serology, Immunology and Biochemistry*, U.S. Government Printing Office, Washington, DC, Unit 9, 1983.

Vandenberg, N., and McMahon, K., Evaluation of OneStep® ABAcard® p30 and Hematrace® Tests for use in Forensic Casework, 17th International Symposium on the Forensic Sciences, Wellington, New Zealand, 2004.

Zweidinger, R.A., Lytle, L.T., and Pitt, C.G., Photography of Bloodstains Visualized by Luminol, *J. Forensic Sci.*, 18, 296, 1973.

Luminol

Albrecht, H.O., Uber die Chemiluminescenz des Aminophthalaurehydrazids. *Z. Phys. Chem. (Leipzig)*, 136:321–330, 1928.

Ballou, S.M., Moves of Murder. *Journal of Forensic Sciences*, 40(4):675–680, 1995.

Cresap, T.R., et al. 1995. The Effects of Luminol and Coomassie Blue on DNA typing by PCR. Presented at the 47th annual American Academy of Forensic Sciences Meeting, Seattle, WA, 1995.

Della Manna, A., and Montpetit, S., A Novel Approach to Obtaining Reliable PCR Results from Luminol Treated Bloodstains. *Journal of Forensic Sciences*, 45(4):886–890, 2000.

Duncan, G.T., Seiden, H., Vallee, L., and Ferraro, D., Effects of Superglue, Other Fingerprint Developing Agents, and Luminol on Bloodstain Analysis. *Journal Assoc. Anal. Chem.*, 69(4): 677–680, 1986.

Erdey, L., Pickering, W.F., and Wilson, C.L., Mixed Chemiluminescent Indicators. *Talanta*, 9:371–375, 1962.

Fregeau, C.J., Germain, O., and Fourney, R.M., Fingerprint Enhancement Revisited and the Effects of Blood Enhancement Chemicals on Subsequent Profiler Plus™ Fluorescent Short Tandem Repeat DNA Analysis of Fresh and Aged Bloody Fingerprints. *Journal of Forensic Sciences*, 45(2):354–379, 2000.

Gaensslen, R.E., *Sourcebook in Forensic Serology, Immunology, and Biochemistry*, Superintendent of Documents, U.S. Government Printing Office, Washington, DC, 1983.

Gill, S.K., New Developments in Chemiluminescence Research. *Aldrichimica Acta*, 16:59–61, 1983.

Gimeno, F.E., Fill Flash Color Photography to Photograph Luminol Bloodstain Patterns. *Journal Forensic Identification*, 39:305–306, 1989.

Gimeno, F.E., and Rini, G.A., Fill Flash Photo Luminescence to Photograph Luminol Blood Stain Patterns. *Journal Forensic Identification*, 39:149–156, 1989.

Gleu, K., and Pfannstiel, K., Uber 3-Aminophthalsaure-hydrazid. *J. Prakt. Chem.*, 146:137–150, 1936.

Grispino, R.R.J., The Effect of Luminol on the Serological Analysis of Dried Human Bloodstains. *Crime Laboratory Digest,* 17(1):13–23, 1990.

Grodsky, M., Wright, K., and Kirk, P.L., Simplified Preliminary Blood Testing. An Improved Technique and Comparative Study of Methods. *J. Crim. Law Criminol. Police Science*, 42:95–104, 1951.

Gross, A.M., Harris, K.A., and Kaldun, G.L., The Effect of Luminol on Presumptive Tests and DNA Analysis Using the Polymerase Chain Reaction. *Journal of Forensic Sciences,* 44(4):837–840, 1999.

Gundermann, K.D., Chemiluminescence in Organic Compounds. *Angewandte Chemie* (International Edition), 4:566–573, 1965.

Hochmeister, M.N., Budowle, B., and Baechtel, F.S., Effects of Presumptive Test Reagents on the Ability to Obtain Restriction Fragment Length Polymorphism (RFLP) Patterns from Human Blood and Semen Stains. *Journal of Forensic Sciences,* 36(3):656–661, 1991.

Hochmeister, M.N., Budowle, B., Sparkes, R., Rudin, O., Gehrig, C., Thali, M., Schmidt, L., Cordier, A., and Dirnhofer, R., Validation Studies of an Immunochromatographic 1-Step Test for the Forensic Identification of Human Blood. *Journal of Forensic Sciences* (44)3:597–601, 1999.

Huntress, E.H., Stanley, L.N., and Parker, A.S., The Preparation of 3-Aminophthalhydrazide For Use in The Demonstration of Chemiluminescence. *J. Amer. Chem. Soc.*, 56:241–242, 1934.

Laux, D.L., Effects of Luminol on the Subsequent Analysis of Bloodstains. *Journal of Forensic Sciences,* 36(5):1512–1520, 1991.

Lytle, L.T., and Hedgecock, D.G., Chemiluminescence in the Visualization of Forensic Bloodstains. *Journal of Forensic Sciences,* 23:550–555, 1978.

McGrath, J., The Chemical Luminescence Test For Blood. *British Medical Journal,* 2:156–157, 1942.

Niebauer, J.C., Booth, J.B., and Brewer, B.L., Recording Luminol Luminescence in its Context Using a Film Overlay Method. *J. Forensic Ident.,* 40(5):271–278, 1990.

Proescher, F., and Moody, A.M., Detection of Blood By Means of Chemiluminescence. *J. Lab. Clin. Med.,* 24:1183–1189, 1939.

Specht, W., The Chemiluminescence of Hemin: An Aid For Finding and Recognizing Blood Stains Important For Forensic Purposes. *Angewante Chemie,* 50:155–157, 1937.

Tamamushi, B., and Akiyama, H., Mechanism of the Chemiluminescence of 3-Aminophthalhydrazide. *Zeitschrift Fuer Physikalische Chemic,* 38:400–406, 1938.

Thornton, J.I., and Maloney, R.S., The Chemistry of the Luminol Reaction: Where to From Here? *Calif. Assoc. Crim. Newsletter,* Sept:9–16, 1985.

Thornton, J.I., and Murdock, J.E., Photography of Luminol Reaction in Crime Scenes. *Criminol.,* 10(37):15–19, 1977.

Weber, K., Die Anwendung der Chemiluminescenz des Luminols in der gerichtlichen Medizin und Toxicologie. I. Der Nachweis von Blutspuren. *Z. Gesamte Gerichtl. Med.,* 57:410–423, 1966.

White, E.H., The Chemiluminescence of Luminol. In McElroy, W.D., and Glass, B. Eds., *Light and Life,* Johns Hopkins Press, Baltimore, MD, 1961, pp. 183–195.

White, E.H., and Bursey, M.M., Chemiluminescence of Luminol and Related Hydrazides: The Light Emission Step. *Journal of the American Chemical Society,* 86:941–942, 1964.

Zweidinger, R.A., Lytle, L.T., and Pitt, C.G., Photography of Bloodstains Visualized By Luminol. *Journal of Forensic Sciences,* 18:296–302, 1973.

Chemical Enhancement of Latent Blood Impressions

Bergeron, J., Development of bloody prints on dark surfaces with titanium dioxide and methanol, *Journal of Forensic Identification,* No. 53, 2003.

Bodziak, W.J., *Footwear Impression Evidence,* 2nd ed., CRC Press, Boca Raton, FL, 2000.

Eversdijk, M., and Gelderman, R., Shedding Fresh Light at the Scene of the Crime, *Information Bulletin for Shoeprint/Toolmark Examiners,* Vol. 6, No. 1, March 2000.

Eversdijk, M., Gelderman, R., Nieuw Licht op de Plaats Delict, *Modus,* Jaargang 8, No. 4, September 1999.

Grodsky, Wright, M.K. and Kirk, P.L. Simplified Preliminary Blood Testing. An Improved Technique and Comparative Study of Methods, *Journal of the American Institute of Criminal Law and Criminology,* Vol. 42, 1951.

Manual of Fingerprints Development Techniques, Police Scientific Development Branch, Home Office Police Policy Derectorate, 2nd ed., 1998.

Sears, V.G., Prizeman, T.M., Enhancement of Fingerprints in Blood—Part1: The Optimization of Amido Black, *Journal of Forensic Identification,* Vol. 50, No. 5, 2000.

Spence, L., Asmussen G., Spectral Enhancement of Leucocrystal Violet Treated Footwear Impression Evidence in Blood, *Forensic Science International,* No. 132, 2003.

Teeuwen, A.B.E., van Barneveld, S., Drok, J.W., Keereweer, I. Limborgh, J.C.M., Naber, W.M., and Velders, T., *Enhancement of Footwear Impressions in Blood, Forensic Science International,* No. 95, 1998.

U.S. Department of Justice, Federal Bureau of Investigation, Processing Guide for Developing Latent Prints.

Appendix A
Trigonometric Tables—Sine and Tangent Functions

Degrees	Sine	Tangent	Degrees	Sine	Tangent
0.0	.0000	.0000	46.0	.7193	1.036
1.0	.0175	.0175	47.0	.7314	1.072
2.0	.0349	.0349	48.0	.7431	1.111
3.0	.0523	.0524	49.0	.7547	1.150
4.0	.0698	.0699	50.0	.7660	1.192
5.0	.0872	.0875	51.0	.7771	1.235
6.0	.1045	.1051	52.0	.7880	1.280
7.0	.1219	.1228	53.0	.7986	1.327
8.0	.1392	.1405	54.0	.8090	1.376
9.0	.1564	.1584	55.0	.8192	1.428
10.0	.1736	.1763	56.0	.8290	1.483
11.0	.1908	.1944	57.0	.8387	1.540
12.0	.2079	.2126	58.0	.8480	1.600
13.0	.2250	.2309	59.0	.8572	1.664
14.0	.2419	.2493	60.0	.8660	1.732
15.0	.2588	.2679	61.0	.8746	1.804
16.0	.2756	.2867	62.0	.8829	1.881
17.0	.2924	.3057	63.0	.8910	1.963
18.0	.3090	.3249	64.0	.8988	2.050
19.0	.3256	.3443	65.0	.9063	2.145
20.0	.3420	.3640	66.0	.9135	2.246
21.0	.3584	.3839	67.0	.9205	2.356
22.0	.3746	.4040	68.0	.9272	2.475
23.0	.3907	.4245	69.0	.9336	2.605
24.0	.4067	.4452	70.0	.9397	2.748
25.0	.4226	.4663	71.0	.9455	2.904
26.0	.4384	.4877	72.0	.9511	3.078
27.0	.4540	.5095	73.0	.9563	3.271
28.0	.4695	.5317	74.0	.9613	3.487
29.0	.4848	.5543	75.0	.9659	3.732
30.0	.5000	.5774	76.0	.9703	4.011
31.0	.5150	.6009	77.0	.9744	4.332
32.0	.5299	.6249	78.0	.9781	4.705
33.0	.5446	.6494	79.0	.9816	5.145
34.0	.5592	.6745	80.0	.9848	5.671
35.0	.5736	.7002	81.0	.9877	6.314
36.0	.5878	.7265	82.0	.9903	7.115
37.0	.6018	.7536	83.0	.9925	8.144
38.0	.6157	.7813	84.0	.9945	9.514
39.0	.6293	.8098	85.0	.9962	11.43
40.0	.6428	.8391	86.0	.9976	14.30
41.0	.6561	.8693	87.0	.9986	19.08
42.0	.6691	.9004	88.0	.9994	28.64
43.0	.6820	.9325	89.0	.9998	57.29
44.0	.6947	.9657	90.0	1.000	0.000
45.0	.7071	1.000			

Appendix B
Metric Measurements and Equivalents

Decimal Prefixes and Multiples for Area/Length, Mass Weight, and Volume

Prefix	Symbol	Multiple	Magnitude	Common Name
yotta	Y	10^{24}	1 000 000 000 000 000 000 000 000	heptillion
zetta	Z	10^{21}	1 000 000 000 000 000 000 000	hextillion
exa	E	10^{18}	1 000 000 000 000 000 000	quintillion
peta	P	10^{15}	1 000 000 000 000 000	quadrillion
tera	T	10^{12}	1 000 000 000 000	trillion
giga	G	10^{9}	1 000 000 000	billion
mega	M	10^{6}	1 000 000	million
kilo	K	10^{3}	1 000	thousand
hecto	h	10^{2}	1 00	hundred
deca	da	10^{1}	10	ten
unit*	—	—	1	one
deci	d	10^{-1}	0.1	tenth
centi	c	10^{-2}	0.01	hundredth
milli	m	10^{-3}	0.001	thousandth
micro	μ (mu)	10^{-6}	0.000 001	millionth
nano	n	10^{-9}	0.000 000 001	billionth
pico	p	10^{-12}	0.000 000 000 001	trillionth
femto	f	10^{-15}	0.000 000 000 000 001	quadrillionth
atto	a	10^{-18}	0.000 000 000 000 000 001	quintillionth
zepto	z	10^{-21}	0.000 000 000 000 000 000 001	hextillionth
yocto	y	10^{-24}	0.000 000 000 000 000 000 000 001	heptillionth

* unit = meter (m) for area/length, gram (g) for mass/weight, and liter (L) for volume (capacity)

Metric Equivalents
Length Measurements
Some common multiples in powers of 10 include:

kilometer (km) = 103 meters
decimeter (dm) = 10^{-1} meters
centimeter (cm) = 10^{-2} meters
millimeter (mm) = 10^{-3} meters

micrometer (μm)* = 10^{-6} meters
nanometer (nm) = 10^{-9} meters
angstrom (Å) = 10^{-10} meters

1 millimeter (mm) = 1000 micrometers (μm)* = 1,000,000 nanometers = 10,000,000 angstroms (Å)
1 angstrom (Å) = 0.1 nanometers (nm) = 0.0001 micrometers (μm) = 0.00000001 centimeters (cm)
10,000 angstroms (Å) = 1000 nanometers (nm) = 1 micrometer (μm) = 0.0001 centimeters (cm)
10,000 micrometers (μm) = 10 millimeters (mm) = 1 centimeter (cm) = 0.01 meters (m)

* Micrometer (μm) may appear as the older designation, micron (μ)

1 millimeter (mm) = 0.1 centimeter (cm)
10 millimeters (mm) = 1 centimeter (cm)
10 centimeters (cm) = 1 decimeter (dm) = 100 millimeters (mm)
10 decimeters (dm) = 1 meter (m) = 1000 millimeters (mm)
10 meters (m) = 1 dekameter (dam)
10 dekameters (dam) = 1 hectometer (hm) = 100 meters (m)
10 hectometers (hm) = 1 kilometer (km) = 1000 meters (m)

Area Measurements

100 square millimeters (mm^2) = 1 square centimeter (cm^2)
10,000 square centimeters (cm^2) = 1 square meter (m^2) = 1,000,000 square millimeters
100 square meters = 1 are (a)
100 ares = 1 hectare (ha) = 10,000 square meters
100 hectares = 1 square kilometer (km^2) = 1,000,000 square meters

Volume Measurements
Some common multiples in powers of 10 include:

1 liter (L) = 10^3 milliliters (mL)
1 milliliter (mL) = 10^{-3} liters (L)
1 microliter (μL) = 10^{-6} liters (L)
1 microliter (μL) = 0.001 milliliter (mL) = 0.000 001 liter (L)
1000 microliters (μL) = 1 mL = 0.001 liter (L)
1,000,000 microliters (μL) = 1000 milliliters (mL) = 1 liter (L)
10 milliliters (mL) = 1 centiliter (cL) = 0.01 liter (L)
10 centiliters (cL) = 1 deciliter (dL) = 100 milliliters (mL) = 0.1 liter (L)
10 deciliters = 1 liter (L) = 1000 milliliters (mL)
10 liters = 1 dekaliter (daL)
10 dekaliters = 1 hectoliter (hL) = 100 liters (L)
10 hectoliters = 1 kiloliter (kL) = 1000 liters (L)

Cubic Measurements

1000 cubic millimeters (mm^3) = 1 cubic centimeter (cm^3) = 1 milliliter (mL) (liquid)
1000 cubic centimeters (cm^3) = 1 cubic decimeter (dm^3) = 1,000,000 cubic millimeters (mm^3)
1000 cubic decimeters (dm^3) = 1 cubic meter (m^3) = 1 stere = 1,000,000 cubic centimeters (cm^3) = 1,000,000,000 cubic millimeters (mm^3)

Mass/Weight Measurements
Some common multiples in powers of 10 include:

kilogram (kg) = 10^3 grams (g)
centigram (cg) = 10^{-2} grams (g)
milligram (mg) = 10^{-3} grams (g)
microgram (μg) = 10^{-6} grams (g)
nanogram (ng) = 10^{-9} grams (g)

Appendix B

picogram (pg) = 10^{-12} grams (g)
1 milligram (mg) = 1000 micrograms (μg) = 1,000,000 nanograms (ng) = 1,000,000,000 picograms (pg)
1 picogram (pg) = 0.001 nanograms (ng) = 0.000001 micrograms (μg) = 0.000000001 milligrams(mg)
10 milligrams (mg) = 1 centigram (cg)
10 centigrams (cg) = 1 decigram (dg) = 100 milligrams (mg)
10 decigrams (dg) = 1 gram (g) = 1000 milligrams (mg)
10 grams (g) = 1 dekagram (dag)
10 dekagrams (dag) = 1 hectogram (hg) = 100 grams (g)
10 hectograms (hg) = 1 kilogram (kg) = 1000 grams (g)
1000 kilograms (kg) = 1 metric ton (t)

Useful Metric Measurement Conversions to English System

centimeters × 0.40 = inches
meters × 3.28 = feet
meters × 1.09 = yards
kilometers × 0.62 = miles
grams × 0.035 = ounces
kilograms × 2.20 = pounds
liters × 2.10 = pints
liters × 1.06 = quarts
liters × 0.264 = gallons (U.S.)
milliliters × 0.034 = fluid ounces

Useful English System to Metric Measurement Conversions

inches × 2.54 = centimeters
feet × 0.30 = meters
yards × 0.91 = meters
miles × 1.60 = kilometers
ounces × 28.3 = grams
pounds × 0.45 = kilograms
pints × 0.47 = liters
quarts × 0.95 = liters
gallons (U.S.) × 3.8 = liters
fluid ounces × 29.6 = milliliters (cubic centimeters)

Temperature Conversions

Degrees Celsius (C) = {Degrees Fahrenheit (F) − 32} × 0.555
Degrees Fahrenheit (F) = {Degrees Celsius (C) × 1.8} + 32

Fahrenheit Degrees°	Celsius Degrees°
−40°	−40°
0°	−17°
25°	−4°
32°	0°
50°	10°
75°	24°
100°	38°
125°	52°
150°	66°
175°	79°
200°	93°

Appendix C
Scene and Laboratory Checklists

CRIME SCENE WORKSHEET--PROPERTY   PAGE 1 of _____   LAB # _____

CASE REFERENCE: _____ Sex __   Race __   Age __
Victim: _____   Suspect: _____
R&I: _____   Cross Ref To: _____
Agency: __ MPD Homicide   __ SCSD Detect.   __ _____
Case Coordinator: _____   Transported By: _____
Others Present at Exam: _____
_____

Date: _____   TIME: Left Lab: _____   *Exam Began:* _____
                               Exam Completed: _____   Returned to Lab: _____

==
Address: _____   Property of: _____

__ **Residence:** House ____   Apt _____   Duplex _____   Townhouse _____   Trailer _____
__ **Commercial:** Retail store _____   Office _____
__ **Other:** _____

==
Site of Exam: _____   Veh. Prop. of: _____
Make: _____   Model: _____   Year: _____
Lic #: _____   State: _____   Expires: _____   Renewal #: _____
VIN #: _____   From: Dash __   Driver's Door ____
Main Body Color: _____   Top Color (if different): _____
Type of Top: Full Vinyl __   Landau __   Metal __   Doors: 2D __   4D ___
Body Style: Sedan __   Truck _____   St. Wagon _____   Hatchback _____   Van _____   Minivan _____
Front Seat: Color: _____   Material: Vinyl __   Cloth _____   Cloth c Vinyl ____
               Style: Bench ____   Bench c Center Arm Rest _____   Bucket _____   60/40 _____
Back Seat: Color: _____   Material: Vinyl __   Cloth _____   Cloth c Vinyl ____
               Style: Bench ____   Bench c Center Arm Rest _____   Bucket _____   60/40 _____
Floorboard: Color: _____   Material: Vinyl __   Carpet _____
Front Floormats: Color: _____   Material: Vinyl __   Carpet _____   Carpet c Vinyl _
Rear Floormats: Color: _____   Material: Vinyl __   Carpet _____   Carpet c Vinyl _
==

TEST CONTROLS:		SPECIMENS RETURNED TO LAB:
Date & Time:	_____	☐ Specimens returned to lab
Phenolp	Tech	
STMP	Tech	☐ No specimens returned to lab

Appendix C

CRIME SCENE TESTING & COLLECTION WORKSHEET

LAB # _____ Page ___ of ___

STAIN	Location or Description	Phenolp.	NOT Collected	Collected As: Lift	Neg Con ?	Crust	Cutting	Item	Other (Specify)	Spec # assigned at Lab
A										
B										
C										
D										
E										
F										
G										
H										
I										
J										
K										
L										
M										
N										
O										
P										

PAGE _____ OF _____   LAB #: _____

NO SUSPECT STAINS CHECKLIST------VEHICLE INTERIOR
Check areas below that were examined and did not exhibit suspect staining

LEFT FRONT--DRIVER SIDE
- _____ Seat
- _____ Side of seat __ Left __ Right
- _____ Front of seat
- _____ Center Armrest
- _____ Seat Backrest
- _____ Side of seat back __ Left __ Right
- _____ Back of Seat Back
- _____ Floor Mat
- _____ Floorboard
- _____ Running board
- _____ Headliner
- _____ Visor
- _____ Upper portion of Door
- _____ Midsection of Door
- _____ Armrest of Door
- _____ Lower portion of Door
- _____ Door Knob
- _____ Door Lock
- _____ Window
- _____ Window Knob/Control
- _____ Windshield
- _____ Dash
- _____ Steering Wheel
- _____ Gear Shift
- _____ Turn Signal
- _____ Accelerator
- _____ Brake pedal
- _____ Clutch pedal

RIGHT FRONT--PASSENGER SIDE
- _____ Seat
- _____ Side of seat __ Left __ Right
- _____ Front of seat
- _____ Center Armrest
- _____ Seat Backrest
- _____ Side of seat back __ Left __ Right
- _____ Back of Seat Back
- _____ Floor Mat
- _____ Floorboard
- _____ Running board
- _____ Headliner
- _____ Visor
- _____ Upper portion of Door
- _____ Midsection of Door
- _____ Armrest of Door
- _____ Lower portion of Door
- _____ Door Knob
- _____ Door Lock
- _____ Window
- _____ Window Knob/Control
- _____ Windshield
- _____ Dash
- _____ Front Ashtray
  - _____ Empty
  - _____ Contained_____
  - _____ Contents recovered

LEFT REAR SEAT
- _____ Seat
- _____ Side of seat __ Left __ Right
- _____ Front of seat
- _____ Center Armrest
- _____ Seat Back Rest
- _____ Side of seat back __ Left __ Right
- _____ Floor Mat
- _____ Floorboard
- _____ Running board
- _____ Headliner
- _____ Upper portion of Door
- _____ Midsection of Door
- _____ Armrest of Door
- _____ Lower portion of Door
- _____ Door Knob
- _____ Door Lock
- _____ Window
- _____ Window Knob/Control
- _____ Back Window
  - _____ Ashtray
  - _____ Empty
  - _____ Contained_____
  - _____ Contents recovered

RIGHT REAR SEAT
- _____ Seat
- _____ Side of seat __ Left __ Right
- _____ Front of seat
- _____ Center Armrest
- _____ Seat Back Rest
- _____ Side of seat back __ Left __ Right
- _____ Floor Mat
- _____ Floorboard
- _____ Running board
- _____ Headliner
- _____ Upper portion of Door
- _____ Midsection of Door
- _____ Armrest of Door
- _____ Lower portion of Door
- _____ Door Knob
- _____ Door Lock
- _____ Window
- _____ Window Knob/Control
- _____ Back Window
  - _____ Ashtray
  - _____ Empty
  - _____ Contained_____
  - _____ Contents recovered

PAGE ____ OF ____   LAB #: _____

Trigonometric Point of Origin Worksheet

STAIN	STAIN WIDTH	STAIN LENGTH	ANGLE OF IMPACT	DISTANCE (D) TO 2D PT OF ORIGIN	(D) X (TAN)

Number of Stains = _____   Sum of (D) X (Tan) [_____]

$$\text{Calculated 3-D Pt of Origin} = \frac{\text{Sum of (D) X (Tan)}}{\text{Number of Stains}} = \underline{\hspace{3cm}}$$

Appendix D
Biohazard Safety Precautions

Attention to biohazard safety is, all too often, limited to the medical community. Many of the tasks performed during the investigation and prosecution of criminal activity by police officers, investigators, forensic scientists, and attorneys also present the potential for contact with many biohazards.

The ways in which exposure may occur and the types of exposure that may occur are more innocuous than they are in a traditional medical setting. This requires that the personnel that handle forensic evidence have an understanding of the general rules of biosafety and be able to apply these guidelines and policies to the diverse situations that can arise in criminal investigations.

This outline will present a basic understanding of the occupational statutes, knowledge about bloodborne pathogens, and some general guidelines geared toward overall biohazard safety. The best defense against biohazard exposures is applying this information along with a using a healthy dose of common sense.

Universal Precautions Policy

The U.S. Occupational Safety and Health Administration defines Universal Precautions as follows: "All blood and certain other body fluids are considered potentially infectious for hepatitis B virus, hepatitis C virus, human immunodeficiency virus (HIV), and other bloodborne pathogens." Bloodborne pathogens are defined as pathogenic microorganisms that are present in human blood, human blood components, and products made from human blood that can cause disease in humans. This means that any blood and certain other body fluids are considered to be potentially infectious.

Body fluids that require the application of universal precautions are as follows:

1. Blood
2. Semen
3. Vaginal fluid
4. Tissue (unfixed)
5. Synovial fluid
6. Amniotic fluid
7. Peritoneal fluid
8. Spinal fluid

9. Pleural fluid
10. Pericardial fluid
11. Any body fluid that is visibly bloody

Universal precautions are necessary when you know a person is infected. If an individual is known to be infected with a bloodborne pathogen, such as HIV, this information should be transmitted to each individual who might be exposed to body fluids from an individual known to be infected. However, it is not always possible to identify the source individual or to know their infectious status. This is particularly true with body fluids found at crime scenes or body fluids found on pieces of evidence. Exposure may occur through skin (especially if it is broken); eyes; mucous membranes, including the nose, mouth, and respiratory tract; and parenteral contact (subcutaneous or intravenous contact).Therefore, a policy of universal precautions is recommended *AT ALL TIMES*.

Human Immunodeficiency Virus

HIV is a retrovirus associated with acquired immunodeficiency syndrome (AIDS). HIV has a core that contains ribonucleic acid (RNA) and enzymes and an outer covering of lipids and proteins. Viruses are not capable of reproduction without infecting a host cell. The HIV binds to the surface of the helper T-cells, a type of white blood cell belonging to the host, to reproduce. The RNA and enzymes of HIV are released into the host cell, which begins copying the viral RNA into double-helix deoxyribonucleic acid (DNA). This double-helix DNA, which now contains the RNA from the HIV, is then incorporated into the DNA of the host cell. In this way, the host cell is deceived into producing more and more virus particles. As HIV gradually depletes the number of helper T-cells, the immune system of the host is rendered virtually powerless.

The host is left vulnerable to other viruses, bacteria, fungi, protozoa, and cancers that would ordinarily not be capable of infecting an individual with an intact immune system. Body fluids that can transmit HIV are blood, semen, vaginal secretions, and possibly breast milk.

However, HIV has been isolated from saliva, tears, urine, cerebrospinal fluid, amniotic fluid, tissue of infected persons, and experimentally infected nonhuman primates. Although the virus has been isolated in all of the aforementioned body fluids, only blood, semen, vaginal secretions, and possibly breast milk have been implicated in the actual transmission of the virus.

Survivability of HIV has been estimated under the following conditions:

1. Liquid, room temperature: 15 days (minimum)
2. Liquid, refrigerated: Much longer than 15 days
3. Dried, room temperature: 3 to 13 days
4. Dried, frozen: Much longer than 13 days

Hepatitis

The general term viral hepatitis includes five types of the disease:

1. Hepatitis A (HAV): Formerly called "infectious hepatitis" contracted through fecal–oral routes.

2. Hepatitis B (HBV): Formerly called "serum hepatitis" contracted through nonintact skin or punctures such as needle sticks.
3. Hepatitis C (HCV): Formerly called "non-A, non-B hepatitis" contracted through nonintact skin and punctures such as needle sticks.
4. Delta hepatitis (HDV)
5. Non-A, non-B, non-C hepatitis: A collective name for all other hepatitis viruses contracted through nonintact skin or punctures such as needle sticks. A human host is 100 times more likely to contract hepatitis than HIV. Although HIV has received more attention, the risk of contracting HBV from the handling of biological materials is actually much greater.

When the source blood is known to be infected with both HBV and HIV, the likelihood of contracting HBV is 100 times greater than the likelihood of contracting HIV. The HBV is known to survive for much longer periods, even in the dried state, than HIV.

Basic Rules for Biohazard Safety

It is essential that universal precautions be observed at all times, and exposure to biohazards should be minimized by eliminating the known routes of exposure by reducing or getting rid of the contact route to the body from the scene, autopsy room, laboratory work space, supplies, and equipment whenever possible. Contact routes include the following:

1. Direct oral route via blood or other body fluids entering the mouth
2. Indirect oral route via smoking or placing pencils in the mouth
3. Ocular route via splashing infectious material into the eyes or rubbing the eyes
4. Inoculation route via puncture wounds caused by needles or other sharp objects
5. Respiratory route via direct respiratory contact or aerosol production

Hands should be washed with warm water and a germicidal soap immediately following contact with infectious materials and after completion of all procedures, including the removal of protective gloves.

Minimizing Biohazard Contamination at the Crime Scene

Establish a "Clean Area" as a working space, and keep this area uncontaminated. Remember that gloves, equipment, and the like that have been taken into the crime scene may be a source of contamination.

1. Leave all supplies and equipment in the "Clean Area" until needed.
2. Organize and assemble equipment to minimize contamination. Take only necessary equipment and supplies into the scene because everything that enters the scene will become potentially contaminated.
3. Determine the type of personal protective equipment (PPE) needed in terms of the quantity of biological fluids present and whether they are in the wet or dry state relative to the tasks to be performed.

The scene should be assessed initially by questioning knowledgeable individuals who have already entered the scene and performing an evaluation walk-through of the scene. During the evaluation walk-through, minimize the number of personnel entering the scene and wear at least protective shoe covers and gloves.

Guidelines for the Selection of the Proper Personal Protective Equipment

1. Protective gloves should be worn as a minimal precaution.
2. If there is a potential for fluid splash, wear face or eye shields.
3. If there is a potential for aerosolization, wear face masks.
4. If there is a potential for clothing contamination, add protective suits and shoe covers.
5. Replace gloves or other PPE when they become visibly soiled, deteriorated, or torn.
6. Do not wash or disinfect surgical or examination gloves for reuse.
7. Remove gloves and other PPE prior to leaving the work area.
8. If cuts or dermatitis are present, use gloves in cases where contact may only be indirect.
9. Prepare a "Discard Location" for disposal of contaminated protective gear and supplies when the scene is exited.
10. Secure the disposal bag to prevent accidental spills.
11. Prepare bleach or other means of decontamination. Household bleach diluted 1:10 with water (approximately 1/4 cup of bleach in a quart of water) must be prepared fresh each day. A minimum of 10 min of contact time is necessary.

Precautions for Evidence Collection

1. Be aware of syringes, needles, knives, razors, nails, broken glass, or sharp metals in the area.
2. When searching confined and blind spaces do not reach in before checking with a flashlight or mirror.
3. Use caution when collecting biological specimens to avoid overt contamination.
4. Change gloves between each item of evidence collected to prevent cross contamination of the evidence.
5. Use masks and eye protection when scraping or cutting dried stains to avoid aerosol exposure.
6. When cutting out stains, use caution with knives and scissors to avoid surface cuts and contamination from the body fluid.
7. Syringes may be used to collect wet stains. Do not use a needle on the syringe unless absolutely necessary because a needle will make the opening smaller and collection more difficult. Needles also present a sharp hazard.
8. Use puncture-resistant containers when packaging sharp items of evidence, and label the outside of the container as "Sharp."
9. Seal evidence packages with tape rather than staples, which present a sharp hazard.
10. Use biohazard labels for evidence items containing blood or body fluids.

Disposables

1. Nonsharp disposables such as paper scales, film boxes, and film canisters may be discarded directly into the collection container.

2. "Sharp" disposables such as scalpels, pens, or pencils to be discarded and broken glassware are first packaged in cardboard, plastic, or another rigid container to prevent punctures, and then they are discarded into the collection container.

Nondisposable equipment and supplies should be decontaminated.

1. These include ink pens to be used again, markers, crime scene kits, cameras, safety glasses, reusable rulers, and scales.
2. Decontamination is done with diluted bleach or commercial wipes.
3. Decontaminated equipment is then returned to the proper storage position.
4. Cleaning supplies are discarded into the collection container.

Personal Protective Equipment Removal

PPE is removed via a stepwise approach to minimize exposure.

1. Eye covers are removed and decontaminated if reusable, then returned to storage location. Cleaning supplies and disposable PPE then are placed in the discard container.
2. Mask and head coverings are removed with caution to prevent indirect exposure and are placed in the discard container.
3. Shoe covers are removed via the "roll off" technique. Invert the interior, uncontaminated surface over the outer, contaminated surface and pull the shoe cover off. Place in the discard container.
4. Protective suits are also rolled off. Invert the interior, uncontaminated surface over the outer, contaminated surface and pull the suit off. Place in the discard container.
5. Gloves are removed last. Start at the wrist and peel the glove off one hand. Grasp the removed glove in the still-gloved hand. Roll down the top of the remaining glove. Grasp the inner, uncontaminated surface of the glove and peel it off while tucking the first glove inside of the second glove. Place unit in the discard container.

Close the discard container by grasping the bag underneath the rolled-down top. Pull the top of the bag upward to close. Secure the top of the bag with tape, twist ties, or the like to prevent spills.

Wash hands with warm water and a germicidal soap. Vionex wipes, Cal-stat, or Sani-cloth wipes may be used if a handwashing facility is not immediately available. These measures are temporary alternatives but not replacements to the handwashing facility.

When leaving the scene, removal and disposal of contaminated materials, supplies, testing items, and personal protective equipment are the responsibilities of the agency.

Selection of Personal Protective Equipment

An important consideration in biohazard safety is selecting the proper PPE to be used. Selection is based on the best conservative judgment of types of exposure that might occur in any given situation.

When in doubt, use more PPE, rather than less. The decision tree shown in Figure D.1 will assist in the selection of the proper PPE to be used.

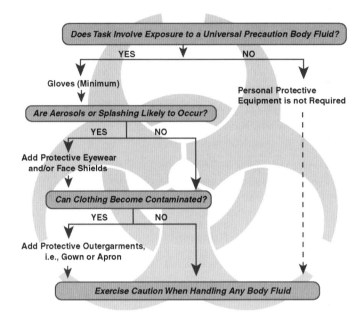

Figure D.1 Personal protective equipment (PPE) decision tree.

References

1. 29 CFR 1910.1030, *Bloodborne Pathogen Standard*, U.S. Department of Labor, Occupational Safety and Health Administration, *Federal Register,* Vol. 56, No. 235, December 6, 1991, Rules and Regulations, Part 119110 (Amended) Subpart Z (Amended), pp. 64179–64180.
2. *Morbidity and Mortality Weekly Report,* U.S. Department of Health and Human Services, Public Health Service, Centers for Disease Control, Atlanta, GA, August 21, 1987, Vol. 36, No. 2S.
3. *Guidelines for Prevention of Transmission of Human Immunodeficiency Virus and Hepatitis B Virus to Health-Care and Public Safety Workers.* "A Response to P.L. 100-607 The Health Omnibus Programs Extensions Act of 1988," U.S. Department of Health and Human Services, Centers for Disease Control, Atlanta, GA, February 1989:2.
4. Masters, Nancy E., *Safety for the Forensic Identification Specialist.* Salem, OR: Lightning Powder, Company, Inc., 1995, p. 31.
5. Bigbee, D., *The Law Enforcement Officer and AIDS,* Fifth ed. U.S. Department of Justice, Federal Bureau of Investigation, 1994, p. 5.
6. Masters, Nancy E., *Safety for the Forensic Identification Specialist,* Salem, OR: Lightning Powder, Company, Inc., 1995, pp. 34–35.
7. *Recommendations for Preventions of HIV Transmission in Health-Care Settings, Morbidity and Mortality Weekly Report,* U.S. Department of Health and Human Services, Public Health Service, Centers for Disease Control, Atlanta, GA, 1987; Volume 36, Supplemental no. 2S: p. 49.
8. Bigbee, D., *The Law Enforcement Officer and AIDS,* Fifth ed. U.S. Department of Justice, Federal Bureau of Investigation, 1994, pp. 19–20.

Appendix E
Court Decisions Relating to Bloodstain Pattern Analysis

State

Alabama

Leonard v. State, 551 So.2d 1143 (Ala. App. 1989)
Robinson v. State, 574 So.2d 910 (Ala. App. 1990)

Alaska

Crawford v. Rogers, 406 P.2d 189 (Alaska 1965)
Pederson v. State, 420 P.2d 327 (Alaska 1966)

California

People v. Carter, 312 P.2d 665 (Cal. 1957)
People v. Hogan, 647 P.2d 93 (Cal. 1982)
People v. Clark, 857 P.2d 1099 (Cal. 1993)

District of Columbia

Eason v. United States, 687 A.2d 922 (D.C. App. 1996)

Florida

Castro v. State, 547 So.2d 111 (Fla. 1989)
Cheshire v. State, 568 So.2d 908 (Fla. 1990)
Morris v. State, 561 So.2d 646 (Fla. 1990)
Reichmann v. State, 581 So.2d 133 (Fla. 1991)

Georgia

Droke v. State, 314 S.E.2d 230 (Ga. 1984)
Coleman v. State, 348 S.E.2d 70 (Ga. 1986)
Coleman v. State, 357 S.E.2d 566 (Ga. 1987)
O'Toole v. State, 373 S.E.2d 12 (Ga. 1988)

Idaho

State v. Rodgers, 812 P.2d 1227 (Idaho 1990)
State v. Rodgers, 812 P.2d 1208 (Idaho 1991)
State v. Raudebaugh, 864 P.2d 596 (Idaho 1993)

Illinois

People v. Driver, 379 N.E.2d 840 (Ill. 1978)
People v. Erickson, 411 N.E.2d 44 (Ill. 1980)
People v. Knox, 459 N.E.2d 1077 (Ill. 1984)
People v. Owens, 508 N.E.2d 1088 (Ill. 1987)
People v. Smith, 633 N.E.2d 69 (Ill. 1993)

Indiana

Fox v. State, 506 N.E.2d 1090 (Ind. 1987)
King v. State, 531 N.E.2d 1154 (Ind. 1988)
Hampton v. State, 588 N.E.2d 555 (Ind. 1992)

Iowa

State v. Hall, 297 N.W.2d 80 (Iowa 1980)

Kansas

State v. Satterfield, 592 P.2d 135 (Kan. 1979)
State v. Ordway, 934 P.2d 94 (Kan. 1997)

Louisiana

State v. Graham, 422 So.2d 123 (La. 1982)
State v. Williams, 445 So.2d 1171 (La. 1984)
State v. Ancar, 508 So.2d 943 (La. 1984)
State v. McFadden, 476 So.2d 413 (La. 1985)
State v. Powell, 598 So.2d 454 (La. 1992)
State v. Howard, 626 So.2d 459 (La. 1993)

Maine

State v. Knight, 43 Me. 11 (1857)
State v. Hilton, 431 A.2d 1296 (Me. 1981)
State v. Philbrick, 436 A.2d 844 (Me. 1981)

Maryland

Tirado v. State, 622 2d 187 (Md. 1993)

Massachusetts

Commonwealth v. Sturtivant, 117 Mass. 122 (1875)
Commonwealth v. Capalbo, 32 N.E.2d 225 (Mass. 1941)
Commonwealth v. Gordon, 666 N.E.2d 122 (Mass. 1996)

Minnesota

State v. Malzac, 244 N.W.2d 258 (Minn. 1976)
State v. Fossen, 282 N.W.2d 496 (Minn. 1979)
State v. Merrill, 428 N.W.2d 361 (Minn. 1988)
State v. Norris, 428 N.W.2d 61 (Minn. 1988)
State v. Robinson, 427 N.W.2d 217 (Minn. 1988)
State v. Moore, 458 N.W.2d 90 (Minn. 1990)

Mississippi

Dillard v. State, 58 Miss. 368 (1880)
Jordon v. State, 464 So.2d 475 (Miss. 1985)
Whittinger v. State, 523 So.2d 966 (Miss. 1988)
Fowler v. State, 566 So.2d 1194 (Miss. 1990)

Montana

State v. Mix, 781 P.2d 751 (Mont. 1989)

Nebraska

State v. Thieszen, 560 N.W.2d 800 (Neb. 1997)

New York

Lindsay v. People of the State of New York, 63 N.Y. 143 (1875)
People v. Comfort, 496 N.Y.S.2d 857 (1985)
People v. Murray, 537 N.Y.S.2d 399 (1989)

North Carolina

State v. Simpson, 255 S.E.2d 147 (N.C. 1979)
State v. Willis, 426 S.E.2d 471 (N.C. 1993)
State v. East, 481 S.E.2d 652 (N.C. 1997)
State v. Clifton, 481 S.E.2d 393 (N.C. 1997)

Oklahoma

Farris v. State, 670 P.2d 995 (Okla. 1983)
Clayton v. State, 840 P.2d 18 (Okla. 1992)
Clayton v. State, 892 P.2d 646 (Okla. 1992)
Hogan v. State, 877 P.2d 1157 (Okla. 1994)
Taylor v. State, 889 P.2d 319 (Okla. 1995)

Oregon

State v. Luther, 663 P.2d 1261 (Or. App. 1983)
State v. Proctor, 767 P.2d 453 (Or. App. 1989)

Pennsylvania

Commonwealth v. Goins, 495 A.2d 527 (Pa. 1985)
Commonwealth v. Duffey, 548 A.2d 1178 (Pa. 1988)

Rhode Island

State v. Chiellini, 557 A.2d 1195 (R.I. 1989)

South Carolina

State v. Myers, 391 S.E.2d 551 (S.C. 1990)

Tennessee

State v. Melson, 638 S.W.2d 342 (Tenn. 1982)
State v. Cazes, 875 S.W.2d 253 (Tenn. 1994)

Texas

Guerrero v. State, 720 S.W.2d 233 (Tex. App. 1986)
Cortijo v. State, 739 S.W.2d 486 (Tex. App. 1987)
Lewis v. State, 737 S.W.2d 857 (Tex. App. 1987)
Salas v. State, 756 S.W.2d 832 (Tex. App. 1988)
Speering v. State, 763 S.W.2d 801 (Tex. App. 1988)
Mowbray v. State, 788 S.W.2d 658 (Tex. App. 1990), vacated
Ex Parte Mowbray, 943 S.W.2d 461 (Tex. App. 1996)
Holmes v. State of Texas, 135 S.W.3d 178 (2004)

Virginia

Compton v. Commonwealth, 250 S.E.2d 749 (Va. 1979)

Washington

State v. Perkins, 204 P.2d 207 (Wash. 1949)

Federal

Wilson v. United States, 162 U.S. 613 (1895)
United States v. Mustafa, 22 M.J. 165 (CMA 1986)
United States v. Mobley, 31 M.J. 273 (CMA 1990)
United States v. Hill, 41 M.J. 596 (Army Ct. Crim. App. 1994)
United States v. Holt, 46 M.J. 853 (N.M. Ct. Crim. App. 1997)
Smith v. Anzelone, 111 F.3d 1126 (4th Cir. 1997)

Appendix F
Court Decisions Relating to Presumptive Blood Tests

Presumptive Testing Cases

State

State v. Stenson, 940 P.2d 1239 (Wash. 1997)
Ex Parte Mowbray, 943 S.W.2d 461 (Tex. 1996)
Commonwealth v. Gordon, 666 N.E.2d 122 (Mass. 1996)
Young v. State, 871 S.W.2d 373 (Ark. 1994)*
Palmer v. State, 870 S.W.2d 385 (Ark. 1994)*
Robedeaux v. State, 866 P.2d 417 (Okla. 1993)
State v. Bronson, 496 N.W.2d 882 (Neb. 1993)
People v. Clark, 833 P.2d 561 (Cal. 1992)
Marcy v. State, 823 P.2d 660 (Alaska 1991)
State v. Lord, 822 P.2d 177 (Wash. 1991)
State v. Moody, 573 A.2d 716 (Conn. 1990)*
People v. Coleman, 759 P.2d 1260 (Cal. 1988)
Smith v. State, 492 So.2d 260 (Miss. 1986)
Wycoff v. State, 382 N.W.2d 462 (Iowa 1986)
Riggle v. State, 585 P.2d 1382 (Okla. 1978)
People v. Talbot, 414 P.2d 633 (Cal. 1966)

Federal

U.S. v. Holt, 46 M.J. 853 (N.M. Ct. Crim. App. 1997)*
U.S. v. Burks, 36 M.J. 447 (CMA 1993)*

Luminol Cases

State

State v. Fukusaku, 1997 WL 570980 (Hawaii 1997)*
Ex Parte Mowbray, 943 S.W.2d 461 (Tex. 1996)*

State v. Workman, 476 S.E.2d 301 (N.C. 1996)
State v. Hyde, 921 P.2d 655 (Ariz. 1996)
Jimenez v. State, 918 P.2d 687 (Nev. 1996)
Houston v. State, 906 S.W.2d 286 (Ark. 1995)
Smolka v. State, 662 So.2d 1255 (Fla. 1995)
State v. Zanter, 535 N.W.2d 624 (Minn. 1995)
State v. Campbell, 460 S.W.2d 144 (N.C. 1995)
Diffee v. State, 894 S.W.2d 564 (Ark. 1995)
Dansby v. State, 675 So.2d 1344 (Ala. 1994)
Young v. State, 871 S.W.2d 373 (Ark. 1994)
Palmer v. State, 870 S.W.2d 385 (Ark. 1994)*
Robedaux v. State, 866 P.2d 417 (Okla. 1993)
State v. Bible, 858 P.2d 1152 (Ariz. 1993)
State v. Wages, 623 N.E.2d 193 (Ohio 1993)
Brenk v. State, 847 S.W.2d 1 (Ark. 1993)*
People v. Cooper, 809 P.2d 865 (Cal. 1991)
State v. Gonzales, 783 P.2d 1239 (Kan. 1989)
Lee v. State, 545 N.E.2d 1085 (Ind. 1989)
State v. Lewis, 543 So.2d 760 (Fla. 1989)
Thompson v. State, 768 P.2d 127 (Alaska 1989)
State v. McElrath, 366 S.E.2d 442 (N.C. 1988)
People v. Henne, 518 N.E.2d 1276 (Ill. 1988)
State v. Martin, 740 P.2d 577 (Kan. 1987)
Johnston v. State, 497 So.2d 863 (Fla. 1986)
People v. Hendricks, 495 N.E.2d 85 (Ill. 1986)
Green v. Superior Court, 707 P.2d 248 (Cal. 1985)
People v. Asgari, 196 Cal.Rptr. 378 (1983)
Waterhouse v. State, 429 So.2d 301 (Fla. 1983)
State v. Jones, 328 N.W.2d 166 (Neb. 1982)
State v. Brown, 293 S.E.2d 569 (N.C. 1982)
State v. White, 235 S.E.2d 55 (N.C. 1977)

Federal

U.S. ex rel. Savory v. Lane, 832 F.2d 1011 (7th Cir. 1987)*
U.S. v. Holt, 46 M.J. 853 (N.M. Ct. Crim. App. 1997)*
U.S. v. Hill, 41 M.J. 596 (Army Ct. Crim. App. 1994)*
U.S. v. Burks, 36 M.J. 447 (CMA 1993)*

Benzidine Cases

State

State v. Hyde, 921 P.2d 655 (Ariz. 1996)
Fisher v. State, 827 S.W.2d 597 (Tex. 1992)
Commonwealth v. Anderson, 537 N.E.2d 146 (Mass. 1989)
Clark v. Ellerthorpe, 552 A.2d 1186 (R.I. 1989)
Commonwealth v. Shea, 519 N.E.2d 1283 (Mass. 1988)

Wycoff v. State, 382 N.W.2d 462 (Iowa 1986)
Commonwealth v. Nadworny, 486 N.E.2d 675 (Mass. 1985)
Commonwealth v. Cuneen, 449 N.E.2d 658 (Mass. 1983)
Nathan v. State, 611 S.W.2d 69 (Tex. 1981)
State v. Clark, 423 A.2d 1151 (R.I. 1980)
Commonwealth v. Best, 411 N.E.2d 442 (Mass. 1980)
State v. Milho, 596 P.2d 777 (Hawaii 1979)
State v. Ruybal, 398 A.2d 407 (Me. 1979)
Riggle v. State, 585 P.2d 1382 (Okla. 1978)
State v. Swain, 269 N.W.2d 707 (Minn. 1978)
State v. Boyd, 359 So.2d 931 (La. 1978)
Chandler v. State, 329 A.2d 430 (Md. 1974)
People v. Johnson, 113 Cal.Rptr. 303 (Cal. 1974)
Commonwealth v. Appleby, 265 N.E.2d 485 (Mass. 1970)
Wilbanks v. State, 266 So.2d 609 (Ala. 1968)
Trotter v. State, 377 S.W.2d 14 (Ark. 1964)
People v. Schiers, 324 P.2d 981 (Cal. 1958)
Commonwealth v. Moore, 80 N.E.2d 24 (Mass. 1948)

Federal

Clark v. Moran, 942 F.2d 24 (1st Cir. 1991)
Real v. Hogan, 828 F.2d 58 (1st Cir. 1987)
Carillo v. Brown, 807 F.2d 1094 (1st Cir. 1986)
Marrapese v. State of R.I., 749 F.2d 934 (1st Cir. 1984)
Clark v. Taylor, 710 F.2d 4 (1st Cir. 1983)
U.S. v. Sheard, 473 F.2d 139 (D.C.Cir. 1972)

Orthotolidine/Ortho-Tolidine/o-Tolidine Cases

State

Commonwealth v. Gordon, 666 N.E.2d 122 (Mass. 1996)
Commonwealth v. Burke, 607 N.E.2d 991 (Mass. 1993)
Commonwealth v. Hodgkins, 520 N.E.2d 145 (Mass. 1988)

Federal

United States v. Hill, 41 M.J. 596 (Army Ct. Crim. App. 1994)

Leucomalachite/Leuco-Malachite Green Cases

State v. Peterson, 494 N.W.2d 551 (Neb. 1993)
State v. Lord, 822 P.2d 177 (Wash. 1991)

Hemastix® Case

People v. Coleman, 759 P.2d 1260 (Cal. 1988)

Phenolphthalein Cases

State

State v. Fukusaku, 1997 WL 570980 (Hawaii 1997)
State v. Stenson, 940 P.2d 1239 (Wash. 1997)
People v. Bradford, 939 P.2d 259 (Cal. 1997)
State v. Workman, 476 S.E. 301 (N.C. 1996)
State v. Peterson, 446 S.E.2d 43 (N.C. 1994)
State v. Moseley, 445 S.E.2d 906 (N.C. 1994)
State v. Keel, 423 S.E.2d 458 (N.C. 1992)
Cheshire v. State, 568 So.2d 908 (Fla. 1990)
State v. McElrath, 366 S.E.2d 442 (N.C. 1988)
State v. Prevette, 345 S.E.2d 159 (N.C. 1986)
Wycoff v. State, 382 N.W.2d 462 (Iowa 1986)
Commonwealth v. Costello, 467 N.E.2d 811 (Mass. 1984)
State v. Brown, 293 S.E.2d 569 (N.C. 1982)
Riggle v. State, 585 P.2d 1382 (Okla. 1978)
Castleberry v. State, 522 P.2d 257 (Okla. 1974)

Federal

Clark v. Moran, 749 F.Supp. 1186 (D.R.I. 1990)
United States v. Holt, 46 M.J. 853 (N.M. Ct. Crim. App. 1997)
United States v. Burks, 36 M.J. 447 (CMA 1993)

Glossary of Terminology

Angle of impact The acute or internal angle formed between the direction of a blood drop in flight and the plane of the surface it strikes.

Area of convergence The area to which stains within a bloodstain pattern can be reconstructed on a two-dimensional surface determined by tracing the long axis of well-defined bloodstains within the pattern to a common area.

Area of origin The three-dimensional area to which stains within a bloodstain pattern can be reconstructed in space using the common area of convergence and the angles of impact.

Arterial pattern A pattern resulting from blood exiting the body under pressure from a breached artery.

Back spatter Blood droplets directed back toward the force or energy that caused the spatter, often associated with gunshot wounds of entrance.

Bloodstain A stain on a surface caused by blood.

Bubble ring Vacuoles in bloodstains that form when blood containing air bubbles dries and retains the circular configuration of the original bubble.

Cast-off pattern A pattern usually linear in configuration when blood is released or flung from a blood-bearing object in motion.

Cessation cast-off pattern A pattern resulting from the rapid deceleration of an object wet with blood.

Clot A gelatinous mass formed as the result of a complex mechanism involving red blood cells, fibrinogen, platelets, and other clotting factors.

Drawback effect The presence of blood in the barrel of a firearm that has been drawn back into the muzzle.

Drip pattern A bloodstain pattern created by free-falling drops of blood striking already-existing blood on a surface commonly associated with satellite spatter. The parent stain on the surface is larger than what would be associated with a single free-falling drop and is usually associated with satellite spatter.

Expired blood Blood that has been blown from the nose, the mouth, or a wound in the respiratory system as the result of air flow or pressure.

Flow pattern   A volume of blood on a surface that moves in one or more directions as a result of the influence of gravity.

Forward spatter   Blood droplets directed away from the force or energy that caused the spatter, often associated with gunshot wounds of exit.

Misting   Blood that has been atomized to a fine spray by the application of force, usually associated with gunshot or explosion events.

Parent stain or drop   A bloodstain from which satellite spatter or wave cast-off stains originate.

Passive bloodstains   Stains and patterns whose physical features indicate that they were created without any significant outside force other than gravity and friction.

Projected bloodstain pattern   Spatter created as the result of a force other than impact.

Scalloping   A serrated edge characteristic of bloodstains.

Serum stain   A yellowish stain resulting from the separation of serum from the retraction of a blood clot.

Skeletonized bloodstain   A bloodstain that consists of a darkened outer peripheral rim with the central portion of the stain having been removed by wiping through the partially dry stain. A skeletonized bloodstain is also produced by the flaking of the central portion of a completely dried stain.

Spatter   Blood stains that exhibit directionality and variation in size and are associated with a source of blood being subjected to external force(s).

Spatter pattern   A distribution of individual spatters on a surface that can be traced to a common area of origin.

Spines   An edge characteristic of bloodstains consisting of narrow, elongated projections from the central area of the stain.

Splash   A bloodstain created by a free-falling volume of blood in excess of 1.0 ml onto a surface from a distance of at least 4 in., OR: An altered bloodstain pattern characterized by a preexisting volume on a surface that has been subjected to additional force creating elongated narrow spines such as stepping into blood.

Swipe   The transfer of blood associated with motion onto a nonbloody surface.

Target   A surface on which blood has been deposited.

Terminal velocity   The maximum velocity that a free-falling drop of blood can accelerate in air, determined to be approximately 25.1 feet per second.

Transfer   The deposition of blood onto a surface as the result of contact.

Void   The absence of blood in an otherwise continuous bloodstain pattern that suggests the presence of an intermediate target that may have been removed.

Wave cast-off   A small stain that has originated from a parent stain as the result of the wavelike action of the original drop striking a surface at an angle of less than 90 degrees.

Wipe   An alteration of a preexisting wet or partially dry bloodstain caused by movement through the existing stain.

Index

A

Abrasions, 21, *22*
  bloodstain patterns within the body, 37, *38*
Acceptance of blood pattern analysis within the scientific community, 8
Acquired immunodeficiency syndrome (AIDS), 518
Actual flight paths, 243–244
Adhesion, 53, 392–394
Adler test, 352
Admissibility of bloodstain pattern analysis in courts, 443–449, 479–483, 484
Aging of blood, 180–181, *182*
Airbags, motor vehicle, 323, 325–326, *327*
Air resistance, 242
Albrecht Mechanism, 372–373
Alcohol effects on bloodstains, 55–56
Alcoholism, 15, 17
Altered bloodstains, 69–70, 179
  aging of blood and, 180–181, *182*
  clotting effects, 188–194, 341–342
  diffusion and capillarity of bloodstains and, 179–180
  dilution and, 194–196
  drying of blood and, 182–187
  due to body transport, 333–334, 336
  fire and soot effects, 200–205, *206*
  fly and other insect activity effects, 206–208, *209*
  removal of blood and, 196–200, *201*
  sequencing of, 212–216
  void patterns and, 210–212, *236*, 237–239, *333*, *340*
American Academy of Forensic Sciences (AAFS), 466–467
American Association for the Advancement of Science (AAAS), 472
American Association of Crime Laboratory Directors (ASCLD), 295
American Association of Neurological Surgeons (AANS), 473

Amido Black
  methanol-based, 402–403
  water-based, 398–402
  water-ethanol based, 403–405
Aneurysm, aortic, 16
Angle of impact, 221–223
Anticoagulants, 56–57
Antihuman hemoglobin, 367–368
Aortic aneurysm, 16
Appeals, postconviction. *See also* Legal issues
  admissibility and, 484
  attorney role in, 485
  collateral attack, 486–487
  denial of due process and, 487–489
  expert qualifications and, 484–486
  ineffective assistance of counsel and, 489–496
  newly discovered evidence and, 487
  perjurious expert testimony and, 487
  prosecutorial misconduct and, 487–489
Application procedures
  fixative, 394–395
  rinse solution, 396–398
Aqueous leucocrystal violet (ALCV), 408–411
2% aqueous solution of 5-sulphosalicylic acid, 393–394
Ardesco Oil Co. v. Gilson, 484–485
Arterial bloodstain patterns, 149–154, *155*
Arteries, 11–12, *13*
Arteriovenous malformations, 16
Assailants, blood spatter on, 127–129, 279
  clothing and shoes, 113, *115–117*, 283–294, 343–344, 388–389
  linking to crime scenes, 344–347
Autopsy photographs, 424–425

B

Backhus, Terri, 496
Back spatter, *135*, 136–138

BackTrack software, 241
  case studies, 254–261
  ellipses used in, 246–248
  laboratory study of double blow using, 249–253
Bacterial infections, 44
Balthazard, Victor, 4
Barskdale, Larry, 207
Baylor v. Estelle, 468–469
Beating events, impact spatter associated with, 119–130, *131*
Bell, James Spencer, 306
Benecke, Mark, 207
Benzidine test, 352
Bevel, Tom, 6, 62, 488–489
Biohazard safety precautions. *See* Safety precautions, biohazard
Biological functions of human blood, 41–42
Blood. *See also* Testing of blood
  adhesion, cohesion, and capillarity, 53, 392–394
  aging of, 180–181, *182*
  banks, 56
  biological functions of, 41–42
  circulation of, 11–13
  clotting, 42, 45–47, *48*, 55–56, 70, 188–194, 341–342
  composition of, 42–47
  confirmatory tests for, 361–364
  diffusion and capillarity of, 179–180
  drying of, 182–187
  hemoglobin, 14, 33–34, 45, *47*
    antihuman, 367–368
  impression, fixing, 392–394
  physical properties of, 48–55
  plasma, 14, 42–43
  platelets, 45–47
  Poiseuille's equation and, 49–50
  pools, 86–88
  postmortem aspects of, 32–38
  pressure, 11–12, 13–15
  rate of loss, 13
  red blood cells (RBC), 42, 45
  relative density, 53–55
  ricocheting, 82–83
  smears, 89
  species origin determination for, 364–365
  surface tension, 50–52, 242
  swipes, 89, 92
  tissue factor (TF), 47
  transfer stains, 94–95, *96*
  use in experiments, 56–57
  viscosity, 48–49, 242
  volume, 13–15, 42
  white blood cells (WBC), 41, 44
  wipe patterns, 93
Bloodborne pathogens, 517–518

Blood Drop Dynamics, 221
Blood drops
  distance fallen, 65–67
  drip patterns, 76–77, *78–80*
  drip trails, 77–80, *81–82*
  formation and travel, 59–60
  shape of falling, 61–63
  volume, 63
  volume as a function of source, 63–65
Blood Dynamics, 6
Blood loss
  from abrasions, 21, *22*
  due to injuries, 21–32
  due to medical conditions, 15–21
  from the gastrointestinal tract, 20–21
  from gunshot wounds, 27–32
  from incised wounds, 24–26
  from lacerations, 21–23, *24*
  from the mouth and nose, 16–19
  from the respiratory system, 20
  from the skin, 15–16
  from stab wounds, 26–27
Bloodstain pattern analysis (BPA). *See also* Bloodstains; Reports
  acceptance within the scientific community, 8
  admissibility in courts, 443–449, 479–483
  collection and preservation of evidence for, 294–296
  data synthesis, 431–432
  determining victim's location in, 217–219
  education and training in, 9
  ethical issues, 466–470
  evidence examinations, 428–432
  general processing at crime scenes, 300–304
  historical development of, 3–7
  history of DNA analysis and, 1
  luminescence and, 377–380
  objectives of, 1–2, 217
  requesters of, 421–422
  sample collection, 343–347
  scientific approach to, 2–3
  scientific working group for, 8–9
  sequence for approaching, 422–425
Bloodstain Pattern Analysis in Violent Crimes, 6
Bloodstain Pattern Analysis-Theory and Practice, 6
Bloodstain Pattern Analysis with an Introduction to Crime Scene Reconstruction, 6
Bloodstain Pattern Interpretation, 4, 6
Bloodstains. *See also* Altered bloodstains; Enhancement, chemical
  altered, 69–70
  arterial, 149–154, *155*
  within the body, 34–38, 55–56
  cast-off, 168–177, 256, 329, 377

Index

classification of, 7–8, *10*, 67–70
on clothing, 283–294
determining location of source of, 217–239
diffusion and capillarity of, 179–180
diluted, 194–196
documentation methods, 264–275
expiratory mechanisms, 157–165
fire and soot effects on, 200–205, *206*
measuring, 223, *224*
passive, 68, *69*, 71–97
produced by medical conditions of the venous system, 155–157
removal of, 196–200, *201*
selection of, 224–225
separating contributors of, 344
sequencing of, 212–216, 316–319
on shoes, 113, *115–117*, 288, *290–292*, 388–389
skeletonized, 185, *186*, 212
spatter, 68, *69*
physical characteristics of, 100–107
physical properties of blood relative to, 99–100
transfer, 88–93
Bodies
alterations in transport, 333–334, 336
feet or shoes of, 334–336
fluids requiring application of universal precautions, 517–518
manner of death evidence on, 342–343
mouth and nose of, 334
posture at time of injury, 336–338
processing of, 329–343
removal of, 330–332
self-defense claims and, 339
sequencing injuries to, 339–341
time lapse effect on bloodstains on, 339
use of scales in photographing, 330
victim viability evidence on, 341–342
Body of bloodstain pattern analysis reports, 437–440
Brady v. Maryland, 489
Bruises, 36, *37*
Burden of proof, 454

C

Cancer
lung, 16, *18*
nasal, 16
Capillaries, 12, 45, *50*
Capillarity of bloodstains, 53, 179–180
Carbon monoxide poisoning, 33
Carter, Alfred, 6
Cast-off mechanisms, 169–177, 256, 329, 377
Catalytic color tests, 351–352

Cerebrospinal fluid (CSF), 189
Chain of custody, 454–455
Chemiluminescence tests, 357–358
Circulation of blood, 11–13
Classification of bloodstains, 7–8, *10*, 67–70
Clayton v. State, 487
Clean area establishment at crime scenes, 519–520
Clothing
examination of bloodstained, 283–294
and shoes, 113, *115–117*, 388–389
suspect, 343–344
and victim viability, 341–342
Clotting, blood, 42, 45–47, *48*
altered bloodstains due to, 188–194, 341–342
effect of drugs and alcohol on, 55–56
spatter patterns and, *70*
Coffee ground emesis, 19, 180
Coffey v. Healthtrust, Inc., 470
Cohesion, liquid, 53
Collection and preservation of bloodstain evidence, 294–296
Collisions, motor vehicle, 323
Commonwealth v. Sturtivant, 479, 480–481, 482–483
Composition of blood, 42–47
Conclusions, bloodstain pattern analysis report, 440–441
Confirmatory tests for blood, 361–364
Contact wounds, 29–30
Content of bloodstain pattern analysis reports, 434–437
Convergence, area of
alternative methods for determining, 233–239
basic trigonometry of, 220–224
defining, 217–219
documentation, 226–232
selection of stains for determining, 224–225
Coomassie Brilliant Blue, 412–413
water-methanol based, 413–414
Court Appointed Scientific Experts (CASE), 472–473
Cremation, 204–205, *206*
Crime reconstruction using BPA, 2, 431–432, 485–486
Crime scenes
bloodstain pattern analysis (BPA) processing, 300–304
clean areas, 519–520
data processing, 304–319
doors and locks at, 305–306
examination, 423–424
importance of thoroughly processing, 297–298
indoor, 275–280, *281*
initial review of, 424
investigator's notes and reports, 264–265, 429
knife blocks at, 307–309

lighting at, 315–316
linking accused to, 344–347
medicines and medicine cabinets at, 309–310
motor vehicles as, 282, 319–329
outdoor, 281–282
photographs, 267–274, 330, 426–427
processing of bodies at, 329–343
refrigerators at, 306
report section, 437–438
sample collection, 343–347
sinks and faucets at, 310, *311, 312–313*, 344
sketch/diagram examination, 265–267, 425–428
specific reports, 434
temperature at, 314–315
trash cans at, 311, 313–314, *314*
video, 274, 277
Crossed-over electrophoresis, 366–367
Cross-examination of experts, 459–466
Crowle's Double Stain, 414–416
Crystal Violet, 408–411
Cyanide poisoning, 33
Cytoplasm, 44

D

Daubert standard, 8, 445–449, 484, 496
Daubert v. Merrell Dow Pharmaceuticals, Inc., 445–449
Davis v. Wallace, 477
Demonstrative evidence, 460–462
Depositions, 464–466
Diagrams, crime scene, 265–267, 425–428
Diffusion of bloodstains, 179–180
Dillard v. State, 481–482
Diluted bloodstains, 194–196
Direct examination of experts, 459–466
Directional analysis of bloodstain patterns
    basic theory, 241–242
    impact angles and, 244–246
    locating blood sources for, 242–244
    using BackTrack software, 246–253
    using ellipses, 246–248, *249*
Discovery process, 455–459
Distant gunshot wounds, 31
Distribution and location of spatter, 102–103
DNA analysis, 1, 294–295, 322, 349–350, 392
Documentation. *See also* Evidence
    area of convergence and origin, 226–232
    of bloodstained clothing examinations, 283–294
    bloodstain pattern, 264–275
    of bodies at the scene, 329–343
    collection and preservation of bloodstain evidence for, 294–296
    and examination of bloodstains in the laboratory, 282–283

importance of accurate, 263–264, 275
indoor crime scene, 275–280, *281*
motor vehicle, 282, 319–329
notes, 264–265, 429
outdoor crime scene, 281–282
photographs, 267–274, 380–383, 424–427
sketches and diagrams, 265–267, 425–428
video, 274, 277
Doors and locks at crime scenes, 305–306
Drag patterns, *91*
Drip patterns, 76–77, *78–80*
Drip trails, 78–80, *81–82,* 345–346
Drops, blood
    actual flight paths, 243–244
    distance fallen, 65–67
    drip patterns, 76–77, *78–80*
    drip trails, 77–80, 345–346
    formation and travel, 59–60, 241–242
    free-falling on angular surfaces, 73–76
    free-falling on horizontal surfaces, 71, *72–73*
    impact angles, 244–246
    shape of falling, 61–63
    virtual flight paths, 242–244
    volume as a function of source, 63–65
    volume of, 63
Drug effects on bloodstains, 55–56
Drying of blood, 182–187
Due process, denial of, 487–489
Dyes. *See* Enhancement, chemical

E

Eckert, William G., 6, 200
Education and training in bloodstain pattern analysis, 9
Electrophoresis, 364
    crossed-over, 366–367
Ellipses, analyzing bloodstains using, 246–248, *249*
Emergency medical technicians' notes or deposition, 430
Enhancement, chemical
    Amido Black
        methanol-based, 402–403
        water-based, 398–402
        water-ethanol based, 403–405
    Aqueous Leucocrystal Violet, 408–411
    Coomassie Brilliant Blue, 412–413
        water-methanol based, 413–414
    Crowle's Double Stain, 414–416
    factors to consider before choosing procedure for, 391–392
    fixatives, 392–394
    fixing solution application procedures, 394–395
    Hungarian Red, 405–408
    preliminary procedures, 392

Index

rinse solution application procedures, 396–398
rinsing procedures, 395–396
titanium dioxide suspension, 416–419
Epstein, Barton P., 6, 183
Esophageal varices, 16, 17
Ethical issues, 466–470
Etude Des Gouttes De Sang Projete, 4
Evidence. *See also* Documentation
bodies as, 329–330
burden of proof and, 454
chain of custody, 454–455
collected from motor vehicles, 282, 319–329
collection and preservation of, 294–296
data processing, 304–319
demonstrative, 460–462
discovery process, 455–459
examinations, sequencing, 428–432
physical, 427–428, 438
specific laboratory reports, 434
weight of, 449–450
Exit wounds, gunshot, 31–32
Ex Parte Mowbray, 487–488
Experiments, blood use in, 56–57
expiratory, 167–168
Experiments and Practical Exercises in Bloodstain Pattern Analysis, 6, 183
Experts
depositions, 464–466
direct and cross-examination of, 459–466
ethical issues and, 467–470
malpractice, 470–477
qualifications of, 450–459, 484–486
testimony, 384, 464–466
Expiratory mechanisms, 160, 259–261
bloodstain experiments, 167–168
Explosions, impact spatter associated with, 131–136

F

Fackler, Martin, 132
Federal Bureau of Investigation (FBI), 8
Federal Rules of Civil Procedure
11, 470
26, 457
Federal Rules of Evidence, 445–448
201, 446
702, 445
Fire and soot effects on bloodstains, 200–205, *206*
Fire Investigation, 201
Fischer, William, 6
Fixing blood impressions
with 2% aqueous solution of 5-sulphosalicylic acid, 393–394
general methods for, 394–395
with methanol, 394

Flight Characteristics and Stain Patterns of Human Blood, 4, 5
Flow patterns, passive, 84–85
Fluorescein, 360–361
Fluorescence tests, 357–358
Fly activity on decomposing bodies, 206–208, *209*
Forensic Science International, 207
Forensic Specialties Accreditation Board (FSAB), 452
Forensic Taphonomy: The Postmortem Fate of Human Remains, 207–208
Forward spatter, *135*, 138–144
Frank, Richard S., 474
Free-falling drops
on angular surfaces, 73–76
drip patterns, 76–77
drip trails, 77–80, *81–82*
on horizontal surfaces, 71, *72–73*
Frye standard, 8, 443–445, 447, 484, 496
Frye v. United States, 443–445

G

Gardner, Ross, 6, 62, 89
Gastrointestinal tract, bleeding from the, 20–21
General Electric Company v. Joiner, 449
Global positioning system (GPS), 266
Granulocytes, 44
Graphic methods for determining area of convergence, 234
Gravitational acceleration, 65–66, 242
Gunshot wounds, 27–32
back spatter associated with, *135*, 136–138
detection of blood in barrels of firearms and, 145–147
forward spatter associated with, *135*, 138–144
impact spatter associated with, 131–136
in motor vehicles, 319–320
void patterns and, 211–212

H

Hayes, Augustus, 480
Head, the, lacerations to, 22–23, *24*
Hemastix test, 356, *357*
Heme, 371
Hemoglobin, 14, 33–34, 45, *47*
antihuman, 367–368
Henderson, Carol, 6
Hepatitis viruses, 517, 518–519
Hesskew, Dusty, 488–489
High-velocity impact blood spatter (HVIS), 7–8
Historical development of BPA, 3–7
Holmes v. Texas, 484
Hospital photographs, 424–425

Human immunodeficiency virus (HIV), 517–518
Hungarian Red water-based, 405–408

I

Illes, Mike, 257–261
Impact, angle of, 221–223
Impact angles, 244–246
Impact spatter. *See also* Spatter bloodstains
  back, *135,* 136–138
  beating and stabbing events, 119–130, *131*
  cause of, 119
  detection of blood in barrels of firearms and, 144–146, *147*
  forward, 138–144
  gunshot, explosions, and power tools, 131–136
  in limited space, 257–261
Incised wounds, 24–26
Indoor crime scenes, 275–280, *281*
Ineffective assistance of counsel, 489–496
Injuries
  abrasion, 21, 37, *38*
  gunshot wound, 27–32
  incised wound, 24–26
  laceration, 21–23, *24*
  stab wound, 26–27
Insect activity and bloodstains, 206–208, *209*
Intermediate-range gunshot wounds, 30–31
International Association for Identification (IAI), 452
International Association of Bloodstain Pattern Analysts (IABPA), 4, *6,* 7, 467
Interpretation of Bloodstain Evidence at Crime Scenes, 6, 200, 382
Introductions, bloodstain pattern analysis report, 436–437
Isoelectric focusing (IEF), 365

J

James, Stuart H., 6, 200, 494–495
Jeffreys, Alec, 1
Jencks Act, 456
Jeserich, Paul, 4

K

Kastle-Meyer test, 353
Kirk, Paul, 4
Kischnick, Kersten, 490–491, 492–494
Kish, Paul, 6, 112, 263, 423
Knife blocks, 307–309
Knight, George, 480

Knight, Lydia, 480
Knight, Mary, 480
Kumho Tire Company, Ltd. v. Carmichael, 449

L

Laber, Terry L., 6, 65, 183
Laboratory Manual on the Geometric Interpretation of Human Bloodstain Evidence, 4
Lacerations, 21–23, *24*
Latent bloodstain impressions enhancement. *See* Enhancement, chemical
Legal issues. *See also* Postconviction procedures
  admissibility, 443–449, 479–483, 484
  burden of proof, 454
  chain of custody, 454–455
  direct and cross-examination of experts, 459–466
  discovery process, 455–459
  malpractice, 470–477
  qualifications of experts, 450–459, 484–486
  weight of the evidence, 449–450
Leucomalachite green test, 354
Leukocytes, 41, 44
Lighting at crime scenes, 315–316
Lincoln & wife v. Inhabitants of Barre, 485
Lindsay v. People of the State of New York, 481
Lividity, postmortem, 32–34, *34*
Lohman, James, 491
Long-range gunshot wounds, 31
Low-velocity impact blood spatter (LVIS), 7
Luminescence, interpretation of, 376–377
Luminol
  case studies, 384–389
  collection of samples and, 378–379
  discovery of, 369–371
  effect of substrates on, 378
  effect on subsequent analysis of bloodstains, 379–380
  interpretation of luminescence produced by, 358–359, 376–377
  interpretation of patterns and, 377–380
  oxidation of, 373
  photography, 380–383
  preparation of, 374–375
  reaction, 371–373
  report writing, 383
  testimony, 384
  use of additional presumptive tests with, 377
Lung cancer, 16, *18*
Lymphocytes, 44

M

MacDonell, Herbert Leon, 200, 423, 450
  expert testimony by, 483, 487, 488

Index

research on bloodstain pattern analysis (BPA) by, 4, *5*, 6, 7, 55, 56, 62, 63, 65
Malpractice, expert witness, 470–477
Marbling of the skin, postmortem, 35
Mattco Forge, Inc. v. Arthur Young Co., 476
McPhail's Reagent, 354
Measurement of bloodstains, 223, *224*
Medical conditions, blood loss due to, 15–21
  of the venous system, 155–157
Medicines and medicine cabinets at crime scenes, 309–310
Medium-velocity impact blood spatter (MVIS), 7
Metabolism, 41
Methanol, 394
  based Amido Black, 402–403
Misconduct, expert, 470–477
Moore, Craig C., 254–257
Motor vehicles, documentation and examination of, 282, 319–329
Mouth
  bleeding from the, 16–19
  photography of, 334
Mowbray, Susie, 488–489

N

Nasal cancer, 16
Neutrophils, 44
Newly discovered evidence, 487
Newtonian fluids, 49
Newton's First Law of Motion, 59
Newton's Second Law of Motion, 59–60
Niagara Regional Police Service, 254–257
Nongranulocytes, 44
Non-Newtonian fluids, 49
Nonserum protein analysis, 367–368
Nose
  bleeding from the, 16–19
  photography of, 334
Notes, documentation, 264–265, 429

O

Objectives of bloodstain pattern analysis, 1–2, 217
Ontario Provincial Police, 257–261
Origin, area of
  alternative methods for determining, 233–239
  basic trigonometry of, 219–224
  defining, 217–219
  documentation, 226–232
  methods of determining, 225–226
o-Tolidine test, 353–354
Ouchterlony double diffusion test, 365–367
Outdoor crime scenes, 281–282

P

Pallor of gums, 32
Passive bloodstains
  blood pools, 84–88
  defined, 68, *69*
  drip patterns, 76–77, *78–80*
  drip trails, 77–80, *81–82*
  flow patterns, 83–84
  free-falling drops on angular surfaces, 73–76
  free-falling drops on horizontal surfaces, 71, *72–73*
  saturation stains, 88
  splashed patterns, 80–83
  transfer, 88–93, 94–95, *96*
People v. Clark, 482
People v. Owens, 444–445
Peptic ulcers, 16, *19,* 180
Perjury, 487
Personal protective equipment, 519–522
Petechial hemorrhaging, 35
Peterson, Ivars, 62
Phenolphthalein test, 353, 360
Photography
  autopsy or hospital, 424–425
  of bodies, 330
  crime scene, 267–274, 426–427
  luminol, 380–383
  of mouth and nose of decedents, 334
  scales used in, 330
Physical evidence
  examination, 427–428
  report section, 438
Physical properties of blood, 48–55
  relative to spatter formation, 99–100
Piotrowski, Eduard, 3–4, *124*
Plasma, 14, 42–43
Platelets, blood, 45–47
Poiseuille's Equation, 49–50
Pools, blood, 84–88
Postconviction procedures. *See also* Legal issues
  admissibility, 484
  appeals, 483
  attorney's role in, 483
  collateral attack, 486–487
  denial of due process and, 487–489
  expert qualifications, 484–486
  ineffective assistance of counsel and, 489–496
  newly discovered evidence and, 487
  perjurious expert testimony and, 487
  prosecutorial misconduct and, 487–489
Postmortem aspects of blood in the body, 32–38
Postmortem injuries, 16
Potolski, Steven, 494Posture of decedents at time of injury, 336–338
Power tools, impact spatter associated with, 131–136
Preparation of luminol, 374–375

Preservation and collection of bloodstain evidence, 294–296
Presumptive testing of blood. *See* Testing of blood
Projection mechanisms. *See also* Spatter bloodstains
　arterial, 149–154, *155*
　bloodstain patterns produced by venous system medical conditions, 155–157
　cast-off bloodstain patterns, 168–177
　definition of, 149
　expiratory bloodstain patterns, 157–165
　experiments, 167–168
Prosecutorial misconduct, 487–489
Protein analysis
　nonserum, 367–368
　serum, 364

Q

Qualifications of experts, 450–459, 484–486

R

Raymond, M. A., 62
Records, autopsy or hospital, 425
Red blood cells (RBC), 42, 45
Redsicker, David, 200–201
Refrigerators at crime scenes, 306
Relative density of blood, 53–55
Removal of bloodstains, 196–200, *201*
Reports
　body, 437–440
　conclusion, 440–441
　content, 434–437
　documentation of area of convergence and origin, 226–232
　finished, 441–442
　introduction, 436–437
　luminol, 383
　phrases used in, 435–436
　purpose of, 433–434
　scene-specific, 434
Research on Blood Spatter, 4
Respiratory system, bleeding from the, 20
Responding officer's notes, 430
Rice, L. D., 480
Ricocheting of blood, 82–83
Riechmann, Dieter, 489–496
Ring precipitin test, 365
Rinsing procedures, fixative, 395–396

S

Saccoccio, Marie, 6
Safety precautions, biohazard
　basic rules for, 519
　hepatitis, 517, 518–519
　human immunodeficiency virus (HIV), 517–518
　minimizing contamination at the crime scene, 519–521
　personal protective equipment and, 519–522
　universal precautions policy, 517–518
Sample collection, 378–379
Satellite spatters
　on horizontal surfaces, 110–111, *112*
　on vertical surfaces, 112–117
Saturation stains, 88
Scales used in photographs, 330
Schmitz, A. J., 369
Scientific and Legal Applications of Bloodstain Pattern Analysis, 6
Scientific approach to BPA, 2–3
Scientific Working Groups (SWG), 8–9
Seatbelts, motor vehicle, 323, *325–326*
Secondary mechanisms spatters, 108–109, *110*
Selection of bloodstains, 224–225
Self-defense claims, 339
Separation of contributors of bloodstains, 344
Sequencing
　for approaching bloodstain cases, 422–425
　bloodstains, 212–216, 316–319
　injuries, 339–341
Serology reports, 429. *See also* Testing of blood
Serum protein analysis, 364
Shape
　of falling blood drops, 61–63
　of spatters, 103–106
Shock, 32
Shoes
　of decedents, 334–336
　documentation and examination of, 288, *290–292*
　luminol photographs and, 388–389
　satellite spatters on, 113, *115–117*
Sinks and faucets at crime scenes, 310, *311*, *312–313*, 344
Size range and quantity of spatter patterns, 100–101, *102–103*
Skeletonized bloodstains, 185, *186*, 212
Sketches, crime scene, 265–267, 425–428
Skin
　bleeding from the, 15–16
　marbling, postmortem, 35
Smears, blood, 89
Soil samples, 281–282
Soot and fire effects on bloodstains, 200–205, *206*
Sourcebook, 369
Source of bloodstains, determining
　angle of impact in, 221–223
　area of convergence and, 217–219
　area of origin and, 219, 225–226

basic trigonometry used in, 220–224
documentation in, 226–232
measuring bloodstains and, 223, *224*
recognition of patterns in, 217–219
selection of stains used in, 224–225
Spatter bloodstains, 68, *69*. *See also* Impact spatter; Projection mechanisms
   definition of, 99
   distribution and location of, 102–103
   horizontal surfaces satellite, 110–111, *112*
   mechanisms association with the creation of, 107, *108*
   physical characteristics of, 100–107
   physical properties of blood relative to, 99–100
   satellite, 110–117
   secondary mechanism, 108–109, *110*
   shape of, 103–106
   on shoes, 113, *115–117*
   significance of identification of, 108
   size range and quantity of, 100–101, *102–103*
   target surface texture effect on, 106–107
   vertical surfaces satellite, 112–117
Species origin determination for blood, 364–365
Splashed bloodstain patterns, 82–83
Stab wounds, 26–27, 344–345
   impact spatter associated with, 119–130, *131*
State of Florida v. Riechmann, 489–496
State of Ohio vs. Samuel Sheppard, 4
State v. Halake, 453
State v. Hall, 482
State v. Knight, 479–480, 483
Surface tension, 50–52, 242
Sutton, T. Paulette, 6
Swipes, blood, 89, 92

T

Takayama test, 363–364
Tardieu spots, 35
Target surface texture effect on spatter appearance, 106–107
Teichmann test, 362–363
Temperature at crime scenes, 314–315
Terminal velocity, 65–67
Testimony. *See* Experts
Testing of blood
   Benzidine, 352
   catalytic color, 351–352
   comparison, 356–358
   confirmatory, 361–364
   DNA analysis and, 1, 294–295, 322, 349–350, 392
   fluorescein, 360–361
   Kastle-Meyer, 353
   leucomalachite green, 354
   luminol, 358–359, 369–389

nonserum protein analysis, 367–368
o-Tolidine, 353–354
Ouchterlony double diffusion, 365–367
phenolphthalein, 353
protocol, 350
ring precipitin, 365
serum protein analysis, 364
species origin determination, 364–365
Takayama, 363–364
Teichmann, 362–363
tetramethylbenzidine, 355–356
using chemiluminescence and fluorescence, 357–358
Tetramethylbenzidine test, 355–356
Thrombocytes, 45–47
Time lapses, 339
Tissue factor (TF), 47
Titanium dioxide suspension, 416–419
Tomash, Craig, 201
Transfer bloodstains, 92, *93–97*
Transient evidence at crime scenes, 316
Trash cans, 311, 313–314, *314*
Trigonometric method for determining area of convergence, 233–234
Trigonometry of right triangles, 220–224
Tuberculosis, 16

U

Uber Entstehung, Form, Richtung und Ausbreitung der Blutspuren nach Hiebwunden des Kopfes, 3
Ulcers, peptic, 16, *19*, 180
United States v. Mustafa, 445
Universal Precautions Policy, 517–518
U.S. Occupational Safety and Health Administration, 517

V

Varicose veins, 15–17
Venous system medical conditions, bloodstain patterns produced by, 158
Video documentation, 274, 277
Virtual flight paths, 242–244
Viscosity, blood, 48–49, 242
Void patterns, 210–212, *236*, 237–239, *333, 340*
Volume, blood, 13–15

W

Waste removal and blood, 41
Water-based Amido Black, 398–402
Water-based Hungarian Red, 405–408
Water-ethanol based Amido Black, 403–405

Water-methanol based Coomassie Brilliant Blue, 413–414
White blood cells (WBC), 41, 44
Wilson v. United States, 482
Wipe patterns, 90
Wonder, Anita, 6
Wounds
   beating, 119–130, *131*
   gunshot, 27–32, 319–320
      back spatter associated with, *135,* 136–138
      detection of blood in barrels of firearms and, 145–147
      forward spatter associated with, *135,* 138–144
      void patterns and, 211–212
   incised, 24–26
   stab, 26–27, 119–130, *131,* 344–345